Silicon-on-Sapphire Circuits and Systems

About the Author

Eugenio Culurciello received the Laurea (M.S.) degree in electronics engineering from the University of Trieste, Italy, in July 1997. His M.S. thesis work was developed at Johns Hopkins University with Professor Ernst Niebur. He joined professor Andreas G. Andreou in the laboratory in January 1998 as a graduate student. He received a second M.S. degree in electrical and computer engineering from Johns Hopkins University, Baltimore, Maryland. In September 2004, he received a Ph.D. degree in electrical engineering at Johns Hopkins University. In July 2004, Professor Culurciello joined the Department of Electrical and Computer Engineering at Yale University as an assistant professor. He established Yale's e-Lab, a VLSI laboratory focused on extending human abilities to interact with the environment through technological advancements. By using emergent fabrication technologies, e-Lab research focuses on the design of biomedical instrumentation and sensory processing circuits to extend human senses, cognition, and health in the same ways that cellular phone and the Internet enlarged human communication capabilities, knowledge, and information retrieval.

His research interests include low-power, mixed-mode VLSI systems with applications to vision sensory systems, compressed sensing and efficient communication systems, address-event communication, implantable sensors, telemetry sensors, biomimetic sensors, and applications in sensor networks, SOI and SOS circuit design, models of SOI MOSFETS, optoelectronic devices, SOI analog-to-digital conversion, SOI RF circuits, radiation-tolerant SOI design, and isolation amplifiers.

Silicon-on-Sapphire Circuits and Systems

Sensor and Biosensor Interfaces

Eugenio Culurciello

New York   Chicago   San Francisco
Lisbon   London   Madrid   Mexico City
Milan   New Delhi   San Juan
Seoul   Singapore   Sydney   Toronto

The McGraw·Hill Companies

Cataloging-in-Publication Data is on file with the Library of Congress.

McGraw-Hill books are available at special quantity discounts to use as premiums and sales promotions, or for use in corporate training programs. To contact a representative please e-mail us at bulksales@mcgraw-hill.com.

Silicon-on-Sapphire Circuits and Systems

1 2 3 4 5 6 7 8 9 0   DOC/DOC   0 1 4 3 2 1 0 9

ISBN   978-0-07-160848-0
MHID   0-07-160848-6

The pages within this book were printed on acid-free paper.

Sponsoring Editor	Copy Editor	Composition
Taisuke Soda	Susan Ginger	International Typesetting and Composition
Acquisitions Coordinator	**Proofreader**	
Michael Mulcahy	Nigel O'Brian, International Typesetting and Composition	**Art Director, Cover** Jeff Weeks
Editorial Supervisor		
David E. Fogarty	**Production Supervisor**	
Project Manager	Richard C. Ruzycka	
Harleen Chopra, International Typesetting and Composition		

For my wife, Kyoung-Soo Lee

Contents

Acknowledgments

The material presented in this book is the result of a decade of work of a multitude of people. All the work presented in this book could not have been accomplished without the help of many great people that have accompanied me in my academic career. In this public acknowledgment, I would like to express my most deep sentiments of gratitude to my thesis advisor, professor Andreas G. Andreou, for his teachings and encouragement, which has culminated in the writing of this book.

I would like to acknowledge all the students and colleagues from Yale e-Lab that have been involved: Zhengming Fu, Farah Laiwalla, Pujitha Weerakoon, Wei Tang, Evan Joon-Huyk Park, Dongsoo Kim, Shoushun Chen, and Hazael Montanaro. Under my supervision, they have all contributed to advance the field of silicon-on-insulator mixed-signal microsystem design.

Many thanks go to all the folks I have interacted and worked with during my PhD study at Johns Hopkins University: to professor Ralph Etienne-Cummings, professor Gert Cauwenberghs and Kim Strohbehn, collaborating with you enriched me and allowed me to explore fields I would have not known otherwise. I express my gratitude to all my PhD graduate students colleagues and postdoctorate fellows: David Goldberg, Philippe Pouliquen, Francisco Tejada, Jennifer Blain, Miriam Adlerstein, Alyssa Apsel, Pamela Abshire, Mark Martin, Zaven Kalayjian, Zhaonian Zhang, Pablo Mandolesi, Pedro Julian, Joshua Cysyk, Paul Giedraitis, Lavida Cooper, Thiago Valladares Sabino Teixeira, Julius Georgiou, Jie Zhou, Viktor Gruev, Mattew Clapp, Francesco Tenore, Ralf Philippe, Roman Genov. Milutin Stanacevic, and Shantanu Chakrabartty.

Thanks to my wife Kyoung-Soo Lee, your moral support and care made this all possible. You are very special to me.

Finally, much gratitude goes to my parents. You have brought me up in this world and have tried to direct me to the best. Every day that goes by I carry within me a piece of you, and I am trying to use it to better this world.

Preface

This book introduces a flavor of silicon-on-insulator (SOI) technologies called silicon-on-sapphire (SOS), describing the fabrication process, the advantages, and the basic devices available for circuit design.

The physical differences in the fabrication of the SOS devices and the insulating substrate make this process quite different from a standard bulk process. The insulating substrate, the floating body and the different thermal properties of the sapphire give rise to characteristics that have to be fully mastered to allow for the design of high-performance circuits.

The objective of this book is to introduce the design of microsystems in SOI and SOS technologies. The structure of the book is organized so that readers can become accustomed to the SOI/SOS fabrication process, starting from the physical constraint of fabrication and moving to device characterization, modeling, and, ultimately, to large-scale microsystems.

The book begins by introducing the SOS fabrication process, its history, and main features. We then describe SOS MOSFET device models and parameters, passive components, basic device characteristics, simple models, CAD models, continuous models. We introduce SOS amplifiers and basic analog circuits, as single-stage amplifiers, differential pairs, and analog switches. Then, SOS linear analog circuits, as operational amplifiers, switched capacitor circuits, and comparators. We describe SOS digital circuits, and, in particular, inverter characteristics.

We also describe mixed-signal SOS analog to digital conversion systems, in particular, resistive and capacitive DAC, ADC architectures, SAR ADC, and Sigma-Delta ADC. We report on SOS photodetectors topologies and performance. We discuss the design of SOS image sensors, with both analog and digital outputs. We introduce SOS biosensor system interfaces, such as voltage and current noise, low-noise voltage amplifiers, low-noise current amplifiers, and sampled and continuous time architectures. We report on SOS advanced analog circuits and systems, such as isolation amplifiers, temperature sensors, and band-gap references.

EUGENIO CULURCIELLO
Yale University, New Haven

Silicon-on-Sapphire Circuits and Systems

CHAPTER 1

The Silicon-on-Sapphire Fabrication Process

1.1 Introduction

Silicon-on-insulator technology (SOI) is one of the available technologies to fabricate integrated circuits. Traditional circuit manufacturing technologies employ a conductive and doped silicon wafer, and, for this reason, are often referred as "bulk complementary metal–oxide semiconductor (CMOS) processes." Bulk CMOS devices and circuits are manufactured on the silicon surface.

SOI technologies use a silicon–insulator–silicon substrate in place of bulk substrates. This kind of substrate allows to reduce the parasitic capacitances for each device and therefore provide improvements in performance.

This chapter discusses a type of SOI called silicon-on-sapphire (SOS), describing its fabrication process, its advantages, and the basic devices available. The objective of this chapter is to familiarize the reader with the SOI/SOS fabrication process, starting from the physical constraint of fabrication, moving to device characterization, and, finally, introducing the design of SOI/SOS circuits and systems.

SOI fabrication processes are discussed in Sec. 1.3 of this chapter. The SOS wafer is introduced in Sec. 1.4, while its fabrication process is described in Sec. 1.5. Section 1.6 presents an overview of SOS metal-oxide-semiconductor field effect transistors (MOSFETs) and their advantages. Section 1.7 discusses in detail SOS MOSFETs in strong and weak inversion and offers preliminary statistical data from an ensemble of different dies. Section 1.7 ends with a useful comparison between data and simulation of SOS MOSFETs.

1.2 Why SOI?

Silicon-on-insulator technology is a key player in the manufacturing of the next generation of high-speed high-performance microprocessors and radio frequency (RF) communication circuits. The SOI technology allows for higher-speed circuits and microprocessors while, at the same time, lowering the power demands of these high-performance components.

In a bulk-based CMOS chip, all fabricated transistors and devices possess a significant parasitic capacitance due to coupling with the substrate. In order to control the devices' behavior, the controlling terminals (gates in MOSFETs) and their substrate parasitic capacitance are charged and discharged at high operational frequencies. The charging of the control gate and its parasitic companion requires time and causes the transistors on the integrated circuit to heat up. The production of heat limits the speed at which microchips can operate. For this reason, integrated microprocessors have poor yield rates above a few gigahertz and, with current densities and heat dissipation capabilities, are not expected to attain future speeds above 5 GHz. SOI integrated circuits do not possess a conducting substrate, which eliminates much of the parasitic capacitance of the controlling terminal of the devices and allows an SOI integrated circuit to operate at higher speeds and lower temperatures. With SOI technology, the gate area of MOS transistors can be fabricated with minimal parasitic capacitance. A low parasitic capacitance circuit will allow faster transistor operation. As transistor latency drops, the speed at which the transistor can process instructions increases, and therefore both high-end microprocessors and RF industries can improve their devices year after year.

International Business Machines (IBM) was one of the first companies to use SOI to fabricate microprocessors (Buchholtz et al., 2000; Anderson et al., 2001). Advanced Micro Devices (AMD) has also been manufacturing in 130-nm, 90-nm, and 65-nm nodes and dual and quad core processors in SOI since 2001 (Pan et al., 2007). Freescale Semiconductors has also been manufacturing PowerPC microprocessors in 180-nm, 130-nm, 90-nm, and 65-nm lines (Bo et al., 2006). In recent years, SOI processors have become even more widespread, due to the introduction of the SOI Cell processor into the popular Playstation 3 platform (Pham et al., 2006). Laboratory testing shows that SOI-based processors feature a 20–25% improvement in transistor switching time compared to bulk CMOS counterparts (Park et al., 1999). SOI has also proven to be useful for low-power applications, since SOI-manufactured processors require an average of 40–50% less power than CMOS ones. This is, again, due to the reduction of undesired stray capacitance.

Another key benefit of SOI is a reduction in soft error rates (SER). Soft errors refer to data corruption caused by cosmic rays and natural radioactive background signals. SER will be an important issue as microprocessors scale to smaller die sizes and lower voltages. SOI chips demonstrate a significant reduction in SER-related issues, even for integrated circuits with a large die area (such as microprocessors).

At the same time, SOI and SOS analog circuits can benefit from the insulating substrate and provide higher performance at lower power draw. In this and the next few chapters of this book, we focus on the advantages of SOS technology for analog and mixed-signal microsystem design. The next generation of mobile-targeted circuits and systems will demand a combination of high-performance digital and analog circuits. This book, by providing a fresh perspective on these new technologies, seeks to offer a fundamental contribution to the successful development of SOI and SOS circuits.

1.3 The SOS Fabrication Process

Silicon-on-insulator is a very-large-scale integration (VLSI) fabrication process in which a thin layer of silicon is deposited on the top of an insulating material (Kuo and Su, 1998). SOS, in which the insulating material is synthetic sapphire, was one of the first SOI processes available. In an SOI process, active metal–oxide–semiconductor (MOS) devices and other passive devices are fabricated in and on the thin layer of silicon.

In SOS, a thin layer of silicon is grown on top of a sapphire substrate, as can be seen in the bottom portion of Fig. 1.1. In this figure, the main difference between bulk CMOS, SOS, and SOI are clearly visible. At the top of Fig. 1.1 we show the cross-section of a typical deep-submicron CMOS process. In this cross-section the example circuit is a CMOS inverter. Notice the large number of layers needed to design NMOS and PMOS devices and the need of wells to insulate them from each other and from the conductive substrate. Also notice the required contacts to the well to keep them in reverse bias with respect to the substrate and reduce parasitic conduction. On the other hand the SOS process cross-section, presented in Fig. 1.1, shows a clear advantage in simplicity and reduction of fabrication masks needed to design the same CMOS inverter circuit. SOS MOSFETs are obtained on a ultrathin (100 nm) film of silicon grown on top of a sapphire substrate. The sapphire substrate itself is grown artificially. The bottom cross-section of Fig. 1.1 shows the same CMOS inverter implemented in an SOI process. SOI uses a buried silicon dioxide layer to separate the top thin film of silicon and obtain similar benefits and reductions in manufacturing steps that SOS provides.

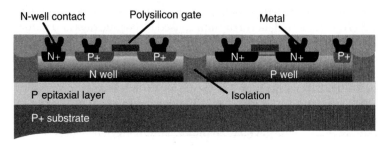

Bulk CMOS Process

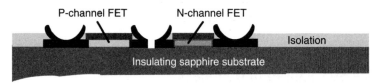

Peregrine SOS Process

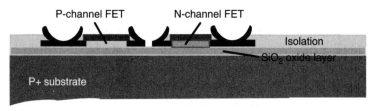

SOI Process

FIGURE 1.1 Comparison of the CMOS wafer cross-section (top), an SOS CMOS wafer (middle), and an SOI CMOS wafer (bottom). The circuit in each of these cross-sections is a CMOS inverter (Peregrine, 2008a).

Historically, the SOS process has been utilized and developed for its property of radiation tolerance (or radiation hardness), the ability to withstand environments with high radiation (Cristoloveanu and Li, 1995). The thin film of silicon is not thick enough to interact with ionizing radiation, and therefore the system is less susceptible than other systems to single-event charged particles upsets. In addition, the insulating layer or substrate shields the thin film devices from radiation-generated hole–electron pairs. In a bulk process, the effect of radiation is severe in the operation of active MOS devices, because the carriers generated in the bulk substrate by ionizing radiation can significantly affect the device characteristics and, moreover, induce latch-up or even permanent damage. The insulating substrate in SOI/SOS eliminates worrisome latch-up conditions. Because there is no substrate, the parasitic bipolar junction transistors (BJTs) are not present, and therefore the problem of latch-up does not arise.

Because of their property of radiation tolerance, SOI and SOS are very popular in military and space applications. Until recently, however, the low yield in the fabrication of VLSI circuits using these processes have prevented them from being successful in the commercial market, restricting them to corner niches in the space and military sectors, where yield and cost are not an issue because of the low numbers of components needed.

1.3.1 Advantages of SOS

From a design point of view, the main advantage of SOI/SOS is the insulation between MOS active devices and the substrate. This insulation frees the devices from the parasitic capacitances in connection to the substrate, increasing the quality factor of passive components and the performance of active MOS devices. The lack of these parasitics to the substrate implies a lack of stray capacitances. Smaller parasitic capacitances reduce the burden on devices and allow transistors to operate faster, as their capacitive load is reduced. Circuit instabilities due to undesired capacitive feedback are also significantly reduced. In the case of MOSFET devices, the feedback capacitance between input and output is due only to the gate capacitance in SOI/SOS.

The substrate isolation gives SOI/SOS a density advantage during fabrication. SOI/SOS devices can be spaced more closely than bulk processes, since the latter need to reverse bias PN junctions to isolate portions of the circuits. This is evident in Fig. 1.1, where a cross-section of a typical bulk process is compared with a SOS process. The circuit in the figure is a CMOS inverter. A typical example of bulk reverse biased isolation is the N-well in a P substrate. The spacing between active regions and wells severely limits density in bulk complementary MOS (CMOS).

Recently, SOI processes have been more and more widely adopted, especially since IBM started manufacturing microprocessors in SOI [PowerPC model 750, year 1999 and 2000 (Buchholtz et al., 2000)]. Such commercial success has been encouraged by the recent development of a commercially available foundry of SOS CMOS technology by Peregrine Semiconductor Corp. of San Diego, CA (Peregrine, 2003). Peregrine is a leading supplier of high-speed mixed-signal integrated circuits (ICs) for wireless and fiber-optic communications. Peregrine's product family is based on its patented ultrathin silicon (UTSi) CMOS wafer fabrication process. A cross-section of a CMOS inverter in this process is given in Fig. 1.1. UTSi ICs have substantial advantages—lower cost, lower power consumption, higher levels of integration, and superior RF performance-in comparison to ICs that are fabricated in competing high-performance mixed-signal processes. Peregrine products include RF MOS switches, high-performance phase-locked loops (PLL), MOSFET mixers, monolithic digital step attenuators, RF prescalers, line drivers and receivers, and frequency synthesizers (Stuber et al., 1998; Megahed et al., 1998).

1.3.2 Disadvantages of SOS

One disadvantage of the SOS process is the lower heat conductivity of the sapphire substrate—typically one-half the conductivity of silicon. Also, a monolith crystal wafer of sapphire is about twice as expensive to fabricate as a silicon wafer. However, the number of fabrication masks is reduced in SOS, because of the absence of some of the bulk CMOS process layers, such as wells, trenches, and other isolation structures.

The cost of developing SOS circuits is constantly decreasing as the number of clients increases and demand rises. With recent trends and increasing demand for low-power low-voltage circuits, SOS currently is and will continue to play an important role in the future of VLSI, as will become apparent in the sections on device characteristics and circuits that follow.

1.4 SOS Wafer Manufacturing

The development of a high-yield SOS process requires many technological advances, some of which are described in this section and in Fig. 1.2. Much of SOS technology was developed in the 1970s, when the process became popular. Sapphire was chosen as a substrate element because of its relative availability and its silicon-compatible properties, in particular its thermal mismatch and its crystallographic parameters (Cristoloveanu and Li, 1995).

An SOS wafer is prepared by cutting a single crystal sapphire (Al_3O_2) approximately $60°$ to the c axis (Kuo and Su, 1998). The sapphire crystal is typically obtained using Czockralsky growth and edge-defined film-fed growth. The sapphire surface has to be carefully finished with subsequent steps of polishing and annealing before it can receive the silicon film.

The sapphire substrate is carefully polished and cleaned ultrasonically before being inserted in the growth apparatus. The single crystal sapphire is placed in a chemical vapor deposition (CVD) chamber,

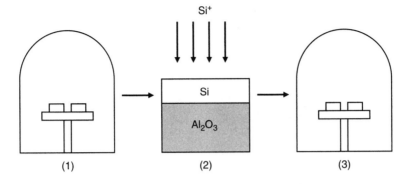

FIGURE 1.2 Fabrication of an SOS wafer: (1) CVD deposition of thin film of silicon on the sapphire substrate at about 900°C, (2) ion implantation of Si+ for amorphization at 300°C, (3) anneal and CVD deposition of silicon at 900°C.

flushed with hydrogen, and heated to 1000–1200°C using an RF generator. When thermal equilibrium is reached, the chamber is filled with silane (SiH_4) and silicon deposition begins.

Initially, only small hemispherical silicon islands grow on the sapphire substrate. These islands gradually enlarge until a thin film of about 20 nm is formed. Due to the crystallographic mismatch of sapphire and silicon, this thin film grows with a high-defect density. Aluminum deposits are also a source of contamination before coalescence, and therefore this initial growth has to occur at high rate (2 μm/min). Subsequent growth is slower to reduce the defect density. When the desired silicon film thickness has been reached, the silane flow is stopped. The thickness of the deposited silicon layer in the Peregrine SOS process is 100 nm.

The wafers are then slowly cooled in hydrogen to anneal and crystallize the surface. The most damaging step is the postgrowth cooling phase, when differential contraction occurs and causes strong compression of the silicon film. Silicon thermal expansion is $3.8 \times 10^{-8}/°C^{-1}$, whereas, it is $9.2 \times 10^{-8}/°C^{-1}$ for sapphire (Al_3O_2).

In the past, low yield has been due to the poor quality of interface between the silicon and the sapphire. Imperfection due to the interface matching of the crystalline structures causes failure lines in the deposited silicon. Peregrine Semiconductors recently developed (1990) a technique to obtain almost perfect silicon in sapphire interfaces, thus bringing the yield of circuits to production acceptable levels. The purity of the SOS interface in Peregrine wafers can be seen in Fig. 1.3. In this figure, the Peregrine wafer is compared to a conventional SOS wafer, where the density of defects is higher (Reedy, 1982);

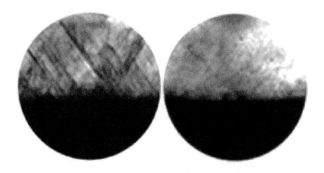

FIGURE 1.3 Micrograph of an unoptimized SOS wafer (left) and Peregrine SOS wafer (right) defects at the silicon–sapphire interface. Notice that Peregrine wafer's defect density is lower than conventional SOS wafer's defect density (?).

(Garcia et al., 1988). The reason is that the sapphire crystal used by Peregrine is rhombohedral and therefore different from the face-centered cubic silicon crystal (Cristoloveanu and Li, 1995). The defect density in a conventional SOS process can reach values as high as 1 M planar faults/cm and 10^9 line defects/cm^2.

A recently developed technique that reduces defect density is called solid phase epitaxy (SPE). In SPE, silicon ions are implanted in the thin silicon film to render it amorphous. The implant of silicon ions on the sapphire surface has to occur at very low energies to minimize damage to the substrate. The implant resets the damaged interface; following a 900°C anneal, the bottom silicon layer is crystallized once again. SPE and CVD can occur at the same time to finely control the thickness of the silicon layer (Reedy et al., 1983).

The use of SPE technique substantially reduces the defect density by acting on the compression strain, by suppressing the transition layer, and by reducing the density of microtwins, stacking faults, and interface traps. The surface can be enhanced with a subsequent silicon implant and epitaxial regrowth.

The electrical characteristics of the SOS wafer are greatly influenced by the fabrication process. In particular, lateral stress, the silicon–sapphire transition layer, and deep inhomogeneities play a dramatic role in altering the silicon energy bands. The twisting of the valence and conduction bands because of fabrication stresses give SOS peculiar transport characteristics. In the case of (100) SOS wafers, the compression stress causes the kx and ky ellipsoids to become more populated with electrons than the kz ellipsoid (which is normal to the silicon surface). Other effects on mobility are due to surface scattering when the doping profiles are high. In addition, self-heating is a

Process	NMOS mobility [cm^2/V s]	PMOS mobility [cm^2/V s]
HP 0.5 μm	431.50	145.38
AMI 0.5 μm	467.79	152.85
SOS 0.5 μm	230.00	134.43

TABLE 1.1   Table of extracted low-field mobility of P and N MOSFETs in three 0.5-μm processes offered by MOSIS (MOSIS, 1999)

common problem of SOS and SOI devices, due to the lower thermal conductivity of the substrates and thus the removal of excess heat. As a result, the effective mass of the electron increases and decreases for holes. The effects visible in the devices include increased hole mobility and reduced drift mobility for electrons (up to 25% as reported in Cristoloveanu and Li, 1995).

Table 1.1 offers an example of the different properties of SOS technology, reporting a comparison of low-field mobility of P and N MOSFETs in three 0.5-μm processes offered by MOSIS (MOSIS, 1999).

In Table 1.1, the NMOS values of low-field mobility are almost half of the values obtained in comparable bulk CMOS process with the same feature size of 0.5 μm.

1.5 Peregrine SOS Process Features

Peregrine offers several SOS processes (see Peregrine, 2003) with a variety of configuration. Some processes are optimized for digital logic, some offer RF quality passives, some are optimized for radiation-hard environments, and some for GHz operation and fast RF circuits. Throughout this book, we have used the Peregrine process named with code FC. The FC process has a single poly layer and three metal layers. The FC process's main advantage is that its third metal layer (metal-3 or metal-thick) is a thick metal layer (3 μm thick). All metal layers are made of aluminum. Metal insulator metal (MIM) capacitors are available between the second metal layer (metal-2) and the top metal layer. This allows the production of very-high-quality factor (Q) passive components such as capacitors and inductors. Capacitors and inductors have a high-quality factor because the effect of parasitic capacitance is very limited. Inductors made in the top metal-thick layer have low series resistance and, therefore, high-quality factor (Q). Resistors also benefits from the insulating substrate because the distributed parasitic capacitance is eliminated. In bulk CMOS processes, large resistors have very-low-quality factors and reduce bandwidth due to the distributed capacitance to the substrate.

The FC process's nominal operating voltage is 3.3 V, and the maximum allowed voltage is 3.6 V.

1.5.1 Local Isolation

Isolation of SOS devices is obtained with a technique usually referred as LOCOS (local isolation of silicon). Figure 1.1 shows a cross-section of a NMOS and PMOS in close vicinity. Between the two devices, the silicon thin film is etched away and a thick field oxide is grown to keep the transistors isolated.

The isolation is obtained with a dual-step LOCOS isolation (Cristoloveanu and Li, 1995). The first step consists of masking the active region of the silicon thin film with silicon nitride. Then field oxide is grown in the field region, and then etched away to produce these rounded active regions (Fig. 1.1). The silicon between the active regions is then removed with reactive ion etching (RIE) all the way to the sapphire substrate. A second local oxide deposition is then applied so that the isolation field oxide has the same thickness of the silicon thin film. The result is the LOCOS isolation in Fig. 1.1. This techniques reduces bird's beak and silicon filaments under the active region. These undesired effects lower the isolation of devices because of encroachment near the sapphire.

1.5.2 Fully Depleted Devices

In the Peregrine SOS, because of the thickness of the silicon thin film (100 nm), the depletion region under the gate reaches the sapphire substrate when the gate is at a zero potential. The devices obtained in this kind of silicon thin film are usually referred as *fully depleted* (FD). FD devices thus have a body that is completely void of majority carriers. This is in contrast to *partially depleted* (PD) devices, where a small layer of majority carriers is present in the body when the gate is biased at a zero potential. Figure 1.4 shows the difference between fully depleted and partially depleted MOSFET devices. Majority carriers in FD devices are absorbed by the source and drain regions.

Fully depleted devices have several advantages over partially depleted devices (Cristoloveanu and Li, 1995). First, the transconductance of the devices is higher because of the better gate control over the silicon thin film. Since the device is already fully depleted with the gate at zero potential, any increase of the gate voltage will induce conduction in the channel. In a PD device, the gate voltage first has to remove the excess majority carrier in the body before contributing to form a channel. Second, fully depleted devices do not suffer from floating body effect.

One adverse effect of the accumulation of majority carriers in the device body is that the threshold voltage becomes a function of the

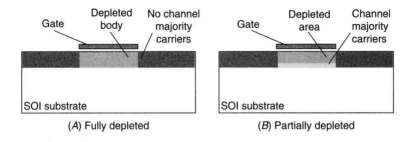

FIGURE 1.4   Comparison between a fully depleted (FD) and partially depleted (PD) SOI/SOS MOSFET. (*A*) FD devices have no majority carriers in the body. (*B*) PD devices have accumulation of majority carriers in the body, which affects the threshold voltage.

gate voltage. This is typical in PD devices, but appears also in FD devices at high-drain voltages. This effect is not desired because a threshold is not uniquely defined in these conditions. In addition, a floating body gives rise to parasitic bipolar and hot-electron effects. These effects can give rise to a kink effect in the drain current of the device, as will be explained in Chap. 2.

1.6   MOS Devices in the Peregrine SOS Process

In this section, we introduce Peregrine SOS MOS devices in the 0.5-μm FC process. Peregrine MOS devices are small silicon islands fabricated above the sapphire substrate, isolated by LOCOS field oxide. The minimum gate length for Peregrine SOS devices is 0.25 μm in the newest GA process and 0.5 μm in the original FC process. Gate length increments are multiple of 0.1 μm.

Figure 1.1 (middle) shows a cross-section of the NMOS and the PMOS of a CMOS inverter in SOS. The inverter does not require a well, and therefore the devices can be closely spaced for high density. The thin layer of silicon where the device channel resides has a nominal thickness of 100 nm. Gate and metal layers are deposited on top of the thin silicon layer. NMOS and PMOS devices are differentiated by the silicon implant PLDD and NLDD and the gate polysilicon implant.

There are three available types of implant for both NMOS and PMOS devices. P-channel devices can be implanted with regular threshold (RP), low threshold (PL), and intrinsic (IP) implants. Having both depletion (low or zero threshold) and accumulation mode (higher

MOS	Description	MOS Type	V_T (V)	Application
RN	Regular V_t	N-channel	0.65	Digital, low-leakage
NL	Low V_t	N-channel	0.14	High-perform. digital, RF
IN	Intrinsic V_t	N-channel	−0.2	High-perform. digital, RF
RP	Regular V_t	P-channel	−0.6	Digital, low-leakage
PL	Low V_t	P-channel	−0.25	High-perform. digital, RF
IP	Intrinsic V_t	P-channel	−0.03	High-perform. digital, RF

TABLE 1.2 Summary table of Peregrine SOS transistors types, threshold voltages, and typical applications.

threshold) is one of the advantages of the isolation properties of the silicon on sapphire technology. In bulk CMOS, accumulation devices cannot be obtained without separate wells because of the lack of isolation in the commonly shared substrate.

N-channel devices can be implanted with regular threshold (RN), low threshold (NL), and intrinsic (IN) implants. Regular-threshold devices are the accumulation devices typical of bulk CMOS; their threshold voltage is targeted to about 0.7 V. Low-threshold devices are accumulation devices with a voltage threshold of about 0.3 V. Intrinsic devices are depletion MOS devices, with a voltage threshold of about 0 V. The minimum-size intrinsic devices can produce μA level currents with no bias voltage. A summary of the transistor types, their thresholds, and their typical usage is given in Table 1.2.

For digital applications, the main tradeoff between transistor types is the magnitude of leakage drain current with the gate voltage set to zero versus the switching speed. Another tradeoff is low-voltage operation versus switching speed, especially for lower-threshold devices. Figures 1.5 and 1.6 show measured transistor characteristics for the different N-channel and P-channel transistors, respectively. Looking at Fig. 1.5 for N-channel devices, it is evident that at zero-gate voltage, regular threshold devices offer only leakage subthreshold currents on the order of less than 1 pA. At zero-gate voltage, low-threshold devices offer tenths of μA of subthreshold current, while an intrinsic device gives hundreds of μA, since the transistor is turned on and the drain current is substantial. The same kind of behavior is visible in Fig. 1.6, but the leakage levels of the P-channel devices are lower than the N-channel ones because of the difference in doping implants. For analog application, the different threshold can be used for a large variety of circuit topologies. Low-threshold transistors allow stacking of devices at low operational voltages. Intrinsic devices can act as

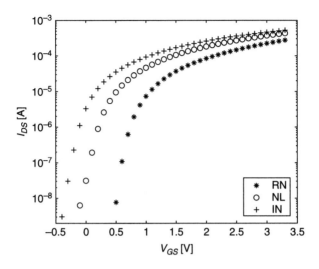

FIGURE 1.5 Measured drain current of 5 × 5 μm N-channel SOS devices: regular threshold (star), low-threshold (circle), and intrinsic (plus) with the drain voltage V_{DS} set at 3 V.

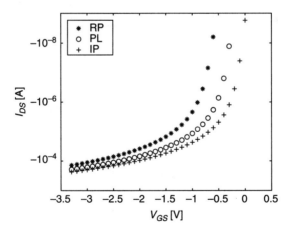

FIGURE 1.6 Measured drain current of 5 × 5 μm P-channel SOS devices: regular threshold (star), low-threshold (circle), and intrinsic (plus) with the drain voltage V_{DS} set at 3 V.

unbiased current sources or can operate at very low supply voltages. A few examples of circuit topologies using Peregrine SOS transistors will be given throughout the entire book and in the chapters that follow.

In summary, the negative impact of lower-threshold voltage transistors is a higher standby current for CMOS application. The benefits of lower-threshold transistors include:

- Improved transition frequency (f_t) due to increased carrier mobility as a result of lower channel doping
- Improved digital switching speed, especially at lower power supply voltages
- Low or zero headroom reduction across transistors
- Reduced kink effect in the N-channel transistors, since the lower threshold transistors have depleted bodies

The kink effect degrades the device characteristics for analog applications; it will be discussed in Chap. 2. Lower-threshold transistors are recommended for high-speed, RF, and analog applications. RN and RP transistors are recommended for digital logic and low-leakage applications.

Compared to bulk CMOS devices, SOS MOSFETs have the advantage of much reduced body effect, and lower short and narrow channel effects (Colinge, 1997). The back-gate effect in SOS devices is virtually zero, so no modulation of the threshold is due to the substrate bias. A short-channel effect can change the threshold voltage by up to four times in bulk CMOS, but less than 50% in SOS devices. Drain-induced barrier lowering, or DIBL, is also lower in SOS MOSFETs as compared to bulk CMOS devices. All these effects are an important feature of SOS and SOI devices. A larger current drive capability can be obtained when body effects are removed, giving SOS devices an additional advantage in speed when compared to bulk CMOS MOSFETs.

1.7 SOS MOS Characteristics for Analog Design

Device characteristics are a very important starting point for state-of-the-art analog design in SOS. Such characteristics can be used to evaluate the best operating region for each transistor in the circuit, and to assess the constant (DC) current levels at each voltage setting. The characteristics also give visual insight to high-order differential parameters, such as the small signal models and the device gain.

We provide here a set of device characteristics collected from SOS MOS devices of various types. We measured above- and below-threshold characteristics, as well as transconductance parameters. Additional characteristics include device current leakage and matching data.

1.7.1 Transistor Transconductance Parameter

The transconductance of a MOS transistor quantifies the drain current variation with a gate-source voltage variation while keeping the drain-source voltage constant. This is an important parameter for the design of analog circuits because it defines a gain profile for each device. In addition, the MOS transistor is a natural transconductor because of the dependence of the input gate voltage on the output drain current. The MOS intrinsic transconductance g_m is defined in Eq. (1.1), where V_{GS} and V_{DS} are the DC bias voltages.

$$g_m = \left[\frac{\partial i_{DS}}{\partial v_{GS}} \right]_{V_{GS}, V_{DS}=const} \tag{1.1}$$

A first-order model for the hand calculation of g_m can be obtained in the three regions of operation: subthreshold, ohmic, and saturation. Each model approximates the transconductance in each region of operation, reducing the complexity of the formulation. Equations (1.2) and (1.3) are the simplified analytical form of g_m, respectively, in the ohmic region and in the saturation region. These equations are derived from Eq. (1.1) and the drain current relation in each operating region.

$$g_m = \mu_0 C_{OX} \frac{W}{L} V_{DS} \tag{1.2}$$

$$g_m = \mu_0 C_{OX} \frac{W}{L} (V_{GS} - V_{th}) \tag{1.3}$$

In Eqs. (1.2) and (1.3), V_{th} is the gate voltage threshold, μ_0 is the carrier mobility, and C_{OX} is the gate capacitance. In weak inversion, MOS transconductance is given by Eq. (1.4).

$$g_m = I_D \frac{k}{V_T} \tag{1.4}$$

where V_T is the thermal voltage, k is the MOS gate capacitance (C_{OX}) divided by itself plus the substrate or device body (C_{dep}), as from Eq. (1.5).

$$k = \frac{C_{OX}}{C_{OX} + C_{dep}} \tag{1.5}$$

Normalized transconductance is given by the ratio of g_m and the drain current (g_m/I_D). In the subthreshold, the normalized transconductance is given by Eq. (1.6).

$$\frac{g_m}{I_D} = \frac{k}{V_T} \tag{1.6}$$

Normalizing the transconductance makes it clear to determine which region of the MOSFET offers the maximum gain as a proportion of the drain current. This is particularly useful for low-power design, where it is important to both reduce the drain current and maximize the transconductance gain. The normalized transconductance of a MOS reaches its maximum value in the subthreshold region of weak inversion, where the MOS shows an exponential relationship between gate voltage and drain current. In weak inversion, the MOSFET the normalized transconductance reaches the theoretical limit of a BJT device, or a transconductance of $1/V_T$ ($g_m = k/V_T$).

The maximum theoretical normalized transconductance in the MOS transistor equals the BJT case if k equals 1. In this case, the normalized transconductance maximum is just above 40 [V^{-1}] at room temperature. Note that, by measuring the transistor's normalized transconductance with precision, it is possible to estimate the MOS transistor's k parameter.

Equations (1.2) and (1.4) can be used for above-threshold design of a first-order circuit topology, but they fail to help in the low-voltage region of weak inversion. Also, they do not take into account short- and narrow-channel effects and channel nonuniformities.

A set of measured data of SOS MOSFETs can help the designer identify operational region and the expected current gain of the device. We measured SOS MOSFET DC characteristics and computed the transconductance. We measured the relationship between the drain current and the gate voltage of the device. We computed the transconductance g_m by numerical methods on the MOS raw characteristics. In order to precisely compute first-order derivative of the MOS characteristics these have to be measured accurately. The measurements of the drain current were taken with increments of gate voltage of $1\ mV$ and with a drain voltage V_{DS} of 3 V. This allows numerical methods of integration to have the required computational precision without overestimating the transconductance values. This data was smoothed with a running average filter to reduce noise.

Figure 1.7 is a plot of the normalized transconductance of a set of RN SOS MOS transistors. The device sizes were (width, length): 1.2×0.8, 2.5×2.5, 25×5, and 25×10 μm. Figures 1.9 and 1.11 are a plot of the normalized transconductance of a set of, respectively, NL and IN SOS MOS transistors. The device sizes were (width, length): 1.2×0.5, 2.5×2.5, 25×5, and 25×10 μm. Equivalently, Figs. 1.8, 1.10, and 1.12 are plots of the normalized transconductance of a set of P-channel SOS MOS transistors, respectively, RP, PL, and IP. The device sizes were

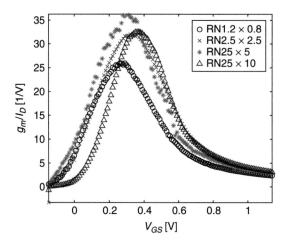

FIGURE 1.7 Measured transconductance parameters for a set of SOS RN MOSFETs. The legend correlates different curves to transistor sizes. The drain voltage is set at 3 V.

(width, length): 1.2×0.5, 2.5×2.5, 25×5, and 25×10 μm. The RN MOSFETs in Fig. 1.7 show a peak normalized transconductance at a gate voltage of 0.2–0.4 V. The RP MOSFETs in Fig. 1.8 show a peak normalized transconductance at a gate voltage of 0.2–0.4 V. The NL

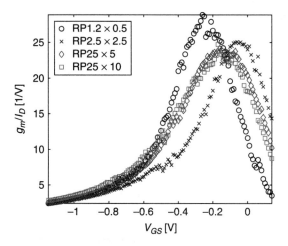

FIGURE 1.8 Measured transconductance parameters for a set of SOS RP MOSFETs. The legend correlates different curves to transistor sizes. The drain voltage is set at 3 V.

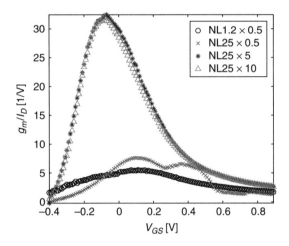

FIGURE 1.9 Measured transconductance parameters for a set of SOS NL MOSFETs. The legend correlates different curves to transistor sizes. The drain voltage is set at 3 V.

MOSFETs in Fig. 1.9 show a peak normalized transconductance at a gate voltage of 0 V. The PL MOSFETs in Fig. 1.10 show a peak normalized transconductance at a gate voltage of 0 V. The IN MOSFETs in Fig. 1.11 show a peak normalized transconductance at a gate voltage

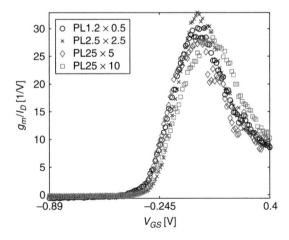

FIGURE 1.10 Measured transconductance parameters for a set of SOS PL MOSFETs. The legend correlates different curves to transistor sizes. The drain voltage is set at 3 V.

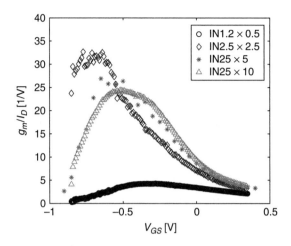

FIGURE 1.11 Measured transconductance parameters for a set of SOS IN MOSFETs. The legend correlates different curves to transistor sizes. The drain voltage is set at 3 V.

of −0.6 V. The IP MOSFETs in Fig. 1.12 show a peak normalized transconductance at a gate voltage of −0.2 V.

Notice that the normalized transconductance data presented here is not flat in the subthreshold region as suggested by the theory and, in

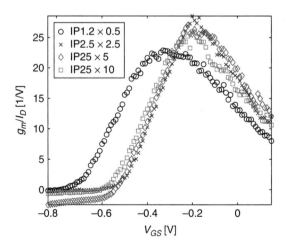

FIGURE 1.12 Measured transconductance parameters for a set of SOS IP MOSFETs. The legend correlates different curves to transistor sizes. The drain voltage is set at 3 V.

particular, Eq. (1.6). Besides, the limitation of the leakage currents of each device, which reduce the normalized transconductance to zero at very-low gate voltages, the subthreshold current become progressively higher than their derivative, resulting in a decrease of the normalized transconductance at low gate voltages. The normalized transconductance data presented here are from multiple devices in different dies, so some mismatch in the response needs to be accounted for. These curve nevertheless provide substantial insight on which devices can be used for specific applications. One final comment is that in Figs. 1.9 and 1.11, the shorter-length devices result in much lower transconductances than larger-length devices. This is due to the short-channel effect and is expected. Intrinsic and low-threshold transistors also have larger leakage current, so their normalized transconductance is reduced.

1.7.2 Modeling the Subthreshold Region

For the design of low-power low-voltage circuits, the subthreshold characteristics of the MOS devices is of great importance and interest. This is because the low current produced by the device in a weak inversion keeps the circuit operating with limited supply current and very-low power consumption. The supply voltage can also be reduced, since the drain currents are small, and this allows the entire circuit to run on an even more stringent power budget. In addition, the exponential characteristic of the subthreshold regime in a MOS is very desirable in many linear applications (Sinencio and Andreou, 1998). Trade-offs and advantages of subthreshold operation of SOS MOS will be discussed in this and the following chapters.

Like conventional bulk MOS transistors, SOS MOSFETs exhibit drain currents that, below threshold, are an exponential function of the gate voltage for a given DC voltage sweep. There are various device physical models to describe this relationship, but we chose to model the drain current with Eq. (1.7).

$$I_D = \frac{W}{L} I_0 e^{\frac{k V_{GS}}{V_T}} e^{\frac{(1-k)V_{BS}}{V_T}} \tag{1.7}$$

W and L are the width and length of the active region of the transistor, V_{GS} refers to the gate-source voltage (equal to the potential of the gate since the source is tied to the ground potential), and V_T is the thermal voltage, equal to kT/q. Since the bulk terminal is not present in SOS MOSFETs, the V_{BS} voltage can be thought as the voltage between the device's body and the source of the device (assume that it is connected to the ground voltage reference). In SOS, the body of the device is a floating one. If the device is fully depleted, the body voltage will not be influenced by the other terminals and the dependance on

V_{BS} can be included in the base current term I_0. Therefore, Eq. (1.8) is used to quantify the exponential behavior only as a function of the gate voltage.

$$I_D = \frac{W}{L} I_0 e^{\frac{k V_{GS}}{V_T}} \tag{1.8}$$

Using Eq. (1.8), we have extracted the parameters I_0 and k for each of the different transistor types and for all transistor sizes. These parameters were obtained fitting the data collected on a variety of MOS transistors with different sizes and for N- and P-channel devices. The drain voltage was kept at 3 V and the gate voltage was varied. The device sizes were (width, length): 1.2×0.8 (N-channel), 1.2×0.5 (P-channel), 2.5×2.5, 25×0.8 (N-channel), 25×0.5 (the P-channel), 25×5, and 25×10 μm. The resulting optimized extracted data is given in Table 1.3 for N-channel devices and in Table 1.4 for P-channel

Transistor	Estimated		Optimized	
	I_{n0}	K_n	I_{n0}	K_n
RN 25 × 10	1.0000E-15	9.9000E-01	8.3206E-16	9.9390E-01
RN 2.5 × 2.5	1.7000E-14	8.2000E-01	1.7178E-14	8.1884E-01
RN 1.2 × 0.8	3.0000E-12	7.0000E-01	4.5128E-12	6.6688E-01
RN 25 × 0.8	3.0000E-15	1.3000E+00	4.3345E-15	1.2704E+00
RN 25 × 5	2.0000E-14	9.9000E-01	4.3735E-14	9.3998E-01
NL 1.2 × 0.5	1.0500E-06	1.3000E-01	1.0611E-06	1.3076E-01
NL 25 × 0.5	9.8000E-07	1.7000E-01	1.0317E-06	1.6274E-01
NL 2.5 × 2.5	6.1000E-08	3.7000E-01	5.6224E-08	3.6635E-01
NL 25 × 10	6.4000E-09	7.9000E-01	6.6946E-09	7.1629E-01
NL 25 × 5	5.0000E-09	3.7000E-01	4.6304E-09	7.7419E-01
IN 2.5 × 2.5	5.3000E-04	5.5000E-01	5.7891E-04	5.5900E-01
IN 1.2 × 0.5	7.9000E-06	1.0500E-01	7.9352E-06	1.0606E-01
IN 25 × 0.5	Erratic Data	Erratic Data	Erratic Data	Erratic Data
IN 25 × 5	1.0500E-04	5.6000E-01	1.5119E-04	5.7836E-01
IN 25 × 10	3.4500E-05	5.8000E-01	3.0982E-05	5.7500E-01

TABLE 1.3 Table of estimated and optimized parameters of the exponential model for SOS subthreshold region of an N-channel device.

Transistor	Estimated		Optimized	
	I_{p0}	K_p	I_{p0}	K_p
RP 2.5 × 2.5	5.8000E-14	−8.0000E-01	7.4495E-14	−7.8301E-01
RP 1.2 × 0.5	1.2000E-13	−8.7000E-01	8.1470E-14	−9.4876E-01
RP 25 × 0.5	1.6000E-10	−5.0000E-01	1.7456E-10	−5.1002E-01
RP 25 × 5	2.0000E-14	7.0000E-01	3.8419E-14	−6.7009E-01
RP 25 × 10	3.0000E-14	−6.1000E-01	2.3773E-14	−6.3195E-01
PL 2.5 × 2.5	2.5000E-10	−8.7000E-01	2.5353E-10	−9.0032E-01
PL 25 × 0.5	8.9000E-08	−4.2000E-01	9.2562E-08	−4.0084E-01
PL 25 × 10	8.8000E-11	−7.8000E-01	6.8887E-11	−8.2088E-01
PL 25 × 5	1.5000E-10	−7.5000E-01	1.5954E-10	−7.7758E-01
PL 1.2 × 0.5	2.5000E-10	−8.7000E-01	2.4350E-10	−8.8315E-01
IP 2.5 × 2.5	3.3000E-08	7.7000E-01	4.2455E-08	−8.0336E-01
IP 1.2 × 0.5	3.5000E-07	−6.5000E-01	4.2641E-07	−6.7694E-01
IP 25 × 0.5	Erratic Data	Erratic Data	Erratic Data	Erratic Data
IP 25 × 5	2.8000E-08	−7.7000E-01	1.8893E-08	−7.3508E-01
IP 25 × 10	5.5000E-08	−7.8000E-01	4.8059E-08	−7.4945E-01

TABLE 1.4 Table of estimated and optimized parameters of the exponential model for SOS subthreshold region of a P-channel device.

devices. *Estimated* values are the input to the error minimization function and *optimized* values are the resulting final values. Once the correct interval bounds were chosen, the optimized parameter values were in almost all cases very close to the estimated values. Note that two transistors, the IN and IP 25 × 0.5 μm, returned nonreliable erratic data and will have to be reexamined in future data collection.

To conclude the collection of subthreshold parameters, we present in the following figures some examples of plots using optimized parameter values for all six types of transistors with dimensions $W = 25$ μm, $L = 10$ μm. The drain voltage was kept at 3 V and the gate voltage was varied. The figures represent data in circles and the extracted exponential model superimposed to validate it. Figure 1.13 shows an RN device, Fig. 1.14 shows an RP device, Fig. 1.15 shows an NL device, Fig. 1.16 shows PL device, Fig. 1.17 shows an IN device, and Fig. 1.18 shows an IP device.

Both regular-threshold N and P devices offer an exponential regime with a dynamic range of four decades, from approximately 10^{-11} to 10^{-7}A.

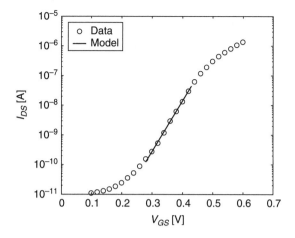

FIGURE 1.13   Subthreshold characteristic of a 25 × 10 μm RN SOS device (circles) with superimposed model (line) extracting the exponential region.

We measured the characteristics of a two regular threshold 1.2 × 5 μm RN and RP transistors with different body potentials. I applied the body potential via a fourth terminal connected to the device's body. While observing the subthreshold characteristics (Fig. 1.19) of

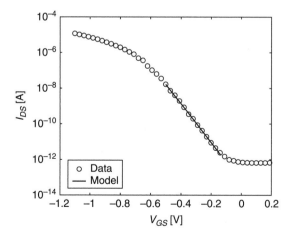

FIGURE 1.14   Subthreshold characteristic of a 25 × 10 μm RP SOS device (circles) with superimposed model (line) extracting the exponential region.

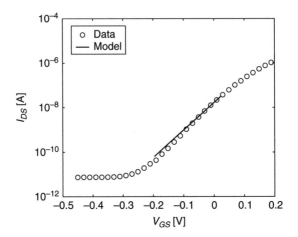

FIGURE 1.15 Subthreshold characteristic of a 25 × 10 μm NL SOS device (circles) with superimposed model (line) extracting the exponential region.

the regular threshold NMOS device biased with drain voltage V_{DS} set at 3 V, we noticed a significant difference in the behavior of the device with bulk floating or at the potential of the source with respect to the case of V_{BS} lower than the source voltage. The curves for V_{BS} floating or zero were significantly different from the curves for $V_{BS} = -0.5$ or -1.

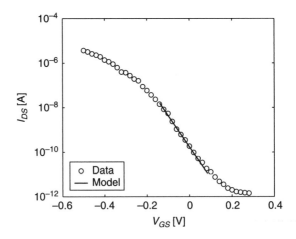

FIGURE 1.16 Subthreshold characteristic of a 25 × 10 μm PL SOS device (circles) with superimposed model (line) extracting the exponential region.

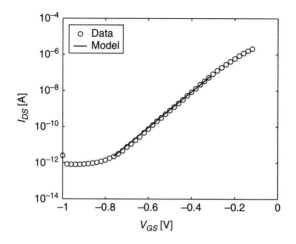

FIGURE 1.17 Subthreshold characteristic of a 25 × 10 μm IN SOS device (circles) with superimposed model (line) extracting the exponential region.

We can explain this phenomenon by recognizing that the device is not fully depleted when the bulk is left floating or no negative voltage is applied. Therefore, the NMOS presents nonzero charge in the bulk region, which lowers the subthreshold slope k. In fact, from Eq. (1.5),

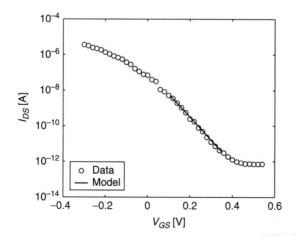

FIGURE 1.18 Subthreshold characteristic of a 25 × 10 μm IP SOS device (circles) with superimposed model (line) extracting the exponential region.

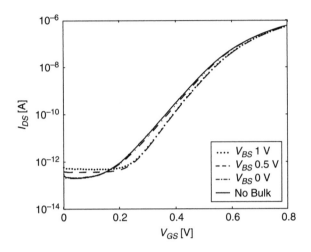

FIGURE 1.19 Subthreshold characteristics of an RN 1.2 × 5 μm MOSFET with $V_{DS} = 3$ V.

higher C_{dep}, caused by residing charges in the bulk, can therefore affect the subthreshold characteristics of the device.

When, on the other hand, an external negative voltage is applied to the bulk, the NMOS device changes characteristics and the subthreshold slope increases. Table 1.5 reports the value of k and the intercept current for all the measured and modeled characteristics.

It is of importance to notice that both types of devices, when fully depleted, do not follow the subthreshold model given in Eq. (1.7). There is no dependency on the bulk voltage V_{BS} on the characteristics. The model thus reduces to the simpler formulation of Eq. (1.8).

This model is valid for V_{BS} zero and floating and also for a large (sufficient to produce full depletion) V_{BS}, given the appropriate set of modeling parameters.

The PMOS transistor, biased at V_{DS} equal to 2 V, also exhibits symmetrical behavior (Fig. 1.20), where the subthreshold characteristics for V_{BS} zero or floating have evident dissimilarities from the ones with a positive applied V_{BS}. Similarly to the NMOS case of Fig. 1.19, the behavior is attributable to the partial depletion of its bulk region.

On the other hand, the PMOS transistor biased at V_{DS} equal to 3 V (Fig. 1.20) showed a dependency on the bulk voltage V_{BS} even for high applied voltages. The dependency affects the subthreshold slope, since the bulk capacitor divider still has an impact at high V_{BS}. An evaluation of the phenomenon will be addressed in future work.

SOS Transistor Type	RN	RP
Transistor Dimensions	1.2 × 5 μm	1.2 × 5 μm
Subthreshold Slope $k(V_{BS} = 0, V_{DS} = 3)$	0.78	0.68
Subthreshold Slope k (estimated at $V_{BS} = 0$)	0.73	0.73
Depletion Depth d_{dep}	72 nm	72 nm
Intercept Current $I_0[A]$ $(V_{BS} = 0, V_{DS} = 3)$	9.0×10^{-15}	-4.5×10^{-16}
Subthreshold Slope $k(V_{BS}$ floating, $V_{DS} = 3)$	0.78	0.68
Intercept Current $I_0[A]$ $(V_{BS}$ floating, $V_{DS} = 3)$	9.7×10^{-15}	-6.0×10^{-16}
Subthreshold Slope $k(V_{BS} = 0.5$ V, $V_{DS} = 2$ V)	0.88	0.84
Intercept Current $I_0[A]$ $(V_{BS} = 0.5$ V, $V_{DS} = 2$ V)	5.5×10^{-16}	-4.0×10^{-19}
Subthreshold Slope k (estimated at $V_{BS} = 0.5$)	0.76	0.76
Depletion Depth d_{dep}	95 nm	95 nm
Operating Temperature	293 K	293 K

TABLE 1.5 Summary of weak inversion parameters for SOS transistors with bulk floating or tied to a fixed voltage.

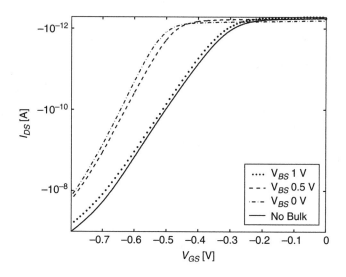

FIGURE 1.20 Subthreshold characteristics of an RP 1.2 × 5 μm MOSFET with $V_{DS} = 3$ V.

Transistor (W × L)	Leakage Current [A]
RN 25 × 5	5.82E-12
RN 25 × 10	9.45E-12
RN 2.5 × 2.5	5.10E-13
RN 1.2 × 0.8	4.34E-11
RN 25 × 0.8	3.53E-12
NL 1.2 × 0.5	5.20E-07
NL 25 × 0.5	1.68E-05
NL 25 × 5	7.98E-12
NL 25 × 10	7.32E-12
NL 2.5 × 2.5	2.16E-11
IN 1.2 × 0.5	1.14E-06
IN 25 × 0.5*	1.12E-04
IN 25 × 5	9.70E-12
IN 25 × 10	7.90E-13
IN 2.5 × 2.5	1.80E-12

TABLE 1.6   Leakage currents for N-channel Peregrine
SOS transistors. The star indicates possible erratic data.

1.7.3 Leakage Data

We measured room-temperature leakage data for SOS N- and P- channel devices. The leakage currents are the minimum currents measured
in the weak inversion region. For different types of devices, the minimum current occurs at different gate voltages, as can be seen in Figs. 1.5
and 1.6. The device sizes were (width, length): 1.2 × 0.8 (N-channel,
for the P-channel 0.5), 2.5 × 2.5, 25 × 0.8 (N-channel, for the P-channel
0.5), 25 × 5, and 25 × 10 μm. Table 1.6 reports a summary of the leakage
currents for N-channel devices and Table 1.7 reports a summary of the
leakage currents for P-channel devices.

1.7.4 Above-threshold Statistical
Characteristics

We computed first- and second-order statistical data for some of the
SOS devices operated in strong inversion. The statistical data was collected across many dies, as opposed to being collected in many devices
on the same die. The following statistical data was collected from a

Transistor (W × L)	Leakage Current [A]
RP 25 × 5	−1.17E-12
RP 25 × 10	−6.50E-13
RP 2.5 × 2.5	−5.20E-13
RP 1.2 × 0.5	−6.10E-13
RP 25 × 0.5	−6.75E-12
PL 1.2 × 0.5	−5.90E-13
PL 25 × 0.5	−1.21E-09
PL 25 × 5	−1.39E-12
PL 25 × 10	−1.39E-12
PL 2.5 × 2.5	−9.30E-13
IP 1.2 × 0.5	−6.50E-13
IP 25 × 0.5*	−6.08E-07
IP 25 × 5	−1.67E-12
IP 25 × 10	−7.10E-13
IP 2.5 × 2.5	−5.50E-13

TABLE 1.7 Leakage currents for P-channel Peregrine
SOS transistors. The star indicates possible erratic data.

set of RNs (Fig. 1.21), RP (Fig. 1.22), IN (Fig. 1.23) and IP (Fig. 1.24) devices. The device size was 10 × 2 μm, measured in the dark with a drain voltage of 3 V. This data was collected using a set of 5 dies. Each solid curve is the average of five devices, and the dashed curves immediately above and below the solid line are the maximum and minimum in the set.

Intrinsic devices give more drain current for the same gate voltage. An IN MOS produces about twice as much drain current at the same gate voltage as the RN MOS. An IP MOS produces about 1.5 times the current of the RP MOS.

Note that in Figs. 1.21 and 1.23 the kink effect is visible for high drain to source voltages. The kink effect is an undesired effect due to the floating body of the devices and hot-electron effects. The kink effect will be examined with greater detail in Chap. 2. The kink effect is visible only in the regular-threshold devices and is virtually not present for low-threshold and intrinsic devices, as can be seen from Figs. 1.22 and 1.24.

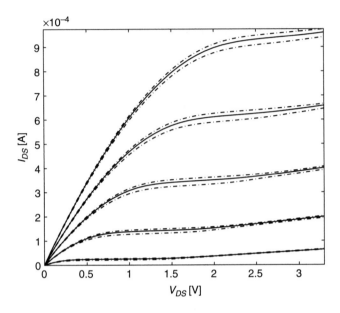

FIGURE 1.21   Above-threshold characteristics of a 10×2 μm Peregrine SOS RN MOSFET. Solid line is the average of five dies, the dotted lines are the maximum and minimum in the set. The drain voltage was set to 3 V.

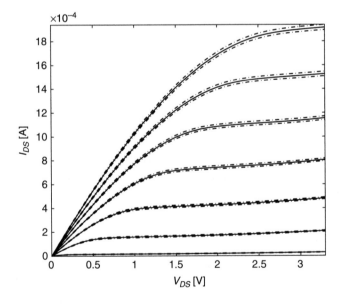

FIGURE 1.22   Above-threshold characteristics of a 10×2 μm Peregrine SOS IN MOSFET. Solid line is the average of five dies, the dotted lines are the maximum and minimum in the set. The drain voltage was set to 3 V.

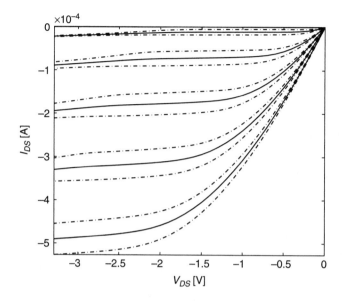

FIGURE 1.23   Above-threshold characteristics of a 10 × 2 μm Peregrine SOS RP MOSFET. Solid line is the average of five dies, the dotted lines are the maximum and minimum in the set. The drain voltage was set to 3 V.

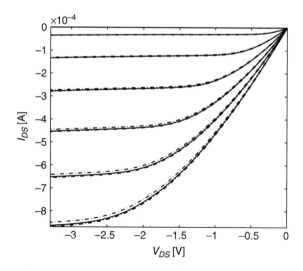

FIGURE 1.24   Above-threshold characteristics of a 10 × 2 μm Peregrine SOS IP MOSFET. Solid line is the average of five dies, the dotted lines are the maximum and minimum in the set. The drain voltage was set to 3 V.

It is also interesting to notice that the intrinsic devices statistics are better than the regular-threshold counterparts. This is due to the fact that regular-threshold devices undergo an additional implant for the threshold that introduces more variability in the device characteristics.

Our data set is not extensive enough to allow us to infer high-order statistical properties of the devices. Since the devices are on different wafers, the matching properties of devices are limited to interdies.

1.7.5 Subthreshold Statistical Characteristics

We computed first- and second-order statistical data for some of the SOS devices operated in weak inversion. The statistical data was collected across many dies as opposed to being collected in many devices of the same die. The following statistical data was collected from a set of RNs (Fig. 1.25), IN (Fig. 1.26), RPs (Fig. 1.27), and IP (Fig. 1.28) devices). The devices' size was 10×2 μm, measured in the dark with a drain voltage of 3 V. This data was collected using a set of five dies. Each solid curve is the average of five devices and the dashed curve immediately above the solid line are the maximum and minimum in the set.

Our data set is not very extensive and does not allow us to infer high-order statistical properties of the devices. Since the devices are

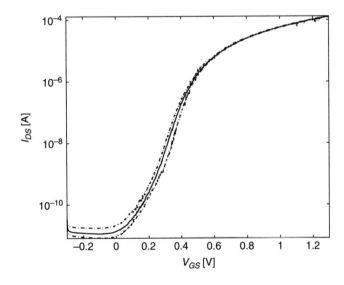

FIGURE 1.25 Subthreshold characteristics of a 10×2 μm Peregrine SOS RN MOSFET. Solid line is the average of five dies, the dotted lines are the maximum and minimum in the set. The drain voltage was set to 3 V.

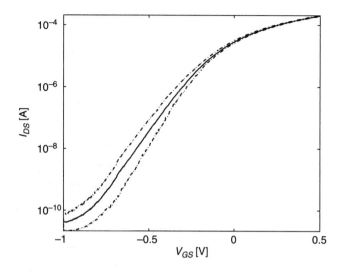

FIGURE 1.26   Subthreshold characteristics of a 10×2 μm Peregrine SOS IN MOSFET. Solid line is the average of five dies, the dotted lines are the maximum and minimum in the set. The drain voltage was set to 3 V.

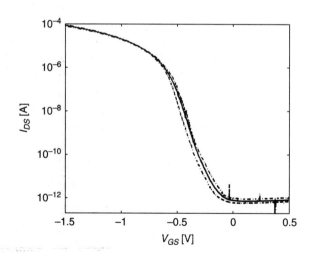

FIGURE 1.27   Subthreshold characteristics of a 10×2 μm Peregrine SOS RP MOSFET. Solid line is the average of five dies, the dotted lines are the maximum and minimum in the set. The drain voltage was set to 3 V.

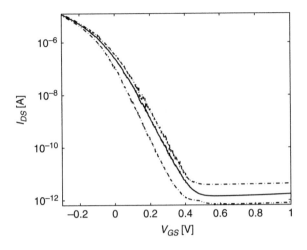

FIGURE 1.28   Subthreshold characteristics of a 10×2 μm Peregrine SOS IP
MOSFET. Solid line is the average of five dies, the dotted lines are the
maximum and minimum in the set. The drain voltage was set to 3 V.

on different wafers, the matching properties of devices are limited to
interdie.

1.7.6 Comparison Between Data and Simulation of SOS MOSFETs

The subthreshold weak-inversion region of MOS transistor has progressively attracted more and more interest in the research and design
community since the late 1980s. The low currents and voltage needed
to operate subthreshold circuits make them a key player in today's
market, which is focused on low-power portable devices. The exponential characteristics of MOS operated in subthreshold also facilitate
the implementation of several linear or log-linear circuits, many of
which became popular for analog filters and analog current-mode
functionals. But the subthreshold region of MOS transistors is often
not very well characterized in today's simulators, which are often targeted to massive digital circuits or standard analog blocks. An incon-
sistency between simulations and fabricated devices can be disastrous
for low-power low-voltage analog designers. An error in the exponential region causes output currents errors of orders of magnitude, because of the high gain of the device. While the agreement between data
and simulation of above-threshold SOS MOSFETs is quite adequate,
simulation of the MOS characteristics in the subthreshold region can
have substantial differences with measured data.

The reason for this disagreement is twofold. First, the model BSIM3.3 level 49 used for simulations can model subthreshold in regular bulk processes, but only to a first-order approximation. BSIM3.3 is useful to calculate leakage currents but not detailed enough for very precise design of low-voltage analog circuits. In particular, the model fails to be precise at very-low current levels and near the threshold. BSIM3.3 is a piecewise model patching different region's models into a single function. This is usually done by means of a spline-based smoothing function, allowing us to calculate the derivatives around discontinuity patch points. The constant shrinking of the features of fabrication processes has also complicated modeling efforts with short- and narrow-channel effects, self-heating, and channel nonlinearities due to fabrication and doping profiles.

Second, BSIM3.3 is designed to model bulk devices and was successively extended to include modeling SOI devices. In case of Peregrine SOS, the model simply ties the bulk voltage of the BSIM3.3 model to a very negative voltage (-15 V for SOS). It is clear that this is a gross approximation that does not take into account the physical differences of the device. Fully depleted devices need to model a floating body; the BSIM3.3 model uses a drive bulk node. Additional errors derive from the barrier lowering of diodes, again due to the floating body and impact ionization. The lack of a substrate contact induces majority carrier accumulations in the lower device body. This effect is not modeled in BSIM3.3.

A possible solution to this problem is to use separate simulators for circuits operating in weak inversion. A good single-region model of the MOS transistor is the EKV model (EPFL, 2004). The EPFL-EKV MOSFET model equations for simulation will be used and described to model SOS MOS transistors in Chap. 2.

A comparison of simulation and measured data in the subthreshold region is provided in Figs. 1.29 to 1.32. Figure 1.29 shows a superimposed plot of both data (collected on a 10×2 μm Peregrine SOS RN MOSFET with V_{DS} set to 3 V) and simulation performed with Cadence Analog Artist using the Spectre simulator and Peregrine SOS BSIM3.3 level 49 provided in the 2004 design kit version 2.8. The data are plotted using circles, the simulation is the solid line. Figure 1.30 shows the same comparison for a 10×2 μm Peregrine SOS IN MOSFET with V_{DS} set to 3 V. Figures 1.31 and 1.32 show the same comparison respectively for a 10×2 μm Peregrine SOS RP and IP MOSFET, again with V_{DS} set to 3 V.

The model fails to equate to the measured data for N-channel transistors. For the RN transistor in Fig. 1.29, the model fails to predict not only the slope of the data but also the shape; more importantly, the region around the threshold is miscalculated by about two orders of magnitude. In case of the IN transistor in Fig. 1.30, the slope of the

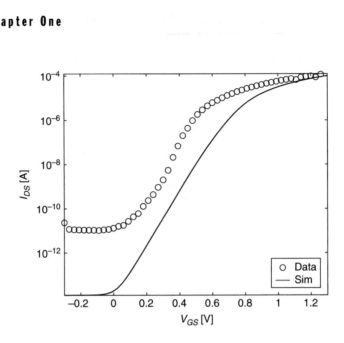

FIGURE 1.29 Comparison of data (circles) and simulation (solid line) of the subthreshold characteristics of a 10×2 μm Peregrine SOS RN MOSFET with $V_{DS} = 3$ V.

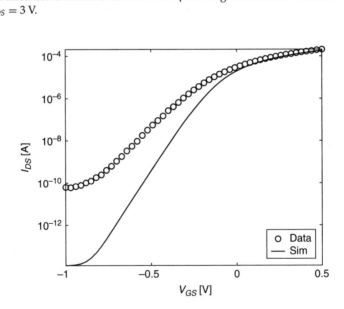

FIGURE 1.30 Comparison of data (circles) and simulation (solid line) of the subthreshold characteristics of a 10×2 μm Peregrine SOS IN MOSFET with $V_{DS} = 3$ V.

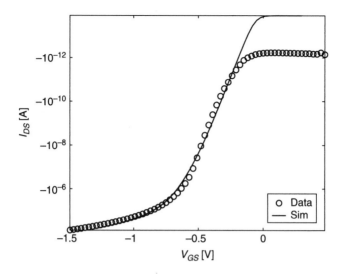

FIGURE 1.31 Comparison of data (circles) and simulation (solid line) of the subthreshold characteristics of a $10 \times 2 \ \mu m$ Peregrine SOS RP MOSFET with $V_{DS} = 3$ V.

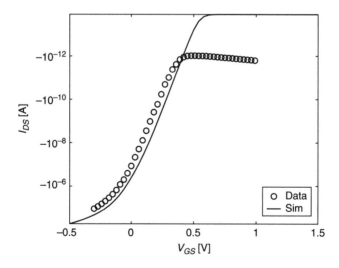

FIGURE 1.32 Comparison of data (circles) and simulation (solid line) of the subthreshold characteristics of a $10 \times 2 \ \mu m$ Peregrine SOS IP MOSFET with $V_{DS} = 3$ V.

curve is miscalculated, producing an error of four orders of magnitude (!) for low current levels.

It is clear that this not an acceptable model for the subthreshold region of a Peregrine SOS N-channel device. On a more positive note, the model of the subthreshold region of both RP and IP SOS devices is very precise and can reliably predict the current levels of the device simulated. Given this set of results, I encourage the designer to use P-channel devices wherever possible if the circuit under development may be influenced by such dramatic errors as we witness in the N-channel case.

1.8 Summary

In this chapter, we introduced SOS CMOS technology. We described the process flow and the main characteristics of the process.

Devices have been described and characterized, both in the weak and the strong inversion. Parameters in all regions of operation have been extracted. This parameter set is useful for the analog designer, as it summarizes the expected performance of the device in the SOS process.

We have discussed some preliminary statistical data and the matching of simulation with measured data in the weak inversion region of operation.

It is hoped that this will convince the reader that the SOS process is indeed the most thoughtful choice for the design of the low-power low-voltage high-performance circuits needed in today's mobile market.

CHAPTER 2

SOS MOSFET Modeling

2.1 Introduction

A commercially available SOS technology is a relatively new VLSI fabrication process and for this reason it has not been fully characterized, as has happened in the last 30 years for most bulk processes. While the models for MOSFETs transistors allow us to design with confidence, the physical dissimilarities in the fabrication of the SOS devices and the insulating substrate cause the SOS MOSFETs to be different from their bulk counterpart. While the devices demonstrate behavior resembling the typical current–voltage behavior of devices in a bulk process, the insulating substrate, the floating body, and the different thermal properties of the sapphire give rise to unconventional characteristics. These nonconventional characteristics have to be fully understood and described to make the design of high-performance circuits possible. The SOS process, with various threshold devices, is the perfect medium for the design of low-power, low-voltage, high-speed circuits. In order to take advantage of the benefits deriving from the technology and produce state-of-the-art circuits, the designer needs to have a good understanding of the advantages and disadvantages of the SOS technology and insight into device operation. This is the focus of this chapter, which offers a comparison of conventional and advanced modeling techniques for SOI processes, and the SOS process in particular.

We begin by reporting first-order models for the SOS MOSFETs in the strong inversion region of DC operation in Sec. 2.2. The model extracts basic device parameters that are very useful for hand calculation and help in the design of circuits by evaluating the current-to-voltage relationship of all typical devices. All models are compared to measured data for validation.

We then present a more advanced unified, closed-form analytical drain current model for partially and fully depleted SOS MOSFETs in Sec. 2.3. The analytical MOSFET model was developed using first-order principles of operation of the device and basic explanation of the physical constraints responsible for hot-carrier effects. The analytical model allows to obtain values of the channel field and to find closed-form solutions for the critical voltages, giving rise to kink effects and output nonlinearities. The reliability of the model was addressed and verified with experimental data. The work presented in Sec. 2.3 allows us to obtain a better understanding of the basic MOS devices in SOS and their limitations due to hot-carrier degeneration and other effects in SOS design.

In addition to giving the textbook model of the SOS transistor, we also provide modeling parameters by means of the EKV model equations in Sec. 2.4. Extracted parameters are useful for a unified single-region model of the SOS transistor. The hot-electron induced kink effect was also incorporated in the EKV model.

We report on a model of bipolar devices in Sec. 2.5. The model is validated against data collected from a fabricated SOS chip. Section 2.5.2 discusses possible application of the SOS process as a BiCMOS process, using a combination of circuit topologies employing available MOSFETs and BJT devices.

Finally, we report on the design and testing of four kinds of nonvolatile memories in the SOS process. We report on the test results of all four devices and demonstrate that MIM-based NMOS floating gate cells can be used to achieve a threshold shift of several volts and no significant decay of threshold voltage after more than hundreds of small retention tests.

2.2 A Model of the SOS MOSFET in Strong Inversion

This section focuses on models of the drain current of SOS MOSFETs operating above threshold. We measured the drain currents of several transistors with varying gate and drain voltages while keeping the source constant. We measured five P-channel and five N-channel transistors; their sizes (width, length) were 1.2×0.8, 2.5×2.5, 25×0.8, 25×5, and $25 \times 10\ \mu m$ for the N-channel transistors and 1.2×0.5, 2.5×2.5, 25×0.5, 25×5, and $25 \times 10\ \mu m$ for the P-channels transistors. There are regular-threshold, low-threshold, and intrinsic SOS for both N and P types, or six types total (RN, NL, IN, RP, PL, and IP, respectively). The sizes of the devices were chosen to represent the most typical transistors used in both digital and analog design. Minimum-size transistors are generally used for digital circuits and can show short- and narrow-channel effects in the drain current versus drain voltage. These effects are interesting from a modeling point of view for estimating the power consumption of large switching logic arrays.

Wide and narrow transistors are useful to model the DC characteristics of buffers and drivers. Large and wide-channel transistors are used for analog design, since their characteristics present less early voltage and nonlinear effects. It is particularly useful to model a square transistor because of its wide use in current sources. The DC characteristic of these transistors are very important for analog design.

2.2.1 Strong Inversion Model

Modeling on the drain current versus gate and drain voltages was conducted with the source tied to the ground. The first approach is an effective model of the transistor with velocity saturation and channel length modulation. It is able to predict with great precision the drain current versus gate and drain voltage relationship in SOS transistors (Rabaey, 1996; Serrano-Gotarredona et al., 1999; Karlsson and Jeppson, 1992).

This model was divided into two operational regions: the linear region for $V_{GS} - V_{TH} > V_{DS}$ and the saturation region for $V_{GS} - V_{TH} < V_{DS}$. The two regions and their respective models are given in Eq. (2.1).

$$I_{DS} = K_x \frac{W}{L} \frac{(V_{GS} - V_{th}) V_{DS} - \frac{V_{DS}^2}{2}}{1 + \theta (V_{GS} - V_{th})} (1 + \lambda V_{DS}) \quad linear$$

$$I_{DS} = \frac{K_x}{2} \frac{W}{L} \frac{(V_{GS} - V_{th})^2}{1 + \theta (V_{GS} - V_{th})} (1 + \lambda V_{DS}) \quad saturation$$

(2.1)

V_{TH} is the threshold voltage of the device in volts, K_x is the transconductance parameter in $[A/V^2]$, θ is the mobility reduction factor due to the vertical electric field, λ is the channel length modulation parameter. W and L are the transistor dimensions of length and width of the gate. I_{DS} is the drain current, V_{GS} is the gate-to-source voltage, and V_{DS} is the drain-to-source voltage.

2.2.2 Strong Inversion Model Results and Discussion

Tables 2.1 and 2.2 report all the measured data for each SOS transistor type. The first column indicates the transistor type and the parameter measured. The parameters previously indicated in the model Sec. 2.2.1 are V_{th}, K_x, λ, and θ. The rows just above the data indicate the size of the transistor. The sizes were (width, length) 1.2×0.8, 2.5×2.5, 25×0.8, 25×5, and 25×10 μm for the N-channel transistors and 1.2×0.5, 2.5×2.5, 25×0.5, 25×5, and 25×10 μm for the P-channel transistors. Transistors include regular-threshold, low-threshold, and intrinsic transistors for both P and N types, or six kinds of transistors

RN	1.2 × 0.8	1	2.5 × 2.5	1	25 × 5	3
Vth	0.45	7.01E-01	0.7	7.05E-01	7.00E-01	7.10E-01
KX	4.30E-05	5.31E-05	6.50E-05	8.14E-05	7.40E-05	9.35E-05
λ	0.03	4.87E-02	0.03	4.36E-02	3.00E-02	3.50E-02
θ		1.51E-01		1.57E-01		1.50E-01
RP	**1.2 × 0.5**	**1**	**2.5 × 2.5**	**1**	**25 × 5**	**1**
Vth	−0.5	−5.20E-01	−0.7	−6.73E-01	−0.7	−6.29E-01
KX	2.60E-05	3.33E-05	3.30E-05	3.58E-05	3.80E-05	4.03E-05
λ	0.04	−5.95E-02	0.035	−4.52E-02	0.03	−1.88E-02
θ		−1.50E-01		−7.20E-02		−4.06E-02
NL	**1.2 × 0.5**	**1**	**2.5 × 2.5**	**1**	**25 × 5**	**1**
Vth	0.28	−1.64E-01	0	5.16E-02	0.07	1.14E-01
KX	4.00E-05	5.11E-05	6.00E-05	8.65E-05	6.80E-05	8.38E-05
λ	0.01	−2.71E-02	0.05	2.01E-02	0.04	6.15E-02
θ		2.13E-01		1.37E-01		1.12E-01
PL	**1.2 × 0.5**	**2**	**2.5 × 2.5**	**1**	**25 × 5**	**1**
Vth	0	−7.75E-02	−0.28	−3.54E-01	−0.28	−3.33E-01
KX	4.30E-05	3.48E-05	3.30E-05	4.33E-05	4.10E-05	5.14E-05
λ	0.02	6.18E-03	0.03	−5.23E-02	0.03	−6.70E-02
θ		−1.28E-01		−1.30E-01		−1.44E-01
IN	**1.2 × 0.5**	**1**	**2.5 × 2.5**	**1**	**25 × 5**	**3**
Vth	0	−2.79E-01	0	−3.65E-01	0.00E+00	−3.26E-01
KX	3.90E-05	5.97E-05	8.00E-05	9.40E-05	1.00E-04	1.14E-04
λ	0.02	−2.08E-02	0.05	−1.33E-02	2.00E-02	9.19E-02
θ		2.58E-01		1.18E-01		1.02E-01
IP	**1.2 × 0.5**	**2**	**2.5 × 2.5**	**1**	**25 × 5**	**1**
Vth	0	1.70E-01	0	−2.03E-01	0	−1.71E-01
KX	3.10E-05	3.98E-05	4.20E-05	5.46E-05	4.00E-05	6.57E-05
λ	0.03	9.78E-03	0.015	−9.81E-02	0.015	−4.04E-02
θ		−1.43E-01		−1.95E-01		−2.32E-01

TABLE 2.1 Summary table of Peregrine SOS transistors. For each transistor, the first column is the model with no velocity saturation. The second column is the model with velocity saturation. The numbers at the right side of the transistor sizes refer to the die number. MOSIS run T09B-SOI05.

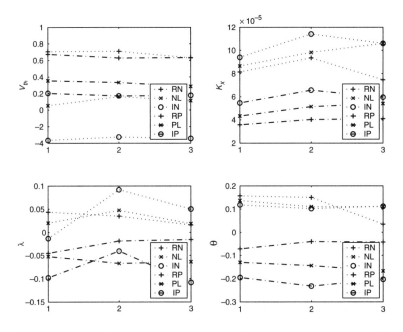

FIGURE 2.1 SOS transistors parameters: a comparative table. Three geometries of transistors (width, length) have been tested: 2.5 × 2.5 μm, 25 × 5 μm, and 25 × 10 μm. These sizes are respectively labeled as 1, 2, and 3 in the figure.

total. The number immediately following the transistor size is the identifier of the particular die where the transistor has been measured. For strong inversion modeling, three dies have been measured, but only some of the results are reported in Table 2.1. For each transistor, the first column is for the model with no velocity saturation (θ is zero). The second column is the model with velocity saturation. Both models have been verified with Matlab scripts computing a minimum of a multi-variable function. The Matlab routine that performs this is *fminsearch*.

Figure 2.1 compares the extracted transistor parameters from Table 2.1. We measured, extracted, and compared three geometries of transistors: 2.5 × 2.5 μm, 25 × 5 μm, and 25 × 10 μm. These sizes are, respectively, labeled 1, 2, and 3 in Fig. 2.1. The chosen geometries are a more reliable representative of the transistor since shorter-channel transistors, such as 1.2 × 0.5 μm and 25 × 0.5 μm, presented second-order velocity saturation and self-heating effects that are beyond the scope of this first modeling effort.

RN	25 × 10	2	25 × 0.8	1
Vth	0.7	6.30E-01	0.45	4.68E-01
KX	7.40E-05	7.46E-05	4.30E-05	6.91E-05
λ	0.03	1.61E-02	0.03	1.03E-02
θ		3.50E-02		2.50E-01
RP	**25 × 10**	**1**	**25 × 0.5**	**1**
Vth	−0.7	−6.33E-01	−0.2	−2.81E-01
KX	3.80E-05	4.10E-05	2.40E-05	4.73E-05
λ	0.03	−1.57E-02	0.03	−3.49E-03
θ		−4.28E-02		−2.97E-01
NL	**25 × 10**	**1**	**25 × 0.5**	**1**
Vth	0.16	1.12E-01	0	−1.59E-01
KX	8.20E-05	1.06E-04	4.00E-05	7.09E-05
λ	0.04	1.87E-02	0.04	−2.91E-02
θ		1.11E-01		3.14E-01
PL	**PL25×10**	**1**	**PL25×0.5**	**1**
Vth	−0.28	−2.86E-01	0	6.80E-02
KX	4.40E-05	5.40E-05	2.70E-05	4.50E-05
λ	0.03	−6.36E-02	0.06	−9.42E-03
θ		−1.67E-01		−3.20E-01
IN	**IN 25 × 10**	**1**	**IN 25 × 0.5**	**weird 1**
Vth	0	−3.43E-01	0	−5.45E-01
KX	1.00E-04	1.06E-04	4.50E-05	8.83E-05
λ	0.02	5.03E-02	0.07	−2.06E-02
θ		1.12E-01		3.92E-01
IP	**IP 25 × 10**	**1**	**IP 25 × 0.5**	
Vth	0	−1.78E-01	0	3.41E-01
KX	4.30E-05	5.96E-05	3.10E-05	4.70E-05
λ	0.015	−1.08E-01	0.03	3.85E-02
θ		−2.03E-01		−2.22E-01

TABLE 2.2 Summary table of Peregrine SOS transistors. For each transistor, the first column is the model with no velocity saturation. The second column is the model with velocity saturation. The numbers at the right side of the transistor sizes refer to the die number. MOSIS run T09B-SOI05.

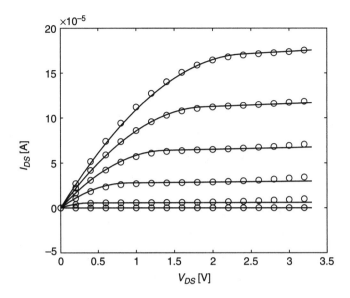

FIGURE 2.2 Modeling results of a 2.5 × 2.5 μm RN SOS with the parameters of Table 2.1. The drain current model (solid line) is plotted over the data (circles) collected. Data for $V_{GS} = 0 - 3.3V$ in 0.5 V steps.

We were able to estimate reliably the threshold voltage and the transconductance of all types of SOS transistors. Parameters λ, θ are also extracted and present more variability depending on the geometry of the transistors.

While reporting all of the results for the modeling effort might be excessive, we report in Fig. 2.2 to 2.7 the modeling results of 2.5 × 2.5 μm SOS MOSFETs. The transistors modeled are type RN in Fig. 2.2, type NL in Fig. 2.3, type IN in Fig. 2.4, type RP in Fig. 2.5, and type PL in Fig. 2.6 and IP in Fig. 2.7. The models predict perfectly the drain current of the device as a function of the drain voltage and gate voltage. In the RN case, (Fig. 2.2) there is some evidence of a hot-electron kink effect visible at high-drain voltages and low-gate voltages. The kink effect is not visible in the other N-channel transistors, NL (Fig. 2.3) and IN (Fig. 2.4). A visible kink effect is also present in the RP transistor (Fig. 2.5) and is still present in the PL transistor (Fig. 2.6). The IP transistor model (Fig. 2.7) is, by contrast, kink free.

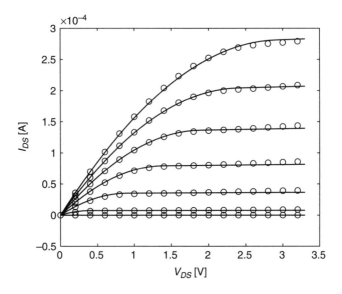

FIGURE 2.3   Modeling results of a 2.5 × 2.5 μm NL SOS with the parameters of Table 2.1. The drain current model (solid line) is plotted over the data (circles) collected. Data for $V_{GS} = 0 - 3.3$ V in 0.5 V steps.

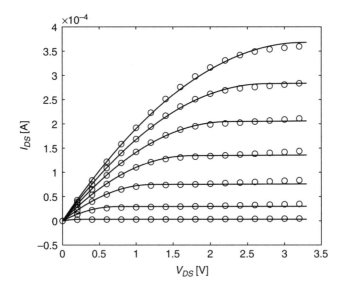

FIGURE 2.4   Modeling results of a 2.5 × 2.5 μm IN SOS with the parameters of Table 2.1. The drain current model (solid line) is plotted over the data (circles) collected. Data for $V_{GS} = 0 - 3.3$ V in 0.5 V steps.

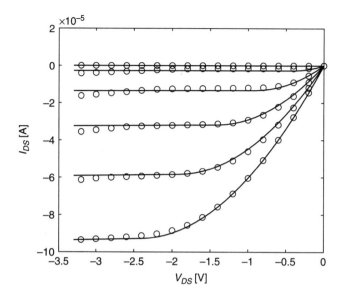

FIGURE 2.5 Modeling results of a 2.5 × 2.5 μm RP SOS with the parameters of Table 2.1. The drain current model (solid line) is plotted over the data (circles) collected. Data for $V_{GS} = 0 - 3.3$ V in 0.5 V steps.

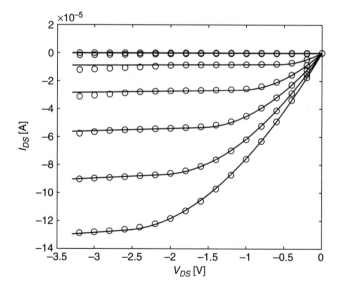

FIGURE 2.6 Modeling results of a 2.5 × 2.5 μm PL SOS with the parameters of Table 2.1. The drain current model (solid line) is plotted over the data (circles) collected. Data for $V_{GS} = 0 - 3.3$ V in 0.5 V steps.

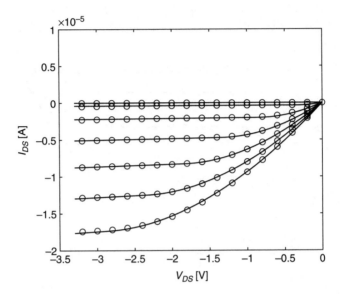

FIGURE 2.7   Modeling results of a 2.5 × 2.5 μm IP SOS with the parameters of Table 2.1. The drain current model (solid line) is plotted over the data (circles) collected. Data for $V_{GS} = 0 - 3.3$ V in 0.5 V steps.

2.3   Hot Electron Effects in SOS MOSFETs

At high values of drain-to-source voltage (V_{DS}) and in saturation, SOI devices typically produce an undesirable deviation from classical MOSFET operation in the saturation region. The drain current exhibits a change in slope for higher V_{DS} due to hot carrier effects (Reggiani, 1985; Lundstrom, 2000). Hot-electron effects generally limit the performance of analog circuits designed in silicon-on-insulator (SOI) technologies. These effects are responsible for many floating-body behaviors and irregularities in SOI devices. The buildup of majority carriers in the device body is mainly responsible for the onset of these effects. A typical drain current affected by hot-electron kink effect is clearly visible in Fig. 2.8. This figure is from real data collected from an RN SOS device of size 1.2 × 5 μm with V_{GS} set to 1 V.

The kinklike behavior in the SOS devices is due to the acceleration of channel carriers. As V_{DS} rises, the pinch-off of the channel causes carriers to group around the source region (Figs. 2.9 and 2.10). In this figure, we suppose the device is pinched-off, with V_{DS} $V_{GS} - V_{th}$. This accumulation generates a high electric field between the channel and the drain of the device. If the field surpasses a critical value, carriers

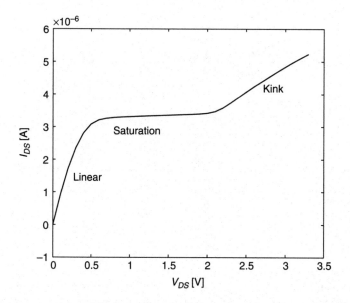

FIGURE 2.8 Explanatory drawing of a MOSFET presenting kink effect for $V_{DS} > 2.2$ V. This figure is from real data collected on a RN 1.2×5 μm SOS device with $V_{GS} = 1$ V.

from the channel itself are accelerated to the point at which they provoke impact ionization as they reach the drain region (Reggiani, 1985).

Impact ionization generates two carriers from a single one at the drain of the device. Generated minority carriers move rapidly up the drain, while their counterpart, the majority carriers, flow into the floating body of the device and accumulate.

Colinge, 1997 reports that P-channel devices are generally free from kink effects since the number of hole-electron pairs generated by energetic holes is much lower than the quantity induced by electrons. Data collected from our sample transistors shows the opposite behavior, where PMOS devices have a much more noticeable kink effect. This is due to the enhancement of the carrier lifetime to match the mobility of P-channel devices to the mobility of N-channel devices. This enhancement simplifies output current matching between devices with different channel and facilitates digital design. In addition, note that the defects of the SOS fabrication process enhance hole mobility at the expense of electron mobility (refer to Chap. 1).

Figures 2.11 to 2.14 show typical characteristics measured for regular-threshold MOSFETs designed in the SOS process by Peregrine

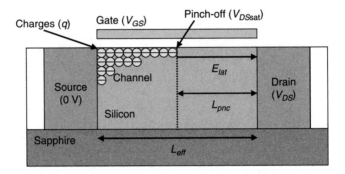

FIGURE 2.9 Electric field and pinching of the channel in a SOS MOSFET. At high V_{DS}, L_{pnc} becomes longer, as a result, the electric field E_{lat} can become large enough to accelerate minority carriers from the channel.

Semiconductors. These MOSFETs were designed with a fourth terminal making contact to the device's body (similar to a bulk contact). Fig. 2.11 is for a 1.2×5 μm NMOS, device, and Fig. 2.12 for a 5×5 μm NMOS device. Figure 2.13 is for a 1.2×5 μm PMOS, device, and Fig. 2.14 for a 5×5 μm PMOS device. The characteristics for both floating-bulk devices (data represented by a plus) and bulk-terminated devices (circle) are also shown. Clearly, in both gender characteristics of the regular-threshold devices, a visible presence of hot-carrier degeneration is present. The PMOS devices exhibit a larger effect, due to the different doping profiles required to match their carrier mobility to that of the NMOS devices.

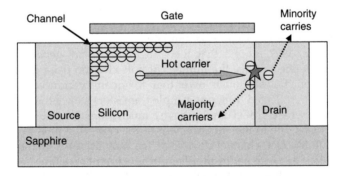

FIGURE 2.10 Hot-carrier generation due to impact at the drain region of a SOS MOSFET–presenting kink effect. Minority carries are swept away by the electric field at the drain, but majority carriers inject in the device body, modulating the threshold.

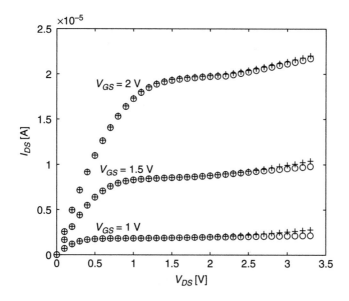

FIGURE 2.11   Hot-carrier effect in regular threshold 1.2 × 5 μm SOS NMOS device with bulk contact (circle) and with bulk floating (plus).

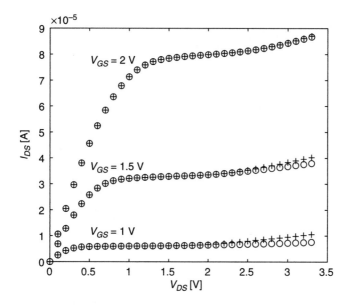

FIGURE 2.12   Hot-carrier effect in regular threshold 5 × 5 μm SOS NMOS device with bulk contact (circle) and with bulk floating (plus).

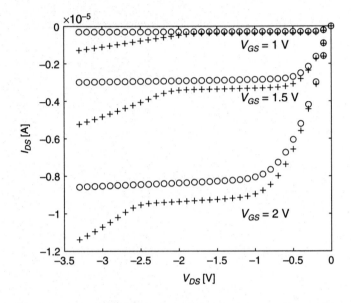

FIGURE 2.13 Hot-carrier effect in regular threshold 1.2 × 5 μm SOS PMOS device with bulk contact (circle) and with bulk floating (plus).

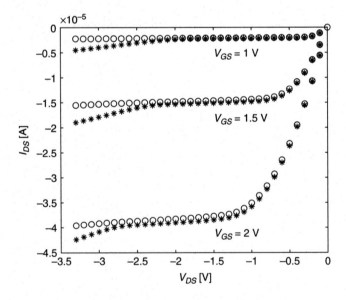

FIGURE 2.14 Hot-carrier effect in regular threshold 5 × 5 μm SOS PMOS device with bulk contact (circle) and with bulk floating (plus).

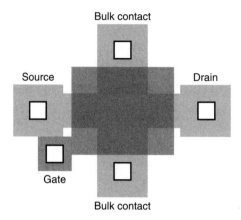

FIGURE 2.15   Layout of the four-terminal SOS transistors. Two body contacts are provided on both sides of the channel. The body contact is a P+ contact for an NMOS and N+ for PMOS. The contact forms a small barrier between N+/P+ and the device body N/P doping.

Providing a contact to the floating body of the device can eliminate the kink effect. This contact allows the draining of majority carriers out of the device body, provided that the resistance of the body is not too large. In this case, in fact, body contact is not completely sufficient to eliminate kink effects. The design of devices furnished with a bulk contact alleviates the kink effect and parasitic lateral bipolar effects typically found in SOI devices. The devices under test did not show parasitic effects due to the presence of a parasitic bipolar. The transistors are provided with two body contacts, on both sides of the channel (Fig. 2.15). This geometry, as shown the data (circle) of Figs. 2.11 to 2.14, suppresses undesired behaviors in the transistors characteristics. Notice that the body contact is much more effective in the PMOS than the NMOS transistor. When a contact is added to the body of the device, because of the difference between the doping of the contact and the body itself, a barrier is formed between the P (body) and P + (body contact) layers. This barrier impedes the drainage of the minority carriers in the NMOS case, but not in the PMOS case. Lowering the body voltage below the ground potential would have helped the NMOS body contact to drain minority carriers, but the body potential was only lowered to the 0V ground potential in Figs. 2.11 and 2.12.

In thin-film SOI devices (Colinge, 1997), the full depletion gives the silicon below the gate an almost infinite resistance, which renders the body contact completely ineffective. Since the devices we tested and measured still produced a significant reduction of the kink effect by using bulk contacts, we conclude that either the devices are not fully depleted or their backsides are not in accumulation.

For the devices to be fully depleted, the maximum depletion region with zero applied voltages, given by $x_{d\,max}$, must be higher than the thickness of the silicon thin film computed in Eq. (2.2).

$$x_{d\,max} = \sqrt{\frac{4\epsilon_{si}\psi_S}{q\,N_A}} \tag{2.2}$$

In our case, $x_{d\,max}$ is about 70 nm at threshold, while the silicon locos thickness is 100 nm (Peregrine Semiconductors FC process). An exception to the full depletion is some thin accumulation or inversion layers at the bottom of the locos in the presence of a large negative or positive bias at the back gate (not provided in the SOI process under consideration). The back interface must be fully depleted as well, otherwise the hot-carrier effect will manifest.

Kink behavior is generally not modeled and not included in circuit simulators. While the kink is not a problem for digital design, where the increase in current makes the circuit faster, for analog circuits it is certainly a worrisome behavior that has to be taken into account. In the design of analog circuits the additional uncertainty in the drain current of the devices makes the design more prone to undesired output mismatch. In conventional circuit blocks such as a differential pair configuration, for example, the mismatch would be a nonlinear output characteristic. In a mirror or current source, the kink effect would change the output current and the steady-state operational point of the circuit in which it is meant to operate. Radio frequency circuits are also affected by higher noise due to the kink effect (Rozeau et al., 2000). Finally, notice that operating the analog circuit with reduced power supplies (1.5–2.5 V) eliminates the kink problems in a high-gate voltage drive, in addition to saving power consumption.

2.3.1 Standard Hot-Electron Effects Models
Several sources in the literature (Sze, 1981; Howes et al., 1990, to cite a few) claim that, in SOI substrates, the kink effect due to hot-carrier degeneration can be attributed to avalanche phenomena. Specifically, the current increase in the drain due to impact ionization generates a positive feedback mechanism. When the energy of the hot carriers is high, avalanche phenomena occur, and the current is multiplied by an exponential factor M. This parameter models the exponential behavior of the device under the influence of high-energy carriers from

the channel. The value of M can be computed by assessing the positive feedback current generated by impact ionization and using Eq. (2.3).

$$M = 1 + A(V_{DS} - V_{DSk}) \exp \left(-\frac{B}{V_{DS} - V_{DSk}} \right) \qquad (2.3)$$

$$I_{DSk} = MI_{DS}, \quad V_{DS} \geq V_{DSk} \qquad (2.4)$$

Equation 2.4 gives an expression for the drain current in the kink region as a function of the drain current obtained with standard models. The multiplicative factor is linearly proportional to the drain-to-source voltage for small V_{DS} after the critical V_{DSk}, and then it assumes an exponential dependence. In the model, A and B are process-dependent parameters, while I_{DS} is the first-order model for a transistor in saturation, defined in Eq. (2.5).

$$I_{DS} = \frac{K}{2}(V_{GS} - V_{th})^2 [1 + \lambda(V_{DS} - V_{DSsat})], \quad V_{DS} \geq V_{GS} - V_{th}$$

$$(2.5)$$

When V_{DS} surpasses the critical value V_{DSk}, the drain current is multiplied by factor M, which models the effect of avalanche in the impact ionization.

The exponential model just described is not able to replicate the kink behavior observed in the SOS devices under consideration. Figures 2.16 and 2.17 show a tentative fit of the conventional avalanche model applied to the set of collected data. The kink effect clearly has a largely linear dependence on the drain-to-source voltage V_{DS}, rather than the exponential behavior described in the literature. This applies for drain voltages V_{DS} in the recommended operational range of up to 3.3 V; for higher-drain voltages an exponential runaway current is expected but has not been recorded in our set of data. This may signify that no avalanche current multiplication occurs in a SOS RP MOSFET. On the other hand, the hot carriers clearly generate a linear increase in the drain current, proportional to their number. Degenerative avalanche effects definitely occur for higher V_{DS} near breakdown levels.

Conventional exponential models rely on the activation of the parasitic BJT transistor. SOS MOSFET characteristics do not present exponential behavior in the kink region. This is due to the inefficiency of the parasitic BJT transistor, which is unable to generate high levels of collector current. This can be seen in Fig. 2.18, where the base current measured from a four-terminal device is not higher than a few hundred picoamperes for any gate voltage (in a 5×1.2 μm RP MOSFET). The base (or bulk) current for an equivalent NMOS device was on the order of a few nanoamperes. Due to the long base, efficiency is low.

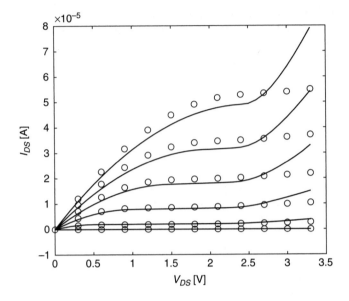

FIGURE 2.16 Standard SOI MOSFET avalanche kink model (solid line) compared to the device data (circles) for a 1.2×5 μm NMOS. Data for $V_{GS} = 0 - 3.3$ V in 0.5 V steps.

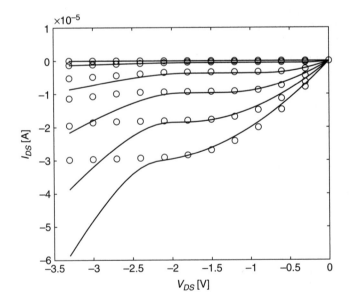

FIGURE 2.17 Standard SOI MOSFET avalanche kink model (solid line) compared to the device data (circles) for a 1.2×5 μm PMOS. Data for $V_{GS} = 0 - 3.3$ V in 0.5 V steps.

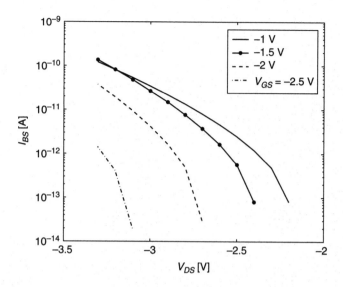

FIGURE 2.18   Bulk current in a four-terminal 1.2 × 5 μm RP transistor in the SOS process.

We extracted the parasitic BJT parameters α_0 to be 0.743 and β about 2 for a four-terminal device 5 × 1.2 μm RP reported in Fig. 2.20 and Sec. 2.5. Similarly, we can show that in Fig. 2.19 the collector current level never exceed the drain current value of the MOSFET.

Additional models show a linear kink for a short V_{DS} range, followed by a second saturation. This is due to hot-electron injection current to the base (body of the device) raising the body voltage and forward biasing the source to the body diode. This in turn increases the recombination current and reduces the body voltage. This mechanism acts as a negative feedback on the drain current and reduces the effect of the kink. SOS MOSFETs do not present similar behavior.

2.3.2 Hot Carrier Generated Kink Effect Modeling of SOS Devices

The generation of hot carriers occurs when the electric field inside the pinched channel surpasses a critical value $E_{lat,c}$. Values of this field are between 10^6 V/m (Rabaey, 1996) and $2 \cdot 10^7$ V/m (Sze, 1990). The value of the critical voltage V_{DSk}, at which the kink effect arises, is determined by the distance between the pinched-off channel and the drain (L_{pnc}). When reaching this value of V_{DS}, electrons from the channel can accelerate through the length of the depleted channel to the

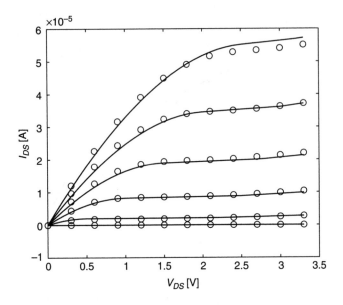

FIGURE 2.19 The proposed SOS MOSFET kink model (solid line) compared to the device data (circles) for a RN 1.2 × 5 μm device. Data for $V_{GS} = 0 - 3.3$ V in 0.5 V steps.

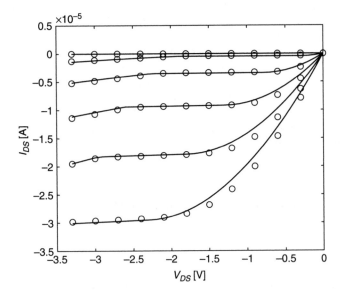

FIGURE 2.20 The proposed SOS MOSFET kink model (solid line) compared to the device data (circles) for a RP 1.2 × 5 μm device. Data for $V_{GS} = 0 - 3.3$ V in 0.5 V steps.

drain area and acquire sufficient energy to provoke impact ionization [Eq. (2.6)].

$$\left|\overrightarrow{E_{lat,c}}\right| \propto \frac{V_{DSk}}{L_{pnck}} \tag{2.6}$$

V_{DSk}, similar to V_{DSsat}, is influenced by V_{GS}. In fact, the channel profile is flattened by higher V_{GS}, when V_{DS} is held constant. When this happens, at higher V_{GS}, the effective critical length L_{effk} is attained later than predicted. This is also explained by Eq. (2.7).

$$\left|\overrightarrow{E_{lat}}\right| \propto \frac{V_{DS} - V_{GS}}{L_{pnc}} \tag{2.7}$$

If V_{GS} increases, so must V_{DS} in order to maintain the same L_{pnc} and the same electric field. In other terms, V_{GS} subtracts from V_{DS} and reduces the pinched-off region of the channel. Fig. 2.9 shows a vertical section of the channel. The channel is influenced by the vertical field imposed by V_{GS} and a longitudinal field imposed by V_{DS}. Note that the lateral electric field is close to zero in the channel, since it is acting as a conductor. The electric field manifests between the channel and the drain in the pinched-off region. Although the lateral electric field is high for the short pinched-off length L_{pcn}, the distance is not enough to accelerate the carrier above the impact ionization energy. Therefore, both a high electric field and a high L_{pcn} are necessary to give carriers sufficient energy to provoke collision ionization at the drain interface. Fig. 2.10 represents the channel during impact ionization near the drain diffusion.

We can also express the energy E_J of a carrier immersed into the pinched-off device electric field using Eq. (2.8).

$$E_J = q \int_l \overrightarrow{E}_{lat} dl \tag{2.8}$$

Carriers in the channel must surpass a critical energy threshold in order to provoke impact ionization at the drain; we call this critical energy E_{Jc}. E_{Jc} is the value of E_J, imposed by some limit values of the pinched-off length l and the lateral electric field. We can, therefore, express the critical drain voltage V_{DSk} voltage, since the voltage scalar is itself defined as the work the field has to perform on a particle to move it from one region to another [Eq. (2.9)].

$$V_{DSk} = \frac{E_{Jc}}{q} \tag{2.9}$$

A value of the electric field inside the channel can be computed by solving the Gauss equation in a small box contour region. On a first approximation, since solving the integral for E_j is very involved and brings excessive complication into the model, we can approximate the value of the lateral electric field at the point where the channel is pinched off with Eq. (2.10).

$$E_{Jc} = \frac{-q \, (2 - \eta_1) \cdot V_{DSsat}}{L_{eff} - L_{pnc}} \tag{2.10}$$

η_1 is a fitting parameter that takes into consideration the error in the approximation of the field and its distribution between the source and drain. E_{Jc} is the lateral electric field from the pinch-off point of the channel just below the gate oxide to the upper right corner of the drain region. L_{eff} and L_{pnc} are, respectively, the effective channel length and the pinched-off length of the channel. V_{DSk} can, therefore, be expressed using Eq. (2.11).

$$V_{DSk} = \int_l \vec{E}_{lat} dl = q \int_l \frac{-(2 - \eta_1) \cdot V_{DSsat}}{L_{eff} - L_{pnc}} dl \tag{2.11}$$

The result of integration is expressed in Eq. (2.12).

$$V_{DSk} = V_{DSk0} + \frac{-(2 - \eta_1) \cdot V_{DSsat}}{L_{eff} - L_{pnc}} L_{pnc} \tag{2.12}$$

Instead of solving a complex set of nonlinear equations for the calculation of L_{pnc} (see Fig. 2.9) as done in the literature (Kuo and Su, 1998), note that the pinched-off region of the channel is simply the depletion region of the drain-to-channel reverse-biased diode. In fact, the pinched-off area is the region that, by definition, contains no inverted minority carriers. While, again, a bidimensional computation of the depletion region would require numerical methodologies, the only important value for our analysis is the value at the interface with the silicon dioxide, where the electric field is confined by the boundary conditions imposed by MOSFET geometry. The depletion region between drain and channel can be computed using Eq. (2.13).

$$L_{pnc} = \sqrt{\frac{\epsilon_{si}}{q} \frac{N_A + N_D}{N_A N_D} (\phi_F - V)} \tag{2.13}$$

Where $V = V_{DS} - V_{GS} + V_{TH}$ is the difference between drain voltage and the pinch-off point of the channel, $\epsilon_{si}, N_A, N_D, \phi_F$ are the permittivity of silicon, the drain and channel implants, and the Fermi level of the drain-to-channel diode, respectively. While the potential at the drain is V_{DS}, the pinch-off voltage is V_{DSsat}. Imposing $V_{DSsat} = V_{GS} - V_{TH}$, a final value for V_{DSk} can be calculated.

V_{DSk0} is the critical voltage for V_{DS} just above threshold. Its value is -1.7 V for a RP and 2.0 V for an RN MOSFET in the SOS process.

The dependence of the critical voltage on V_{GS} can be explained by thinking about the distribution of charge in the channel. Submicron devices in a thin silicon film will have a nonlinear channel shape because of the reduced dimensions. A higher value of V_{GS} (at a fixed V_{DS}) will attract more charges toward the oxide interface, thus effectively stretching the channel toward the source. This in turn decreases the pinched-off portion of the channel. The electric field seen by channel's electrons will then be lower than the critical value. Because of this, at higher V_{GS} the kink appears at higher V_{DS}. That is, V_{DSk} increases with V_{GS}.

When the lateral electric field surpasses the critical value, the carriers accelerated by such fields are able to create impact ionization close to the drain region. This mechanism increases the drain current, and at the same time generates a majority carrier current to the bulk. The minority current influences the bulk potential, effectively modulating the transistor threshold (Cristoloveanu and Li, 1995).

The majority carriers injected into the body can forward bias the source body diode, creating the current I_{BS} in Eq. (2.14).

$$I_{BS} = I_{S0}\left(e^{\left(\frac{V_{BS}}{V_T}\right)} - 1\right) \tag{2.14}$$

The drain current [Eq. (2.15] of the SOS MOSFET can then be calculated using the following relation, given that the device is in saturation and the drain-to-source voltage exceeds the critical voltage V_{DSk}.

$$I_{DS} = \frac{K}{2}(V_{GS} - V_{thk})^2[1 + \lambda(V_{DS} - V_{DSsat})], \; V_{DS} \geq V_{GS} - V_{th},$$

$$V_{DS} \geq V_{DSk} \tag{2.15}$$

Note that the slope of the drain current changes once the drain-to-source voltage reaches the critical value V_{DSk}. This effect is the kink behavior itself. Given the linearity of the kink drain current with the drain-to-source voltage, we can model the kink itself as a change in the threshold of the device. This change is linearly proportional to the drain-to-source voltage V_{DS}, as seen in Eq. (2.16).

$$V_{thk} = V_{th0} - \xi(V_{DS} - V_{DSk}) \tag{2.16}$$

The threshold of the transistor is the shifted down by a quantity proportional to the V_{DS} voltage. The parameter ξ was estimated empirically to be of 0.03 for an NMOS transistor and 0.09 for a PMOS transistor, both 1.2×5 μm channel size. Table 2.3 reports a list of the parameters used in the model and the process variables involved in the above equations.

When contact to the bulk is provided, modulation of the bulk by injection of majority carriers cannot occur. In fact, when the bulk is

SOS Transistor Type	RN	RP
Transistor Dimensions	$1.2 \times 5 \ \mu m$	$1.2 \times 5 \ \mu m$
Threshold Voltage	0.5V	−0.8V
Transconductance	$4.1 \cdot 10^6$	$2.73 \cdot 10^6$
Channel Length Modulation	0.05	−0.02
Thr. Adj. For Floating Body	0.00	0.07
V_{DSk0}	2 V	−1.7 V
Electric Field Parameter η_1	0.5	0.5
Operating Temperature	293 K	293 K
Intrinsic Doping	$1.5 \cdot 10^{16} \ cm^{-3}$	$1.5 \cdot 10^{16} \ cm^{-3}$
Substrate Doping	$6 \cdot 10^{22} \ cm^{-3}$	$6 \cdot 10^{22} \ cm^{-3}$
Channel Doping	$1.7 \cdot 10^{23} \ cm^{-3}$	$1.7 \cdot 10^{23} \ cm^{-3}$

TABLE 2.3   Table of summary for the SOS kink model parameters.

connected to the source or a lower potential, all the majority carriers are collected by the source diffusion. Fig. 2.18 shows the bulk current for a RP SOS MOSFET with four terminals. The currents are exponential and follow the behavior modeled by Eq. (2.14).

The model for the saturation region was the same simple first-order model discussed in the previous section.

2.3.3 Results and Discussion

All transistor characteristics measurements were conducted in an electrically shielded dark chamber. The ambient temperature was not strictly controlled, but it remained around 293 K for the duration of the experiments.

Results from the proposed model are plotted in Figs. 2.19 and 2.20. Figure 2.19 is the proposed SOS MOSFET kink model for a RN 1.2 × 5 μm N-channel device, while Fig. 2.20 is the proposed SOS MOSFET kink model for a RP 1.2 × 5 μm P-channel device. The model predicts with precision both the onset of the kink and its slope for different gate voltages and as a function of the drain voltage.

SOS devices, as can be seen in Figs. 2.11 to 2.14, present not only a kink effect but also a noticeable change in the saturation currents when operated with three terminals, as opposed to the same device operated with four terminals and bulk contact. This effect is due to a threshold change in the devices resulting from the accumulation of majority carriers in the channel. Since the saturation current is higher for a floating-body (three-terminal) MOSFET, the threshold is effectively diminished. Decrease of the threshold voltage V_{th} was measured to be on the order of 0.07 V for a RP SOI transistor, while there was no threshold change for the corresponding NMOS device. The size of both devices was 1.2×5 μm. Threshold change can be attributed to the creation of a charge pocket inside the bulk of the PMOS device. The lower threshold is due to the bulk modulation resulting from to the majority carriers trapped inside the bulk. This modulation manifests itself as a nonzero bulk-to-source voltage V_{BS}. Inverting the floating-body equation, we can calculate the resulting V_{BS} voltage from Eq. (2.17).

$$V_{BS} = \left(\frac{\Delta V_{th}}{\gamma} + \sqrt{|2\phi_F|} \right)^2 - |2\phi_F| \qquad (2.17)$$

Our measurements lead to a V_{BS} of 0.154 V. Using the I_{BS} current relation in Eq. (2.14) We can therefore assume that the I_{BS} current is on the order of 44.48 nA.

A measure of the bulk current in a four terminal PMOS device, as it appears in Fig. 2.18, proves the effectiveness of the bulk contact in the removal of the majority carriers trapped in the body and also provides means to assess their amplitude.

Note again that V_{BS} remained practically zero for the NMOS device. Therefore, we report that no trapping of majority carriers occurs in the bulk of a floating-body NMOS device in the SOS process. Table 2.3 summarizes additional modeling parameters.

We did not measure any current for the regular threshold PMOS device, which means that no appreciable gate injection current was obtained.

In conclusion, the kink effect reported for SOS devices can be modeled purely by a threshold shift. The majority carriers injected from the drain to the channel increase with the onset of hot electrons. This, in turn, generates a current to the bulk. Since the lifetime of the carriers in the bulk is higher than the recombination rate, the bulk voltage V_{BS} rises and results in a effective decrease of the threshold throughout the entire operational region of the device.

2.4 EKV Model and Parameter Extraction

The EKV MOSFET model developed by Enz et al., 1995 and Bucher et al., 1996a,b reports a simple transistor equation (or *single expression*) valid in all regions of operation. The peculiarity of this model is that it preserves continuity of the derivatives with respect to any terminal voltage in the entire range of validity of the model.

The EKV MOSFET model is a compact simulation model built on fundamental physical properties of the MOS structure. For this reason, it is intended for the design and simulation of low-voltage, low-current analog and mixed analog-digital circuits developed with modern submicron processes. Peregrine's SOS process can be modeled with great precision with the EKV model, as the MOS structure is predominant in this process. In fact, additional parasitic capacitive structures are minimized by the isolated substrate and the intrinsic isolation between devices. Therefore, the gate influence on the SOS MOSFET channel will determine the drain current with a simple and reduced set of equations. This set of equations is valid below threshold and offers great precision. This property favors the design of precise low-power circuits, translinear analog circuits, and subthreshold current-mode circuits.

The EVK model also takes into account a recent model of the SOS kink effect, as we developed and explained in Sec. 2.3.2. Due to increased mobility of the carriers because of higher lattice temperature at high currents, this effect is visible in deep saturation in most SOS MOSFETs. We added a simple equation that models the kink effect and solves the problem of unaccounted currents in the design of analog circuit with MOS devices operating in deep saturation.

2.4.1 The EKV Model

We used the EPFL-EKV MOSFET model for the SOS devices under test. This flavor of the EKV model was developed in the Electronics Laboratories, Swiss Federal Institute of Technology (EPFL), Lausanne, Switzerland (Enz et al., 1995; Bucher et al., 1996a,b). The model is formulated as a *single expression*, preserving continuity of higher-order derivatives with respect to terminal voltages. Empowered with the EKV equations, in this section, we will model the DC characteristics of SOS MOSFETs to obtain the drain current as a function of the terminal voltages.

The EPFL-EKV MOSFET DC model used for SOS devices includes equations that quantify the behavior of the device in presence of the following effects:

- basic geometrical and process related aspects as oxide thickness, junction depth, effective channel length, and width
- effects of doping profile, substrate effect
- modeling of weak, moderate, and strong inversion behavior
- modeling of mobility effects due to vertical and lateral fields, velocity saturation

- short-channel effects as channel-length modulation, source and drain charge-sharing (including for narrow channel widths), reverse short channel effect

- modeling of substrate current due to impact ionization (kink effect)

In the EKV model, voltages are all referred to the substrate. In the case of SOS devices, where there is no global substrate, we referred the voltages to a common ground node. For single transistors, this common ground node can be the potential of the body of the device. In the model, V_S, V_D are the intrinsic source and drain voltages, meaning that the voltage drop over extrinsic parasitic resistive elements is supposed to have already been accounted for externally. V_D is the electrical drain voltage and is chosen such that $V_D > V_S$. Body reference allows the model to be handled symmetrically with respect to source and drain, a symmetry that is inherent in common MOSFET layout. V_G is the gate voltage.

The following equations are intended for an N-channel MOSFET. P-channel MOSFETs equations are equivalent, but the polarity of the voltages is inverted prior to computing the current. The P-channel currents will thus result in a negative value.

A set of parameters used in the EKV model used is given in Table 2.4.

Name	Description	Value	Units
COX	Gate oxide cap.	0.00334	F/m^2
V_{TO}	Threshold voltage	varies (-0.5 to 0.8)	V
γ	Body effect	0.711	$V^{1/2}$
ϕ	Channel Fermi potential	0.8	V
KP	Transconductance	varies (30-120)	$\mu A/V^2$
θ	Mobility reduction	0.157	V^{-1}
UCRIT	Longitudinal critical field	7	$V/\mu m$
XJ	S,D Junction depth	0.1	μm
DL	Channel length correction	0.05	μm
DW	Channel width correction	0.1	μm
λ	Depletion length coeff.	varies (2-3)	—
LETA	Short-channel effect	0.3	—
WETA	Narrow-channel effect	0.1	—

TABLE 2.4 Summary table of the EKV model parameters for Peregrine SOS transistors.

COX is given by the ratio of ϵ_{OX}/TOX, where TOX is the thickness of the oxide at the gate. γ is defined in the standard MOS Eq. (2.18). NSUB is the body/channel doping in an SOS device. NSUB in SOS devices is $1.7 \cdot 10^{23}$ $(1/\text{m}^3)$. This value was used to compute parameter γ. V_{TO} is the long-channel threshold voltage defined in the standard MOS Eq. (2.19). V_{FB} is the flat-band voltage. We did not compute VTO in the model; its value was extracted using measured data from SOS devices. ϕ is defined in the standard MOS Eq. (2.20).

$$\gamma = \frac{\sqrt{2q\,\epsilon_{SI}\text{NSUB} \cdot 10^6}}{\text{COX}} \tag{2.18}$$

$$V_{TO} = V_{FB} + \varphi + \gamma\sqrt{\varphi} \tag{2.19}$$

$$\varphi = 2V_T \ln\left(\frac{\text{NSUB} \cdot 10^6}{n_i}\right) \tag{2.20}$$

V_T is the thermal voltage and n_i is the intrinsic concentration of electrons in silicon. Note that ϕ is a parameter-dependent coefficient, since V_T and n_i are also. KP is the transconductance of the device, defined as $\text{KP} = \text{U0} \cdot \text{COX}$. U0 is the mobility of the device at room temperature. KP was extracted from fabricated SOS MOSFET. The value of U0 can be extracted from the value of KP. Variable UCRIT is the longitudinal critical field given by $\text{UCRIT} = \text{VMAX}/\text{U0}$. VMAX is the saturation velocity for the SOS process. We used a value of $7 \cdot 10^6$, given by the Spice SOS model in RN transistors. XJ, or diffusion length of source, drain is 0.1 μm in the SOS process.

The pinch-off voltage V_P is the performing actor in the EKV model. All other equations are standard MOS equations. The V_P voltage corresponds to the channel voltage for which the inversion charge becomes zero in a nonequilibrium state. The V_P voltage depends only on the gate voltage and physical parameters V_{TO}, γ, ϕ. An expression of the pinch-off voltage V_P is given in Eq. (2.21). The n parameter is defined as the weak inversion slope factor. It depends on the same parameters set as V_P and is defined in Eq. (2.22).

$$V_P = V_G - V_{TO} - \gamma\left(\sqrt{V_G - V_{TO} + \sqrt{\psi_{si0}} + \frac{\gamma^2}{2}} - \sqrt{\psi_{si0}} + \frac{\gamma}{2}\right) \tag{2.21}$$

$$n = 1 + \frac{\gamma}{\sqrt{V_P + \varphi}} \tag{2.22}$$

The mobility reduction due to the vertical field inside the device is taken into account by parameter θ. We can compute the equivalent transconductance factor by using Eq. (2.23).

$$\beta = \text{KP}\,\frac{W_{eff}}{L_{eff}}\,\frac{1}{1 + \theta V_P} \tag{2.23}$$

Finally, after all the above-mentioned quantities have been computed, the EKV drain current of the MOSFET is given by Eq. (2.24), where I_F and I_R are, respectively, the forward and reverse currents.

$$I_D = I_F - I_R \tag{2.24}$$

The forward current I_F and reverse current I_R are given by Eq. (2.25). The tail current I_S and the thermal voltage U_t are respectively given in Eqs. (2.26) and (2.27).

$$I_{F(R)} = I_S \cdot \left(\log \left(1 + e^{\frac{V_P - V_{S(D)}}{U_t}} \right) \right)^2 \tag{2.25}$$

$$I_S = 2n\beta U_t^2 \tag{2.26}$$

$$U_t = kT/q \tag{2.27}$$

The SOS EKV model was implemented using the MATLAB scripting language. The model has been verified by computing a minimum of a multivariable function and using the MATLAB routine *fminsearch*.

2.4.2 EKV Model Results and Discussion

A list of EKV model parameters extracted for different types SOS MOSFETs is given in Table 2.5 (Karlsson and Jeppson, 1992). The size of the transistors was $2.5 \times 2.5 \ \mu m$. Three N-type and three P-type MOSFETs were tested, organized by threshold voltage. For each type, the three different thresholds were regular (around 0.7 V), low (0.3 V), and intrinsic (0 V).

We report in Figs. 2.21 to 2.26 the modeling results of $2.5 \times 2.5 \ \mu m$ SOS MOSFETs using the EKV model. The transistor modeled are RN in Fig. 2.21, NL in Fig. 2.22, IN in Fig. 2.23, RP in Fig. 2.24 and PL in Fig. 2.25 and IP in Fig. 2.26. The model predicts perfectly the drain

Type	KN	V_{TO}	λ	θ	KST	KSLOPE
RN	79.5	0.51	2.48	0.16	2.1	0.05
NL	96.5	−0.057	1.17	0.19	2.1	0.05
IN	135.1	−0.36	1.55	0.35	2.1	0.05
RP	34.4	0.49	1.68	0.031	2.2	0.07
PL	40.9	0.18	1.86	0.075	2.2	0.07
IP	56.5	−0.034	2.40	0.17	2.2	0.02

TABLE 2.5 Extracted parameters of a Peregrine SOS Transistor with the EKV model including the kink effect.

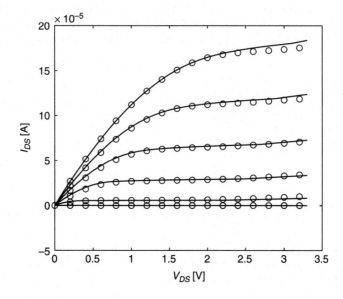

FIGURE 2.21   Modeling results of a 2.5 × 2.5 μm RN SOS with the EKV model parameters of Table 2.5. The drain current model (solid line) is plotted over the data (circles) collected.

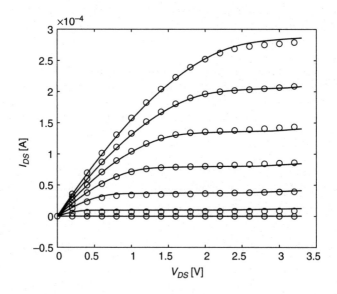

FIGURE 2.22   Modeling results of a 2.5 × 2.5 μm NL SOS with the EKV model parameters of Table 2.5. The drain current model (solid line) is plotted over the data (circles) collected.

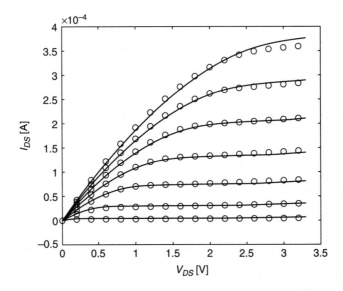

FIGURE 2.23 Modeling results of a 2.5 × 2.5 μm IN SOS with the EKV model parameters of Table 2.5. The drain current model (solid line) is plotted over the data (circles) collected.

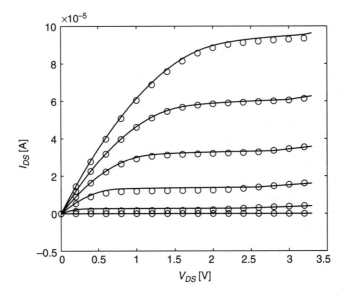

FIGURE 2.24 Modeling results of a 2.5 × 2.5 μm RP SOS with the EKV model parameters of Table 2.5. The drain current model (solid line) is plotted over the data (circles) collected.

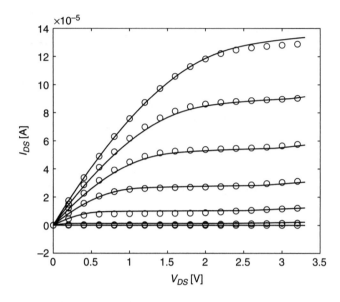

FIGURE 2.25 Modeling results of a 2.5 × 2.5 μm PL SOS with the EKV model parameters of Table 2.5. The drain current model (solid line) is plotted over the data (circles) collected.

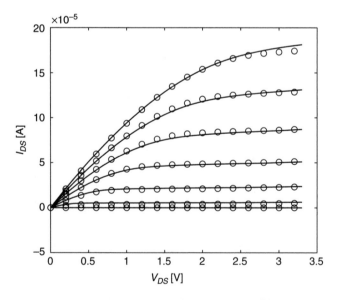

FIGURE 2.26 Modeling results of a 2.5 × 2.5 μm IP SOS with the EKV model parameters of Table 2.5. The drain current model (solid line) is plotted over the data (circles) collected.

current of the device as a function of the drain voltage and gate voltage. In the RN case of Fig. 2.21 there is some evidence of hot-electron kink effect visible at high-drain voltages and low-gate voltages. The kink effect is not visible in the other N-channel transistors NL and IN of Figs. 2.22 and 2.23. A visible kink effect is also present in the RP transistor of Fig. 2.24 and is still present in the PL transistor of Fig. 2.25. The IP transistor model of Fig. 2.26 is kink free.

The kink effect was included in this model. The model equations have been described in Sec. 2.3.2. KRT and KSLOPE are, respectively, the onset of the kink and the kink slope. In Sec. 2.3.2 these two quantities are respectively called ξ and V_{DSk}.

2.5 Models of SOS Four Terminal MOSFETs Operated as Bipolar Transistors

Four terminal devices processed in SOS can also be operated as bipolar junction transistors (BJT). This can be accomplished by using the channel as a base. The devices, while not exhibiting exceptionally beneficial gains, provide a very high dynamic range in the classical exponential input–output characteristic and can therefore be used in specific circuits that take advantage of it. Figure 2.27 shows a plot of the collector current with respect to the collector voltage and varying base voltage. The device shows regular characteristics that resemble the MOSFET ones. The possibility of creating a successful BJT device enriches the process with new possible circuit topologies and circuits. The SOS process can therefore be used effectively as a BiCMOS process when designing low-gain bipolar circuits. Translinear circuits and thermally insensitive sources are a couple of examples of the many possible applications and circuits that can be designed with SOS BJT devices (Sinencio and Andreou, 1998).

The following sections report the modeling of the current gain of BJT devices and an analysis of the recombination in the base region. The model is then compared to the collected data.

2.5.1 Modeling Bipolar Devices in SOS

Identical versions of the four-terminal MOSFET devices just described in the hot-electron kink section can also be operated as a BJT. The lateral BJT is the only possible structure for bipolar SOI (Fig. 2.28). In fact, the SOS 100 nm thin-film silicon does not offer enough space to accommodate the vertical BJT structure.

In conventional optimized BJT devices, the base region has very small dimensions compared to the collector region, especially the

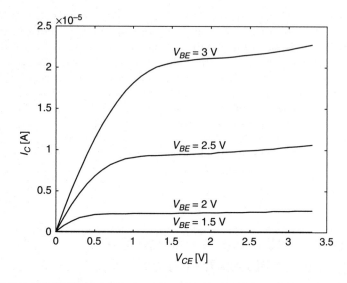

FIGURE 2.27   Characteristics of a 1.2 × 5 μm NMOS operated as NPN BJT. V_{CE} is the drain voltage of the SOS MOSFET, V_{BE} is the bulk to source voltage.

emitter region. The emitter is much more heavily doped than the collector in order to obtain a single-sided PN junction between emitter and base, and therefore reduce recombination in the base itself. In fact, if the emitter doping is higher than the base, the holes current

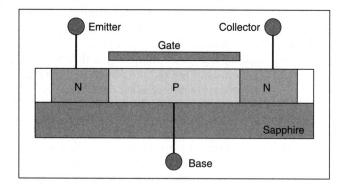

FIGURE 2.28   Bipolar junction transistor layout in the SOS process. The MOS gate is tied to ground. The body of the device is the base terminal, and the drain and source are the collector and emitter, respectively.

can be ignored. The injected (or better emitted) electrons in the base are accelerated towards the collector because of the electric field. The reverse-bias hole current from the base to the emitter can cause recombination of the electrons directed to the collector with the holes present in the base region (Sze, 1981; Rabaey, 1996).

The SOS process is strictly a CMOS process, and the design manual (Peregrine, 2003) does not mention the possibility of obtaining BJT transistors. The performance of bipolar devices cannot be optimized unless specific doping profiles are available. In the SOS process, the channel implant that constitutes the base of the bipolar is doped only to a level of about one-third of the drain and source region that constitute the collector and emitter of the BJT. For this reason, the base-emitter and collector-base junctions are both the double-sided abrupt type. In the double-sided emitter-base junction, the holes current of a NPN BJT cannot be neglected. This partly accounts for the lower current gains measured in the BJT under consideration.

The following subsection reports a model of the BJT device fabricated in the SOS process. Current gain and log-linearity of the device are stressed for promoting the use of these devices in commercial circuits and applications.

Parameter α_0 is the current gain of the device in common base configuration. Parameter β_0 is the current gain in common emitter configuration. These parameters can be related by Eq. (2.28).

$$\alpha_0 = \frac{\beta_0}{1 + \beta_0} \tag{2.28}$$

We estimated the parameter β_0 from the data collected from a 1.2×5 μm NPN device to be on the order of 2.9. This allows to estimate the parameter α_0 from the same set of data to be 0.743. α_0 can be expressed (Sze, 1981) by the following equation:

$$\alpha_0 = \frac{\partial I_{nE}}{\partial I_E} \frac{\partial I_{nC}}{\partial I_{nE}} \frac{\partial I_C}{\partial I_{nC}} = \gamma \cdot \alpha_T \cdot M \tag{2.29}$$

In the above equation, the first right term is defined as the emitter efficiency γ, the second term is the base transport factor α_T, and the last is the collector multiplier factor M. The base transport factor can be estimated with Eq. (2.30).

$$\alpha_T = 1/\cosh\left(\frac{W}{L_B}\right) \tag{2.30}$$

In Eq. (2.30), W is the base width and L_B is diffusion length of minority carriers in the base. L_B can be calculated using process-dependent parameters and physical data (Sze, 1981). Since the base width for the device under test was considerably higher than the diffusion length, parameter α_T was close to unity and therefore can be neglected in

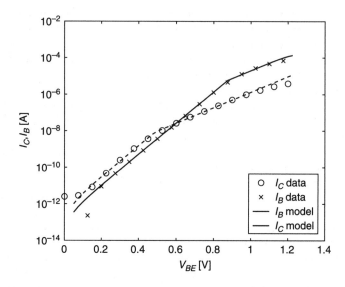

FIGURE 2.29 Modeling results of a 1.2×5 μm NMOS operated as a NPN BJT in SOS. Data is for collector and base currents I_C and I_B.

the estimation of the common base current gain α_0. In addition, since the device under test has a single collector and its geometry was not designed for obtaining intrinsic gain, the M factor has a value of unity.

Therefore, the emitter efficiency γ is the only important parameter in the determination of the gain α_0. Its value can be estimated simply by inverting the relation $\alpha_0 = \gamma \alpha_T M$. The value was estimated to be 0.743 in the NPN transistor under analysis.

The data collected from a 1.2×5 μm NMOS operated as a NPN BJT in SOS shows excellent match of the emitter base–emitter current with the theoretical exponential characteristic in Eq. (2.31). Figure 2.29 reports the modeling results.

$$I_B = I_{B0} \exp \left(\alpha_1 \frac{q V_{BE}}{kT} \right) \tag{2.31}$$

The collector current follows the base current as expressed by Eq. (2.32).

$$I_C = \beta_0 I_B \tag{2.32}$$

On the other hand, the collector current exhibits quite a reduced log-linear region in the voltage to current relation. This can be explained by the observation that when the base to emitter junction is forward biased, its depletion region is the main source of carriers capture. As the

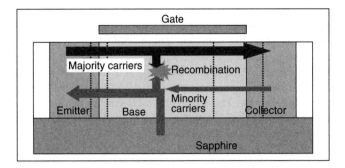

FIGURE 2.30 Current components and recombination in the BJT device. Minority carriers are injected in the base, but given the large size of the base a large portion recombines.

base-emitter voltage increases, so does the depletion region between these two regions. This causes the generation of recombination centers for the majority carriers coming from the emitter and directed to the collector. Recombination of the majority carriers with the base counterparts is sketched in Fig. 2.30. In other words, the base has a high recombination rate for the emitter current, which in turn reduces the effective collector current produced by the device. In this condition there is a forward current of minority carriers in the base. The current density of this forward (F) current follows a nonideal diode relation given by Eq. (2.33).

$$J_F \approx \exp\left(\frac{q\,V_{BE}}{nkT}\right) \tag{2.33}$$

The parameter n varies from 1 to 2 (Sze, 1981) as the recombination reaches equilibrium. We measured a value of n equal to 2 for the device under test, when fitting the data with the above mentioned recombination theory. Recombination-generation processes currents manifest at a biasing voltage as low as 0.5 V, where the quantity $q\,V/kT$ has a value of 19.8.

A high-injection condition, where both drift and diffusion currents must be considered inside the base area of the device, also produce current densities that satisfy the equation above and can therefore be modeled by the same recombination principles. Notice again that the short log-linear region is attributed to the large base width and the double-sided profile of the base. Since the emitter is not degenerately doped with respect to the base doping, the depletion region in the two terminals will have comparable size. In addition, the wide base allows for a high-depletion region before breakdown occurs and for additional recombination due to trapping in the base, especially if the electric field between base and collector is not able to accelerate the minority carriers to the collector. This is not at all an ideal case for a high-performance BJT device, where the base width is kept short to reduce recombination and obtain high-current gain in both common emitter and base configuration. Fig. 2.30 is a lateral view of the fabricated BJT device. The base (channel) contact was provided by means of body contacts. Emitter and collector are the MOSFET drain and source regions. The gate of the device was tied to a fixed negative potential to fully deplete the device.

2.5.2 A Chance for BiCMOS SOS?

Peregrine Semiconductor's SOI process by is capable of being a BiCMOS process, since it is possible to obtain bipolar junction transistors with fair characteristics. Bipolar devices are usually desired for their high gain, something that cannot be obtained using the current process. In radio frequencies, bipolar devices provided high gain and bandwidth compared to CMOS transistors. Note, however, that the SOS process, deprived of parasitics and the influences of the substrate, can provide high bandwidth and can be successfully employed in commercial-radio frequency–operated circuits. However, the need for high-bandwidth devices can be satisfied by SOS MOSFETs (providing transition frequency on the excess of 60 GHz Peregrine, 2003). Nevertheless, useful circuits can be designed when BJTs are available in the same process.

One very important circuit, now part of almost every integrated commercial silicon device, is the thermal voltage–referenced self-biasing circuit. The circuit presented in Fig. 2.31 provides a thermal voltage reference proportional to V_T. When the current in the two branches is biased to operate in the exponential region of the BJT device, the output of the element is proportional to absolute temperature (PTAT).

The use of bipolar devices also provides great benefits to low-voltage low-power integrated circuits like the translinear class (Sinencio and Andreou, 1998). The extended log-linear region of operation for the bipolar devices greatly encompasses the one offered by MOSFETs transistors. Translinear circuits operate in the exponential region of the device to be able to perform high performance and precise analog operations. If the operational range is restricted to the subthreshold characteristic of the MOS transistor, the dynamic range of the circuit is limited. Also, a small deviation from the exponential regime of operation provokes severe distortion of the output result, since the device stops exhibiting the desired input/output characteristic. On the other hand, the use of BJT largely extends the exponential region of operation and allows for higher dynamic ranges of inputs.

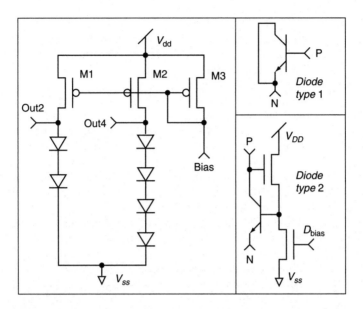

FIGURE 2.31   PTAT circuit. The diodes in the left circuit can be one of the two types on the right.

Finally, recombination or high-injection nonlinearities, which are still governed by exponential functions, give much less output distortion than the one encountered when using MOSFETs.

2.6 SOS Flash Memory Devices

Flash memory is a nonvolatile digital memory that can be electrically erased and reprogrammed. This memory does not need power to retain the data stored. Flash memory store information in an array of floating gate transistors (or cells), each of which traditionally stores one bit of information. Flash memory offers fast read and access times (on the order of 1–10 μs) and better shock resistance than hard disks. These characteristics recently made flash memory very popular for storage on battery-powered devices such as cell-phones, PDAs, and music players.

Several floating-gate devices have been designed in commercially available bulk CMOS processes by research groups (Yang and Andreou, 1994; Martin, 1998; Srinivasan et al., 2005; Hasler, 2005). SOI flash memories have been modeled (Chan et al., 2004; Stanojevic et al., 1997) but not fabricated. In this section, we report on the design and test results of four kinds of nonvolatile memories in a commercially available 0.5 μm SOS CMOS process (Peregrine, 2008a). SOS is a flavor of SOI targeting high-performance analog circuits. We fabricated PMOS and NMOS cells with MOS- and MIM-based floating gates.

Each flash memory cell is similar to a standard MOSFET except that it has two gates instead of just one. These two gates are "control gates" (CG), similar to standard MOS transistors. These control gates are capacitively coupled to a "floating gate" (FG) that is insulated all around by an oxide layer (SiO_2-Ni-SiO_2). The FG resides between the CG and the MOS cell. Because the FG is insulated by the oxide layer, its electrons are trapped, giving it the ability to store analog (multilevel) or digital (two-levels) information. When electrons are present on the FG, they add to the CG electric field, modifying the threshold voltage (V_t) of the cell. During cell readout, a specific voltage applied on the CG will make the MOS drain current change in relation to the V_t of the cell, which is controlled by the number of electrons on the FG.

A NOR flash is programmed by hot-electron injection. When a large voltage is placed on the CG, a strong enough electric field pushes lucky electrons traveling from channel to drain onto the FG, provided they have enough vertical momentum. To erase a NOR flash cell, electron tunneling is used. A large voltage is placed between the CG and the MOS source, which pulls the electrons off the FG. In battery-powered single-supply devices, this high voltage is generated by an on-chip charge pump. SOS charge pumps circuits (Culurciello et al., 2005a) can be modified to program SOS flash memory cells.

2.6.1 SOS Flash Memory Design

Using the SOS process, we fabricated four test structures of floating gates for flash memories. Our design targets NOR flash memories. The SOS process provides six available voltage threshold MOSFETs, three for NMOS, three for PMOS. The thresholds are 0.7, 0.3, and 0V (Peregrine, 2008b). Our devices use high-threshold (0.7 V) standard MOS devices to allow for the largest possible threshold shifts. The SOS MOSFETs are fabricated on a 100-nm-thick silicon layer.

We designed two kinds of floating gates: one based on metal-oxide-semiconductor (MOS) capacitors and one based on metal–insulator–metal (MIM) capacitors. For each kind of floating gate, we have fabricated both NMOS and PMOS test cells. Figure 2.32 reports a schematic caption of the test cells. Figure 2.32(a) shows the MIM floating gates and Fig. 2.32(b) the MOS floating gates. In Fig. 2.33, we show the layout of the MIM-based test structure, while in Fig. 2.34 we report the MOS-based layout. In both figures, the NMOS and PMOS cells are on the right side, while the CG are the two MIM/MOS capacitors on the

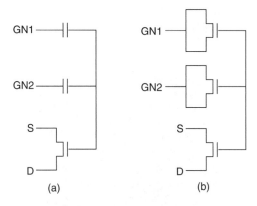

FIGURE 2.32 Schematic of the nonvolatile memories in SOS CMOS. GN1-2 are control gates (CG), and S and D are the source and drain of the MOS memory cell.

left side. The configuration is similar to the schematic of Fig. 2.32. The size of the MOSFETs is $W/L = 2/2$ µm. The MOS used as memory cell was a standard transistor, with no gate to drain/source overlap (layout rules were respected). The capacitance values used for MIM CG is 54 fF, and 6.5 fF (max) for the MOS CG.

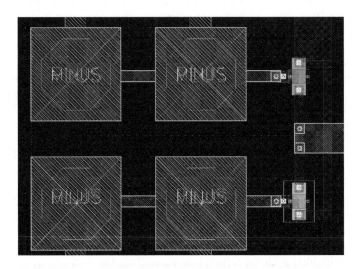

FIGURE 2.33 Layout of the MIM-based nonvolatile memories in SOS CMOS. The MIM capacitor of 54 fF is the CG GN1-2 of Fig. 2.32.

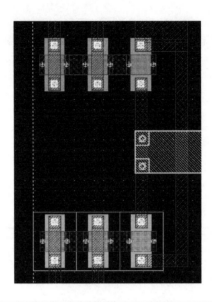

FIGURE 2.34 Layout of the NMOS-based nonvolatile memories in SOS CMOS. The MOS capacitor of 6.5 fF is the CG GN1-2 of Fig. 2.32.

2.6.2 SOS Flash Memories
Experimental Results

We have tested the memory cells using a Agilent 4156A Semiconductor Parameter Analyzer. The devices were packaged in a DIP16 package before evaluation.

MOS $I_{ds}-V_{gs}$ data collected during the programming of the MIM NMOS cell is given in Fig. 2.35. The MIM-based NMOS device reported a 0.4-V voltage threshold (V_T) shift. This shift was achieved by stress biasing of $V_G/V_D = 11/5.5$ V for 10 s, with respect to the initial condition of $V_D = 0.1$ V. Several drain and gate stress voltage configurations were applied ($V_G/V_D = 4/2, 5/2, 6/3, 7/3, 8/4, 9/4.5$, $10/5, 11/5.5$) for a variety of programming times (10 ms, 100 ms, 1 s, 10 s). This result proves that flash memory can be designed in the SOS process. This resulting threshold shift was also similar to what has been obtained with hot-electron injection by other research groups using bulk-CMOS process with a 0.5-μm feature size (Hasler, 2005).

Figure 2.36 reports the threshold shift obtained from Fig. 2.35. Threshold shifts of 0.2–0.4 V were obtained with $V_G/V_D = 11/5.5$ V biasing for 100 ms to 10 s. Various degrees of threshold shifts were also obtained with the settings mentioned above.

By using biasing voltages of $V_g/V_d = 12/6$ V for 2 and 6 s, respectively, the resulting threshold shift is increased by 7 V, as can be

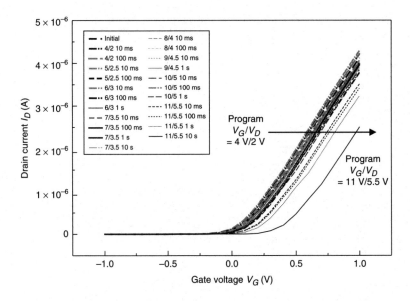

FIGURE 2.35 I_{DS} vs V_{GS} plot changes after programming the MIM-based NMOS nonvolatile memory. Plots varied bottom to top with increasing bias voltages and programming times.

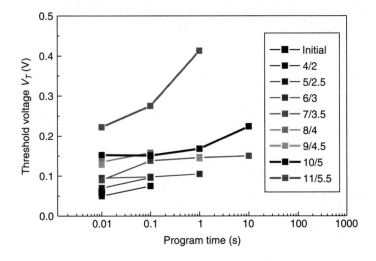

FIGURE 2.36 Threshold voltage (V_T) shifts after programming the MIM-based NMOS nonvolatile memory cell. Plots varied bottom to top with increasing bias voltages and programming times.

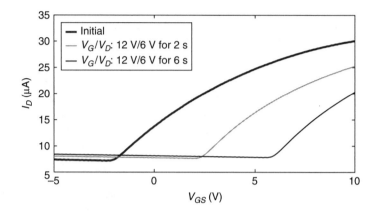

FIGURE 2.37 I_{DS} vs V_{GS} plot changes after programming the MIM-based NMOS nonvolatile memory using biasing voltages of $V_G/V_D = 12/6$ V for 2 and 6, respectively.

see in Fig. 2.37. Notice that in this test, the floating gate was initially charged, and programming shifted the threshold by 7 V. This test is promising because it shows that higher-than-standard programming voltage levels are necessary for programming SOS devices.

Figure 2.38 reports the retention time test of the MIM NMOS memory cell. Good retention is observed in this small scale test (>100 s).

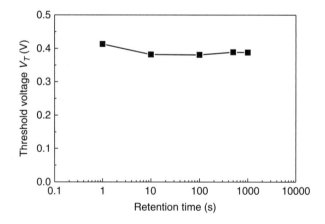

FIGURE 2.38 Data retention of the MIM-based NMOS nonvolatile memory cell.

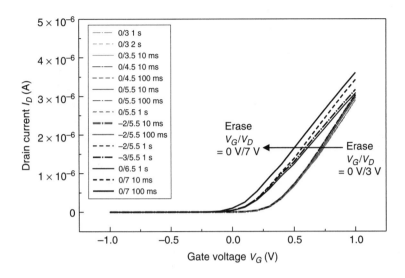

FIGURE 2.39 Erasing of the MIM-based NMOS nonvolatile memory cell. Plots varied from right to left with increasing tunneling biases and erase times.

Given that the quality of silicon dioxide in the SOS technology is not different from 0.5-μm bulk CMOS processes, we do not expect the retention results to differ from the industry standard even in longer-scale retention tests.

Figure 2.39 reports drain current (I_{DS}-V_{GS}) data from the erase test of the MIM NMOS memory cell. This cell can be erased by using $V_G/V_D = 0/7$ V biasing and keeping the drain voltage $V_D = 0.1$ V. In this figure, we report the erase drain currents for various voltages configurations ($V_G/V_D = 0/3, 0/3.5, 0/4.5, 0/5.5, -2/5.5, -3/5.5, 0/6.5\,0/7$) and for various programming times (10 ms, 100 ms, 1 s).

We then proceeded to test the MOS-based NMOS floating gate test memory cell. A significant threshold voltage (V_T) shift was not observed. The programming stress biasing was raised up to $V_G/V_D = 12/6$ V biasing, with respect to the initial conditions with $V_D = 0.1$ V. The measurement was halted at this programming voltages because of junction breakdown concerns. Figure 2.40 shows the I_{DS}-V_{GS} plot changes after programming the MOS-based NMOS memory cell. In this figure, we report the MOS I_{DS}-V_{GS} plot for various stress voltages configurations ($V_g/V_d = 4/2, 6/3, 7/3, 8/4, 10/5\,11/5.5\,12/6$). and for various programming times (10 ms, 100 ms, 1 s, 10 s, 100 s). The data show that the MOS-based NMOS cannot be used as flash memory cells, as threshold voltage shifts were insignificantly small.

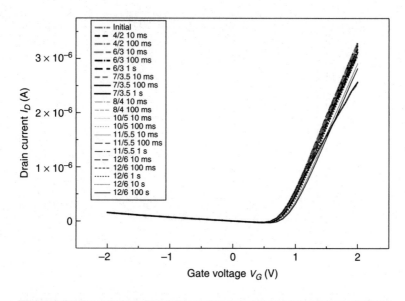

FIGURE 2.40 Programming of a MOS-based NMOS nonvolatile memory cell.

Finally, we proceeded to test the PMOS devices, both MIM and MOS based (Fig. 2.41). Both PMOS devices showed good I_D vs V_G transistor behavior. Unfortunately, no charge injection was observed, and therefore no voltage threshold shift was obtained. This is due to large energy barrier for hole injection in PMOS devices (4.9 eV) as compared to NMOS device (3.1 eV for electrons).

2.6.3 SOS Flash Memories Performance Evaluation and Model

For 0.5-μm bulk CMOS processes, the industry standard flash memory cell has a voltage threshold shift $\Delta V_T = 2.2$ V, obtained with programming voltages $V_G = 11.5$ V and $V_D = 5.5$ V for 3 μs. Our SOS MIM NMOS memory cell was only able to obtain a threshold shift of 0.4 V with a 10-s programming time. On the other hand, with larger programming voltages of $V_G/V_D = 12/6$ V for 2 and 6 s, respectively, we obtained threshold shifts of 7 V. While this is an encouraging result, we developed a model to identify the differences with industry standard in 0.5-μm bulk CMOS processes (Stanojevic et al., 1997).

Figure 2.42 shows a model of the capacitive stack composing the floating gate memory cell. The voltage of the floating gate V_{FG} is given by Eq. (8.15) as a function of the drain and gate voltages.

$$V_{FG} = \alpha_G V_{GS} + \alpha_D V_{DS} \tag{2.34}$$

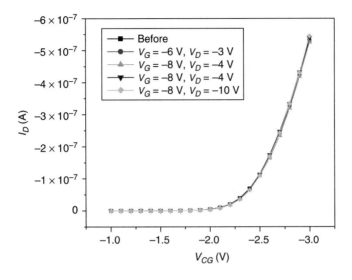

FIGURE 2.41 Programming of MOS-based PMOS nonvolatile memories.

The coupling coefficient α_G of the floating gate voltage is given by Eq. (2.35).

$$\alpha_G = \frac{C_{FC}}{C_{FC} + C_S + C_D + C_B} \tag{2.35}$$

In the SOS process, we have C_{FC} of 6.5 fF for a MOS-based CG and 54 fF for a MIM CG. The value of C_B is 6.5 fF for a $W/L = 2/2\ \mu\text{m}$ transistor, while C_S and C_D are on the order of 0.1 fF (Table 2.6). With

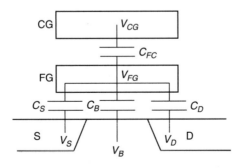

FIGURE 2.42 Model of nonvolatile memories.

Process Technology	0.5 μm SOS
Silicon thickness	100 nm
MOS memory cell size	$W/L = 2/2$ μm
MIM CG capacitance	54 fF
MOS CG capacitance	6.5 fF
MIM NMOS ΔV_T	0.4 V
MOS NMOS ΔV_T	0 V
MIM PMOS ΔV_T	0 V
MOS PMOS ΔV_T	0 V

TABLE 2.6   Summary of SOS FG Devices

these data, we compute $\alpha_G = 0.89$ for the MIM-based floating gate and $\alpha_G = 0.5$ for the MOS-based gate.

Possible reasons for the lower-voltage threshold shift efficiency are:

- unoptimized floating gate coupling voltage
- reduced drain/source junctions and gate overlap
- floating body effect (need modification to coupling coefficient)

First, the stress programming voltages we used were based on the industry standard memories with an $\alpha_G = 0.6$, but this value was different from the SOS devices under test. Second, since the SOS MOS devices were not optimized for flash memories, but for analog circuits, were instead, the overlap of gate and drain region is minimal, reducing the injection and tunneling efficiencies (Cristoloveanu and Li, 1995; Kuo and Su, 1998; Colinge, 1997). Finally, the threshold voltage of a floating-body SOS NMOS is modulated by hot electrons, since holes accumulate in its body (Stanojevic et al., 1997; Chan et al., 2004).

2.7 Conclusions

The newest technological advances in processing large-scale integrated circuits offer a vast number of alternative and new topologies to the designer. SOS promises high speed and reduced noise coupling because of the insulating nature of its substrate. Designers can take advantage of the improved speed and the lack of parasitics to improve their design and simplify it. Multiple threshold transistors allow the designer to optimize standard circuit topologies and invent new ones.

The technology offers even more than it promises in its design manual. Four-terminal MOSFETs not only can operate as the SOS process design manual suggests, they can also encompass standard views and perform self-cancellation of their nonlinearities when operated with an additional bulk terminal. This chapter presented an efficient way to reduce the kink effect in SOI MOSFETs and also offered insights about modeling its behavior and accounting for it during design. Traditional models and more advanced charge models were applied to study and characterize the MOS devices.

In addition, a bipolar transistor can be obtained with the same four-terminal transistor available in the SOS process. When the bulk is operated as the base, the device is able to exhibit a noninsignificant gain. In addition, its very linear exponential behavior is very useful not only for low-power circuits but also for circuits used commonly in integrated systems, such as the PTAT voltage reference.

We also presented four kinds of nonvolatile memories in a commercially available 0.5 μm SOS CMOS process. We tested PMOS and NMOS with MOS- and MIM-based floating gates. We reported on the results of all devices and demonstrate that only MIM-based NMOS floating gate cells can be use to achieve a threshold shift of up to 7 V and retentions of more than a thousand seconds.

In summary, the SOS technology, and SOI in general, offers a variety of nonstandard topologies that can be used to enhance and facilitate the design any traditional building blocks of the VLSI repertoire.

2.8 Data Collection Methods

Measurements from the SOS transistors in this chapter were taken on a probe station using individual SOS multiproject dies containing several test transistors. The measurements were conducted using Keithley 236 Source Measure Units (SMUs). The SMUs were interfaced with a Hewlett Packard Omnibook 800 CT personal computer. The communication between instruments and the computer was implemented using the National Instruments NI-488.2 general purpose interface bus (GPIB). The mathematical tool MATLAB version 6.0.0.88 R12 was used to establish the GPIB interface and monitor the collection of data. A program written in the MATLAB scripting language was used to apply various combinations of gate and drain voltages and to read the SMUs measurements of the resulting drain current. The connections between the device terminals and the SMUs were made using Wentworth Laboratories probes, model PR0195.

CHAPTER 3

Design of SOS Single-Stage Amplifiers and Analog Components

3.1 Introduction

This chapter introduces the basic analog building blocks for SOS circuits and systems. SOS current mirrors and single-stage amplifiers are presented and analyzed in this chapter.

3.2 Analog Characteristics of SOS MOSFETs

When analyzing the performance of analog building blocks, the MOSFETs characteristics are of extreme importance in revealing what can be accomplished and what are the best devices to design a specific circuit block.

In particular, some of the most important parameters for SOS MOSFETs are the output resistance r_o and the transconductance g_m. Both of these quantities are a direct function of the drain current I_D, and thus also a function of the gate and drain bias voltages. The output resistance r_o and the transconductance g_m parameters are defined respectively in Eqs. (3.1) and (3.2).

$$r_o = \left. \frac{\partial I_D}{\partial V_{DS}} \right|_{V_{GS}=const} \tag{3.1}$$

$$g_m = \left. \frac{\partial I_D}{\partial V_{GS}} \right|_{V_{DS}=const} \tag{3.2}$$

These parameters can be obtained from simulations by differentiating the drain current I_D with respect to the gate and drain currents while keeping the other biases constant, as expressed in Eqs. (3.1) and (3.2).

In this section, we report on the simulation results of all the SOS MOSFETs types. We used the following types of transistors: NMOS devices were regular threshold (0.7 V RN), low threshold (0.3 V NL), and intrinsic (0 V IN). PMOS devices were regular threshold (0.7 V RP), low threshold (0.3 V PL) and intrinsic (0 V IP). It is impossible to report all possible combinations of input voltages and devices sizes, therefore here we only report devices with a size of [W, L] of [2 μm, 2 μm], as it is typical for analog transistor to have a minimum length of approximately 4 times the minimum feature size. The biasing configuration we used are: gate voltages of 1 V and drain voltage sweeps from 0 V to 3.3 V for the output resistance simulations, and drain voltages of 3.3 V and gate voltage sweeps from 0 V to 3.3 V for the transconductance simulations. The voltage integration step in these simulations was 0.5 mV. Results for the same channel length but different widths can be computed from the results of these simulations.

Figure 3.1 shows the simulated transconductance parameter g_m for all the SOS NMOS transistors. These data were collected by

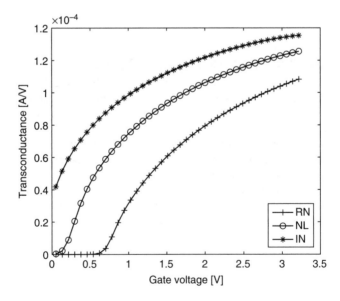

FIGURE 3.1   Transconductance parameter g_m for the SOS NMOS transistors. The NMOS size is 2 μm × 2 μm. The drain voltage V_{DS} was biased at 3.3 V.

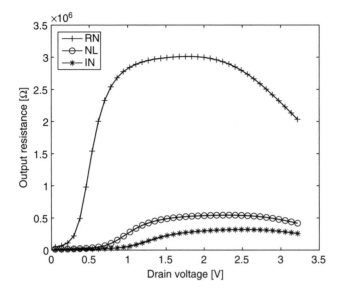

FIGURE 3.2   Output resistance parameter r_o for the SOS NMOS transistors. The NMOS size is 2 μm × 2 μm. The gate voltage V_{GS} was biased at 1 V.

monitoring the drain current I_D and sweeping the gate voltage V_{GS} from 0 V to 3.3 V, with the drain voltage V_{DS} biased at 3.3 V. The transconductance parameter g_m was computed from the the drain current using Eq. (3.2). These data show that the maximum transconductance has a modest increase from type RN to NL to IN. While at high gate voltages, the advantage of one type of transistor versus others is modest, at low gate voltages ($V_{GS} < 1$V), the relative difference in transconductance is quite substantial and is much higher in NL than RN and in IN with respect to NL.

Figure 3.2 shows the simulated output resistance parameter r_o for all the SOS NMOS transistors. These data were collected by monitoring the drain current I_D and sweeping the drain voltage V_{DS} from 0 V to 3.3 V, with the gate voltage V_{GS} biased at 1 V. The output resistance parameter g_m was computed from the the drain current using Eq. (3.1). These data show that the maximum output resistance is significantly higher in RN-type transistors than NL or IN types. RN output resistance in this transistor geometry is 5 times the one on NL and almost 10 times that of IN transistors. This is due to the high conductance of the channels and the fact that a lower threshold means that the channel is easily formed, and thus more conductive. IN transistors are depletion devices, thus the channel is a low-impedance connection between drain and source.

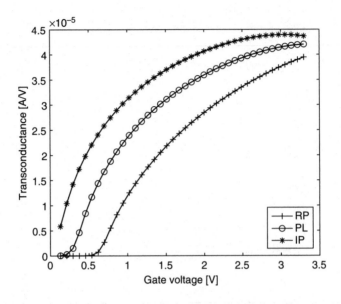

FIGURE 3.3   Transconductance parameter g_m for the SOS PMOS transistors. The PMOS size is 2 μm × 2 μm. The drain voltage V_{DS} was biased at 3.3 V.

Figure 3.3 shows the simulated transconductance parameter g_m for all SOS PMOS transistors. Similar to Fig. 3.1, these data were collected by monitoring the drain current I_D and sweeping the gate voltage V_{GS} from 0 V to 3.3 V, with the drain voltage V_{DS} biased at 3.3 V. The transconductance parameter g_m was computed from the the drain current using Eq. (3.2). These data show that the maximum transconductance has a modest increase from type RP to PL to IP. While at high gate voltages, the advantage of one type of transistors versus the other is modest, at low gate voltages (V_{GS} <1 V), the relative difference in transconductance is quite substantial and is much higher in PL than RP and in IP with respect to PL.

Figure 3.4 shows the simulated output resistance parameter r_o for all the SOS PMOS transistors. Similar to Fig. 3.2, these data were collected by monitoring the drain current I_D and sweeping the drain voltage V_{DS} from 0 V to 3.3 V, with the gate voltage V_{GS} biased at 1 V. The output resistance parameter g_m was computed from the the drain current using Eq. (3.1). These data show that the maximum output resistance is significantly higher in RP-type transistors than PL or IP types. RP output resistance in this transistor geometry is 4 times the one on PL and almost 9 times that of IP transistors. This is due to the

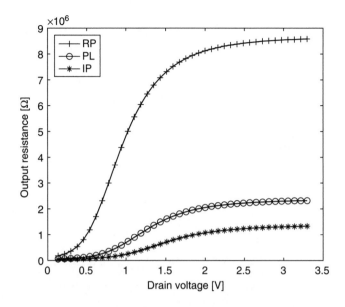

FIGURE 3.4   Output resistance parameter r_o for the SOS PMOS transistors. The PMOS size is 2 μm × 2 μm. The gate voltage V_{GS} was biased at 1 V.

high conductance of the channels and the fact that a lower threshold means that the channel is easily formed and thus more conductive. IP transistors are depletion devices, thus the channel is practically always a low impedance connection between drain and source.

The data presented in this section are for devices with sizes of $[W, L]$ of [2 μm, 2 μm]. When other sizes are used, output resistance is inversely proportional to the channel width W and proportional to the length L. The transconductance parameter is proportional to the channel width W and inversely proportional to the length L. This simple relationship can be used to obtain the parameters for other size transistors. Notice that since there is no body effect in SOS transistors, the transconductance gain and output resistance do not change when transistors are stacked. Short and narrow channel effects significantly modify the results presented above.

Although the results presented here can help the designer to hand calculate parameter and transistor sizes, the difference in bias configuration and the complex dependence of the drain current on all these parameters make it difficult to be useful in all situations. We suggest that designers plot similar curves, just like the last four presented in this chapter, for the type of transistor used in their designs.

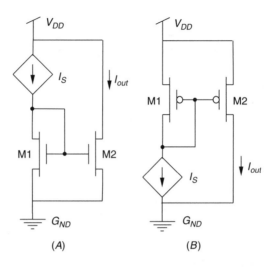

FIGURE 3.5   SOS current mirrors. (A) is an all-NMOS version, and (B) an all-PMOS version. Transistors M1 are the input mirroring transistors, while transistors M2 are the output tranistors.

3.3 SOS Current Mirrors

Current-mirror circuits can copy, multiply, or divide an input current by a fixed parameter. These circuits are one of the most typical components of CMOS circuits and systems because they provide a variety of current references, all dependent on a single input current source. A typical current mirror schematic is given in Fig. 3.5, where both an NMOS—Fig. 3.5(A) and a PMOS—Fig. 3.5(B) current mirror are reported. I_s is the input current of the mirror and I_{out} is the output current. These two currents are related and can be expressed by Eq. (3.3).

$$\frac{I_{out}}{I_s} = \frac{W_2/L_2}{W_1/L_1} \frac{1 + \lambda(V_{out} - V_{DS1,sat})}{1 + \lambda(V_{DS1} - V_{DS1,sat})} \tag{3.3}$$

Equation 3.3 reduces to Eq. (3.4) when both M1 and M2 have identical channel lengths, have a long channel, and the channel-length modulation λ effect can be neglected.

$$\frac{I_{out}}{I_s} = \frac{W_2}{W_1} \tag{3.4}$$

Usually, current mirrors are designed with square large transistors to reduce mismatch between transistors. Layout techniques to reduce mismatch are often employed when a precise ratio of currents is desired. This is often the case in the design of linear circuit components as difference amplifiers, voltage references, operational amplifier, and amplifiers arrays. Connecting in series or parallel the same finger element improves matching (Geiger et al., 1990; Baker et al., 1998; Hastings, 2005).

Mismatch of current mirrors can be attributed predominantly to voltage threshold differences between transistors, although transconductance and other parameters can also contribute. If the threshold of two identical devices changes by a fixed ΔV_{th}, we can, for example, express the difference of transistors M1 and M2 in Fig. 3.5 as $V_{th1} = V_{th} - \Delta V_{th}/2$ and $V_{th2} = V_{th} + \Delta V_{th}/2$. In this case, with both M1 and M2 identical and in saturation, the results of Eq. (3.3) can be expressed by (3.5).

$$\frac{I_{out}}{I_s} = \frac{(V_{GS} - V_{th} - \Delta V_{th}/2)^2}{(V_{GS} - V_{th} + \Delta V_{th}/2)^2} = \frac{\left[1 - \frac{\Delta V_{th}}{2(V_{GS} - V_{th})}\right]^2}{\left[1 + \frac{\Delta V_{th}}{2(V_{GS} - V_{th})}\right]^2} \qquad (3.5)$$

Ignoring the larger powers and second-order exponents, the last form of Eq. (3.5) can be reduced to a the simpler relation in Eq. (3.6). In this equation, one can see that the overdrive voltage can reduce mismatch when increased.

$$\frac{I_{out}}{I_s} = 1 - \frac{2\Delta V_{th}}{V_{GS} - V_{th}} \qquad (3.6)$$

Similar relations can also be computed for subthreshold operation of the transistors, although this region is not usually recommended when high-precision matching is required. The subthreshold region, due to this large current gain, has a poorer mismatch and would require larger transistors to attain the same matching performance. Notice also that the current of mirrors can be arbitrarily reduced by increasing the length of the devices while keeping them operating in the saturation region. This is the preferred way to design low-power current mirrors.

We have simulated six kinds of SOS current mirrors and here report the results and show the difference of operation of each type of transistor. We used the following types of transistors for both M1 and M2: NMOS devices were regular threshold (0.7 V RN), low threshold (0.3 V NL) and intrinsic (0 V IN). PMOS devices were regular threshold (0.7 V RP), low threshold (0.3 V PL), and intrinsic (0 V IP). In simulation, I_s was set to 10 μA, and all MOSFET have the same size $[W, L] = [10$ μm, 1 μm]. The power supply voltage was varied from the entire acceptable range [0 V, 3.3 V] to show the effects on the output of the mirror currents. Figures 3.6 and 3.7 show the output currents I_{out} for NMOS and PMOS types of transistors, respectively.

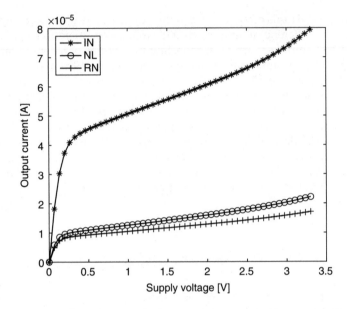

FIGURE 3.6   SOS NMOS current mirrors simulation of the output current I_{out} as a function of the supply voltage V_{DD}. MOSFET sizes were $[W, L] = [10\ \mu m, 1\ \mu m]$. Each MOSFET kind (IN, NL, RN) was used for both M1 and M2.

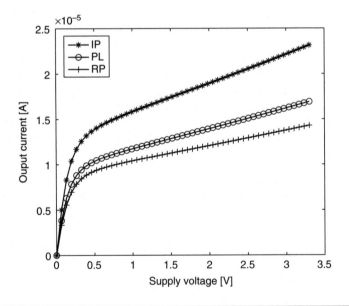

FIGURE 3.7   SOS PMOS current mirrors simulation of the output current I_{out} as a function of the supply voltage V_{DD}. MOSFET sizes were $[W, L] = [10\ \mu m, 1\ \mu m]$. Each MOSFET type (IP, PL, RP) was used for both M1 and M2.

Both PMOS and NMOS regular- and low-threshold transistors provide a change of approximately 50% of the nominal value of (10 μA) of the output current I_{out} when the supply voltage is varied from 1 V to 3.3 V. Intrinsic NMOS and PMOS transistor were not able to mirror the current because the minimum current they can provide with the size used was higher than the source current I_S. These results show that regular threshold MOSFETs provided less variations in the output current as a function of the supply voltage when operated with a supply voltage higher than the voltage threshold. Other kinds of transistors, due to the larger maximum drain current, resulted in larger variations with the supply voltages.

3.4 SOS Supply-Independent Current Reference

When a supply independent current reference is desired, current mirrors are not adequate, due to the large dependance on supply voltages (25%/V, as seen in the previous section). A better alternative is the use of a beta-multiplier reference, as reported in Fig. 3.8. This current reference is composed of two mirrors in feedback, a topology in which both circuits help to stabilize the output current.

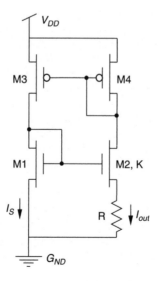

FIGURE 3.8 Schematic of a SOS supply-independent current reference. Two current mirrors in feedback help to stabilize the voltage with respect to the supply V_{DD}. This circuit is called a beta multiplier because the M2 transistor width is K times that on M1. All transistor lengths are supposed to be equal.

This supply-independent current source operates by trying to maintain the same current on both M1 and M2 devices. The gate voltage of transistors M1 and M2 are related by $V_{GS1} = V_{GS2} + I_S R$. If transistor M2 width is a multiplied (by parameter K) version of M1 widths W and the lengths L are the same, then Eq. (3.7) holds. The term "beta multiplier" derives from the fact that $\beta_2 = K\beta_1$, with $\beta = KP\ W/L$ being the device transconductance parameter.

$$V_{GS} = \sqrt{\frac{2I_D}{\beta}} + V_{th} \qquad (3.7)$$

Equation 3.8 expresses the output current I_{out} and is derived from Eq. (3.7), with KP being the transconductance of M1 or M2.

$$I_{out} = \frac{2}{R^2\beta}\left(1 - \frac{1}{\sqrt{K}}\right)^2 \qquad (3.8)$$

We have conducted test simulations of the SOS supply-independent current reference. Transistor sizes in the simulation circuit were, for NMOS: $[W1, L1] = [10\ \mu m, 2\ \mu m]$, $K = 4$, therefore $[W2, L2] = [40\ \mu m, 2\ \mu m]$ and for the PMOS: $[W3,4, L3,4] = [30\ \mu m, 2\ \mu m]$. Each type of MOSFET (IN, NL, RN) was used for both M1 and M2, and (IP, LP, RP) were used for transistors M3 and M4. All transistors in each test were of the same type of threshold. The value of resistor R was set to $R = 77\ k\Omega$. The supply voltage V_{DD} was swept from 0 V to 3.3 V while monitoring the output current. The results are reported in Fig. 3.9.

The simulations show that all types of transistors provide good supply independence, with a variation of the output current of 0.05%/V, which is a factor of 500 better than the current mirrors reported in the previous section. The type of transistor did not make a difference in output variations but only changed the value of the output current due to the difference in transconductance parameters. Notice also that the threshold voltage differences allow intrinsic transistor to be operated from supplies of 0.5 V, low-threshold transistor from more than 0.5 V, and regular thresholds from above 1 V.

3.5   SOS Single-Stage Amplifiers

In this section, we report on the design of single-stage amplifiers in SOS technology. We show what are the differences between the various types of SOS transistors and how to take advantage of the diversity of devices to improve and optimize single-stage analog amplifiers.

In this section we report on common-source amplifiers with diode loads and current source loads, cascoded common-source amplifiers with self-biased current source loads, and source follower amplifiers.

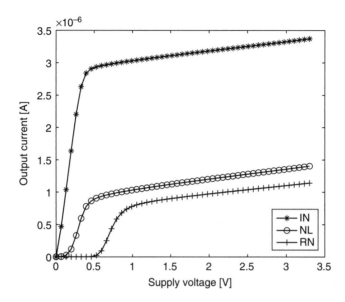

FIGURE 3.9 Simulation of the output current I_{out} of the SOS supply-independent current reference as a function of the supply voltage V_{DD}. Transistor sizes were $[W1, L1] = [10\ \mu m, 2\ \mu m]$, $[W2, L2] = [40\ \mu m, 2\ \mu m]$, $[W3,4, L3,4] = [30\ \mu m, 2\ \mu m]$. Each type of MOSFET kind (IN, NL, RN) was used for both M1 and M2, and (IP, LP, RP) were used for transistors M3 and M4. All transistors in each test were of the same type of threshold.

3.5.1 SOS Common-Source Amplifiers

The common-source amplifier is one of the most popular single-stage amplifiers in CMOS technology. It allows us to amplify an input signal voltage and deliver large output currents. This stage has a large input impedance and a low output impedance, allowing it to decouple input and output while providing current buffering for driving large capacitive and resistive loads.

3.5.1.1 SOS Common-Source Amplifiers with Diode-Connected MOSFET Loads

Figure 3.10 shows a typical common-source amplifier schematic. Transistor M1 in Fig. 3.23 is the input transistor, and transistor M2 is the common-source amplifier load, in this case a diode-connected MOSFET with gate and drain tied together. This is one of the simplest common-source configuration because it is self-biased and does not require an external bias voltage.

Figure 3.10 shows two similar configuration with all NMOS transistors, as in Fig. 3.10(*A*), or all PMOS transistors, as in Fig. 3.10(*B*).

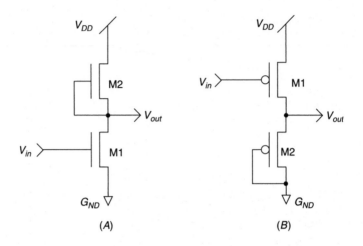

FIGURE 3.10   Common-source amplifiers schematic. (A) is an all NMOS version, and (B) an all PMOS version. Transistors M1 are the input transistor, while transistors M2 are the common-source amplifier load, in this case a diode-connected MOSFET.

The common-source amplifier provides a voltage gain given by Eq. (3.9), where all quantities are small-signal voltages and currents.

$$A_v = \frac{v_{in}}{v_{out}} = -\frac{i_d \frac{1}{g_{m2}}}{i_d \frac{1}{g_{m1}}} = -\frac{g_{m1}}{g_{m2}} \tag{3.9}$$

The results in Eq. (3.9) supposes that the g_m of individual transistors is much lower than their output resistance (r_o), which is the case for SOS MOSFETs if the widths are about equal or higher than the lengths.

In addition, the bandwidth of these amplifiers is given in Eq. (3.10), where A_v(DC) is the low-frequency gain in Eq. (3.9).

$$A_v(f) = \frac{A_v(DC)}{(1 + j\omega\tau_{in})(1 + j\omega\tau_{out})} \tag{3.10}$$

In Eq. (3.10), τ_{in} and τ_{out} are, respectively, input and output time constants. In general, the output time constant prevails, as load capacitances tend to be larger, and sets the first pole in Eq. (3.10). In this case, $\tau_{out} = \frac{g_{m1}}{C_{out}}$ and the amplifier bandwidth f_{3dB} is given by Eq. (3.11), where C_{out} is the output capacitance of the common-source amplifier.

$$f_{3dB} = \frac{g_{m1}}{2\pi C_{out}} \tag{3.11}$$

The SOS process features six kind of transistors for each MOSFET type. We have simulated the AC response of six kinds of

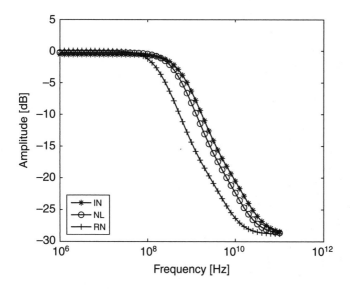

FIGURE 3.11   Comparison of the AC magnitude simulation for SOS NMOS common-source amplifiers based on different transistors (RN, NL, IN). All transistor sizes were $[W, L] = [2 \ \mu m, 2 \ \mu m]$. An output load capacitor of 20-fF was used, with an input DC voltage of 1 V and a power supply of 3.3 V.

common-source amplifiers, each designed with a different kind of transistor, in order to show the differences in the responses. We used the following types of transistors: NMOS devices were regular threshold (0.7 V RN), low threshold (0.3 V NL), and intrinsic (0 V IN). PMOS devices were regular threshold (0.7 V RP), low threshold (0.3 V PL), and intrinsic (0 V IP). We conducted all simulation with input and output capacitive loads of 20-fF. NMOS common-source amplifiers were topologically identical to the one in Fig. 3.10(A), and PMOS amplifier were identical to Fig. 3.10(B).

Figures 3.11 and 3.12 report the values of magnitude and phase AC simulations of SOS NMOS common-source amplifiers. The NMOS simulations were conducted with an input DC voltage of 1 V and a power supply of 3.3 V.

The low-frequency gain was one in all types of amplifiers, as all transistor sizes were $[W, L] = [2 \ \mu m, 2 \ \mu m]$. NMOS devices showed an increasingly higher bandwidth when lower threshold transistors were used. The reason is that lower threshold devices have a slightly lower transconductance g_m but an increased drain current. As can be seen in Sec. 1.6 and Figs. 1.5 and 1.6, the drain current of low-threshold transistors is approximately twice that of regular, and the

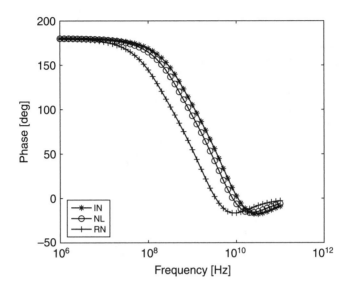

FIGURE 3.12 A comparison of the AC phase simulation for SOS NMOS common-source amplifiers based on different transistors (RN, NL, IN). All transistor sizes were [W, L] = [2 μm, 2 μm]. An output load capacitor of 20-fF was used, with an input DC voltage of 1 V and a power supply of 3.3 V.

drain current of intrinsic is about 2.5 times that of regular transistors. The data in Fig. 3.11 show that NL transistor source followers have a bandwidth (400 MHz) of approximately 2 times the RN transistors (200 MHz), whereas IN transistors amplifiers have a bandwidth of 2.5 times (500 MHz) than RN. These results are consistent with values of g_{m1}, respectively, of 25 μA/V for RN, 50 μA/V for RN, and 65 μA/V for IN using Eq. (3.11).

Figures 3.13 and 3.14 report the values of magnitude and phase AC simulation of SOS PMOS common-source amplifiers. The PMOS simulations were conducted with an input DC voltage of 2 V and a power supply of 3.3 V. The low-frequency gain was one in all types of amplifiers, as all transistor sizes were [W, L] = [2 μm, 2 μm]. Referring to Fig. 3.13, low-threshold LP common-source amplifiers have a bandwidth (290 MHz) that is less than 2 times that of regular threshold RP (180 MHz) and also have intrinsic IP amplifiers with a bandwidth (350 MHz) about twice that of RP ones. These results are consistent with values of g_{m1}, respectively, of 21 μA/V for RP, 35 μA/V for PL, and 45 μA/V for IP using Eq. (3.11).

We also provide simulations of the NMOS common-source amplifier AC magnitude with a supply voltage of 1.5 V in Fig. 3.15. In this case, the input DC voltage was set to 0.8 V. Bandwidth in this case is

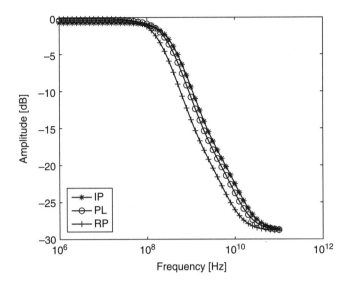

FIGURE 3.13 A comparison of the AC magnitude simulation for SOS PMOS common-source amplifiers based on different transistors (RP, PL, IP). All transistor sizes were $[W, L] = [2 \ \mu\text{m}, 2 \ \mu\text{m}]$. An output load capacitor of 20-fF was used, with an input DC voltage of 2 V and a power supply of 3.3 V.

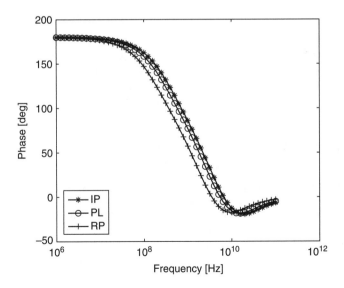

FIGURE 3.14 A comparison of the AC phase simulation for SOS PMOS common-source amplifiers based on different transistors (RP, PL, IP). All transistor sizes were $[W, L] = [2 \ \mu\text{m}, 2 \ \mu\text{m}]$. An output load capacitor of 20-fF was used, with an input DC voltage of 2 V and a power supply of 3.3 V.

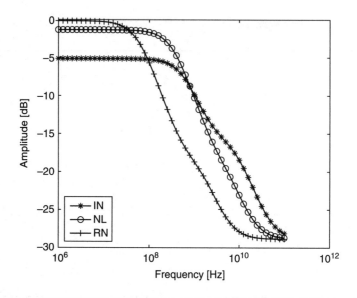

FIGURE 3.15 A comparison of the AC magnitude simulation for SOS NMOS common-source amplifiers based on different transistors (RN, NL, IN). All transistor sizes were $[W, L] = [2\ \mu m, 2\ \mu m]$. An output load capacitor of 20-fF was used, with an DC input voltage of 0.8 V and a power supply of 1.5 V.

60 MHz for RN, 300 MHz for NL and 600 MHz for IN. These results are consistent with values of g_{m1}, respectively, 8 $\mu A/V$ for RN, 38 $\mu A/V$ for RN, and 75 $\mu A/V$ for IN using Eq. (3.11). On the other hand, the DC gain is significantly reduced for the IN case, making it unusable for the application as common source amplifier in this configuration and with a DC input of 0.8 V.

3.5.1.2 SOS Common-Source Amplifiers with Current-Source Load

Common-source amplifiers can also be designed with a current-source load. A schematic representation of this type of common source amplifier is given in Fig. 3.16, where (A) is an NMOS input version, and (B) is a PMOS input version. Transistors M1 are the input transistors, whereas M2 and M3 are current mirrors that reflects the desired bias current I_S. Transistors M2 and M3 are usually the same size, but this is not a requirement. This stage allows accurate control of the bias current to the amplifier, as opposed to accurately sizing the devices. In general, current-source M2 provides a very large output impedance, which in turns gives this common-source amplifier configuration a very large gain.

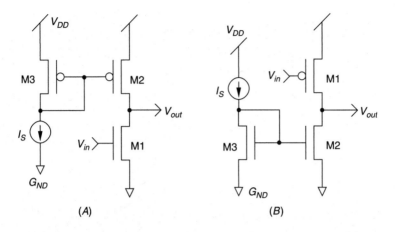

FIGURE 3.16   Schematic of a common source amplifier with a current source load. M1 is the input transistor, and M2 and M3 form a current mirror that biases the stage with a set current I_S. (A) is the NMOS input version, and (B) is the PMOS input version.

The common-source amplifier with current-source load provides a voltage gain given by Eq. (3.12), where all quantities are small-signal voltages and currents.

$$A_v = \frac{v_{in}}{v_{out}} = -\frac{r_{o1}||r_{o2}}{1/g_{m1}} = -g_{m1}(r_{o1}||r_{o2}) \qquad (3.12)$$

We have simulated the AC performance of the common source amplifier with current source load of Fig. 3.16. In these simulations we used a supply voltage of 3.3 V and we set the bias current I_S to 20 μA. We simulated all types of SOS transistors; for example, the amplifier labeled RN has an RN NMOS as input transistor M1, and RP transistor as M2 and M3. All NMOS transistors had a size of $[W, L] = [10$ μm, 2 μm] and all PMOS transistor had a size of $[W, L] = [20$ μm, 2 μm]. All amplifiers were connected to a 20-fF load capacitor at the output. Simulation results are given in Figs. 3.17 and 3.18.

For NMOS amplifiers, the input DC voltage was set to 948, 405, and 60 mV, respectively, for RN, NL, and IN types. The amplifiers are very sensitive to the input DC voltage, and we obtained the values given above from a DC simulation of the operation point that provided the largest gain when the input DC was swept from ground to supply voltage 3.3 V.

Notice that the gain of RN was highest, followed by NL and IN, respectively. The reduction in gain is due to the decrease in output resistance typical of low-threshold and intrinsic transistors, even if

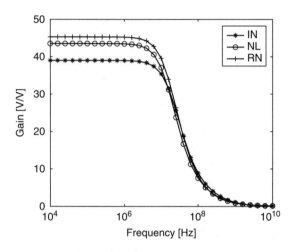

FIGURE 3.17 AC simulations of NMOS input common source amplifier with a current source load. All NMOS transistors had a size of $[W, L] = [10\ \mu m, 2\ \mu m]$ and all PMOS transistor had a size of $[W, L] = [20\ \mu m, 2\ \mu m]$. A 20-fF load capacitor was used at the output.

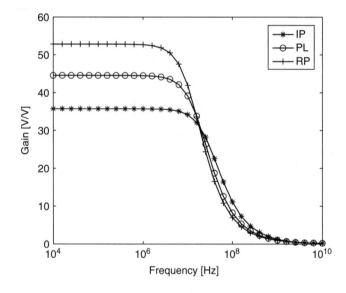

FIGURE 3.18 AC simulations of PMOS input common source amplifier with a current source load. All NMOS transistors had a size of $[W, L] = [10\ \mu m, 2\ \mu m]$ and all PMOS transistor had a size of $[W, L] = [20\ \mu m, 2\ \mu m]$. A 20-fF load capacitor was used at the output.

the transconductance of an M1 increase in NL and IN amplifiers. The decrease in output resistance is due to the larger channel conductance of these devices.

For PMOS amplifiers, the input DC voltage was set to 2.43, 2.8, and 3 V, respectively, for RP, PL, and IP types. The amplifier is very sensitive to the input DC voltage, and we obtained the values given above from a DC simulation of the operation point that provided the largest gain when the input DC was swept from ground to the supply voltage 3.3 V.

Similar to NMOS input amplifiers, notice that the gain of RP was the highest, followed by PL and IP, respectively. Reduction in gain is due to the decrease in output resistance of low-threshold intrinsic transistors, even if the transconductance of M1 increase in PL and IP amplifiers.

The bandwidth of the amplifiers is similar and varies between 100 MHz for RP to 200 MHz for IP, due to the reduction of gain. NMOS input amplifiers have a bandwidth of approximately 150 MHz. The bandwidth is limited by the source current I_S of 20 μA and the load capacitor of 20-fF, which gives a bandwidth of approximately 150 MHz.

The maximum DC gain that we obtained from the simulations in Figs. 3.17 and 3.18 of these stages can be computed by using Eq. (3.12), with g_{m1} being on the order of 100 μA/V for RN, and the output resistance is approximately 400 kΩ, for NL and IN g_{m1} is slightly higher, but the output resistance decreases. The gain of all the stages is nevertheless approximately the same and around 40 V/V.

The AC gain of the amplifier can change by a large amount (one order of magnitude) with the only parameter set by the input DC voltage; therefore, it is challenging to obtain the same results in real fabricated devices and in simulations. When a large gain is desired, using current mirrors and current sources as loads helps the designer significantly, because the load are self-biased. The difficulty of choosing the appropriate input DC voltage, however, makes the design of high-gain stages as common-source amplifiers quite challenging.

From these simulations, we can deduce that a regular threshold transistor as load and an intrinsic or low-threshold input are the best combination that can be obtained in SOS and result in the largest gain.

3.5.2 SOS Cascoded Common-Source Amplifiers

Cascoded common-source amplifiers are a special version of common-source amplifiers that decouple input and output in order to increase the overall voltage gain. A common-source amplifier schematic is shown in Fig. 3.19 for both NMOS and PMOS implementations. Transistor M1 is the input transistor, M2 the current source of the

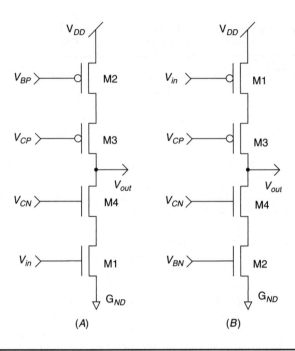

FIGURE 3.19 Cascoded common-source amplifiers schematic. (*A*) is an all NMOS version, and (*B*) an all PMOS version. M1 transistors are input transistors, and M2 transistors are the common-source amplifier load. Transistors M3 and M4 are the cascode transistors.

common-source amplifier, and M3 and M4 are the cascode transistors. M3 and M4 decouple the output impedance of the load from the input transistor, so that the input loading that M1 effectively sees is much lower.

The gain of this stage is increased because input transistor M1 is not loaded directly by the output, instead, it is loaded by the source of M3 (or M4 for the NMOS input), which has a low-input impedance. The gain of the device can be expressed by Eq. (3.13), where g_{m1} is the transconductance of transistor M1 and R_{ocs} is the output resistance. R_{ocs} is the parallel of the output resistance of M3 and M4, given respectively by $g_{m3}r_{o3}$ and $g_{m4}r_{o4}$.

$$A_v = \frac{V_{out}}{V_{in}} = -g_{m1}R_{ocs} \qquad (3.13)$$

We have simulated the AC response of six kinds of cascoded common-source amplifiers, each designed with a different kind of transistor, to show the differences in the responses. The simulation circuit schematic is given in Fig. 3.20, the AC simulation results for

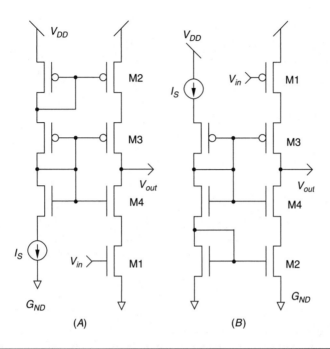

FIGURE 3.20 SOS cascoded common-source amplifiers schematic used for AC simulations. (*A*) All NMOS version. (*B*) All PMOS version. Transistors M1 are the input transistor, and transistors M2 are the common-source amplifier load. Transistors M3 and M4 are the cascode transistors. Transistors on the left side are diode-connected biasing transistors of identical size to the ones on the right side. A current source I_S of 20-μA was used to bias the circuit. A load capacitance of 20-fF was connected to the output.

NMOS are shown in Fig. 3.21 are for NMOS transistors, and for PMOS transistors are shown in Fig. 3.22. We used the following types of transistors: NMOS devices were regular threshold (0.7 V RN), low threshold (0.3 V NL), and intrinsic (0 V IN). PMOS devices were regular threshold (0.7 V RP), low threshold (0.3 V PL) and intrinsic (0 V IP). NMOS cascoded common-source amplifiers were topologically identical to the one in Fig. 3.20(*A*), while PMOS amplifier were identical to Fig. 3.20(*B*). All NMOS transistor sizes were $[W, L] = [10 \ \mu m, 2 \ \mu m]$, while PMOS transistor sizes were $[W, L] = [20 \ \mu m, 2 \ \mu m]$. All simulation we conducted with an output capacitive load of 20-fF. We have used the input DC voltage that resulted in the maximum gain from the amplifier. We have swept the DC input voltage by running a DC simulation and simultaneously looking at the AC gain.

The input DC voltage was set to 940, 400, and 20 mV, respectively, for NMOS RN, NL, IN amplifier type. In the conditions of maximum

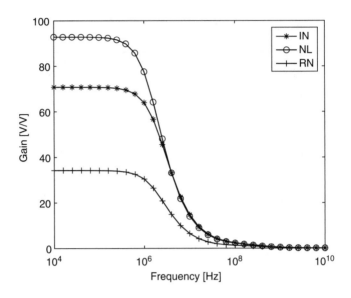

FIGURE 3.21 Comparison of the AC magnitude simulation for SOS NMOS cascoded common-source amplifiers based on different input transistors (RN, NL, IN). All NMOS transistor sizes were $[W, L] = [10\ \mu m, 2\ \mu m]$, while PMOS transistor sizes were $[W, L] = [20\ \mu m, 2\ \mu m]$. The input DC voltage was set to 940, 400, and 20 mV, respectively for RN, NL, IN. The output capacitive load was 20-fF.

gain, referring to Fig. 3.20(A), the gate voltage of M2 is 2.42 V, and the gate voltage for M3 and M4 is 1.55 V for the RN amplifier. These values are 2.8 V and 2.3 V for NL and 3 V and 2.8 V for IN, respectively. The DC gain of the NMOS input amplifiers was 92 for NL, 71 for IN, and 34 for RN. The bandwidths were 3 MHz for NL, 3 MHz for IN, and 2 MHz for RN. The low bandwidths are due to the decrease in the output current: approximately 0.4 μA for NL and IN and 0.3 μA for RN. At these current levels, the output resistance of the cascode stage was on the order of 3–4 MΩ.

The input DC voltage was set to 2.43, 2.8, and 3 V, respectively, for PMOS RP, PL, IP amplifier type. In the conditions of maximum gain, referring to Fig. 3.20(B), the gate voltage of M2 is 952 mV, and the gate voltage for M3 and M4 is 1.4 V for the RP amplifier. These values are 402 mV and 802 mV for PL and 117 mV and 235 mV for IP, respectively. The DC gain of the PMOS input amplifiers was 118 for PL, 63 for IP, and 32 for RP. The bandwidths were 1 MHz for PL, 1.5 MHz for IP, and 3 MHz for RP. The low bandwidths are due to the decrease in the output current: approximately 0.2 μA for PL, 0.3 μA for IP, and 0.4 μA

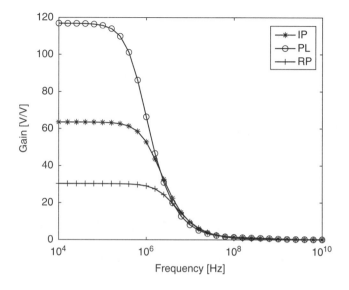

FIGURE 3.22 A comparison of the AC magnitude simulation for SOS PMOS cascoded common-source amplifiers based on different input transistors (RP, PL, IP). All NMOS transistor sizes were $[W, L] = [10\ \mu m, 2\ \mu m]$, while PMOS transistor sizes were $[W, L] = [20\ \mu m, 2\ \mu m]$. The input DC voltage was set to 2.43, 2.8, and 3 V, respectively, for PMOS RP, PL, IP. The output capacitive load was 20-fF.

for RP. At these current levels, the output resistance of the cascode stage was on the order of 3–5 MΩ.

Notice that the gain of the cascoded common-source amplifiers is maximum with the use of low-threshold transistors for both NMOS and PMOS input amplifiers. Low-threshold NMOS input amplifiers reported a gain approximately 2 to 3 times the gain of a regular threshold, whereas intrinsics were approximately twice the gain of a regular threshold. Notice that these gains are higher than the ones obtainable with common-source amplifiers without cascode, as seen in the previous section. The gain with cascode is approximately twice as high with cascoded low-threshold inputs as the one with regular-threshold noncascode amplifiers.

3.5.3 SOS Source-Follower Amplifiers

A source follower amplifier is composed of two transistors where one is the input and one is used as a current source. The output of the source-follower amplifier is a voltage level-shifted version of the input voltage. This stage is used to convert from high impedance to low impedance without the need of gain. The stage is used to decouple

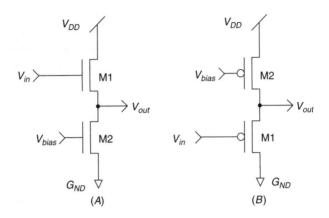

FIGURE 3.23 Source-follower amplifiers schematic. (*A*) All-NMOS version. (*B*) All-PMOS version. Transistor M1 is the input transistor, while transistor M2 is the current source of the source-follower amplifier.

an input from the output while providing a large current source and also a current sink less than the current source of the amplifier.

Figure 3.23 shows two source-follower configurations. An all-NMOS transistors is given in Fig. 3.23(*A*). In this case, both input and output voltages are referred to the supply voltage V_{DD}. An all as PMOS transistor source-follower is given in Fig. 3.23(*B*), where both input and output voltages are referred to ground. Transistor M1 in Fig. 3.23 is the input transistor, and transistor M2 is the current source of the source-follower amplifier. The current sinking capabilities of this stage are limited to the value of the current of these transistors.

The NMOS source-follower amplifier has a gain defined in Eq. (3.14), where $V_{out} = I_d\, r_o$ and $V_{GS2} = I_D/G_{m2}$. From Eq. (3.14), we can deduce that the maximum gain of this stage is one in case that the small-signal output resistance is high. This is usually the case, as this kind of amplifier is used as a buffer.

$$A_v = \frac{V_{in}}{V_{out}} = -\frac{r_o}{r_o + 1/g_{m2}} = \frac{g_{m2}r_o}{g_{m2}r_o + 1} \qquad (3.14)$$

Assuming that the gain of the device is 1, the large signal characteristics of these amplifier is given by Eq. (3.15), where $V_{th,M1}$ is the threshold voltage of the input transistor M1. Refer to Fig. 3.23(*A*) for details. The reason for the behavior in Eq. (3.15) is that transistor M1 is operating in saturation only when the input is higher than one voltage threshold.

$$V_{out} = V_{in} - V_{th,M1} \qquad (3.15)$$

The amplifier bandwidth has a time constant τ_{out} identical to the common-source amplifier and given by Eq. (3.11), where C_{out} is the output capacitance of the source-follower amplifier. This equation holds when the load capacitance is higher than the amplifier internal capacitances.

We have simulated six kinds of source-follower amplifiers implemented with different SOS MOSFET devices. All transistor sizes were $[W, L] = [2 \mu m, 2 \mu m]$. We used the following types of transistors: NMOS devices were regular threshold (0.7 V RN), low threshold (0.3 V NL), and intrinsic (0 V IN). PMOS devices were regular threshold (0.7 V RP), low threshold (0.3 V PL) and intrinsic (0 V IP). The results are given in Figs. 3.24 and 3.25 for the NMOS and PMOS transistors, respectively. The simulations were conducted by sweeping the input voltage V_{in} and with a supply voltage $V_{DD} = 3.3$ V. The bias voltages for the M2 gate transistors were 0.7 V for RN transistors, 0.3 V for NL, and 0 V for IN. The PMOS bias voltages of M2 gate were 2.6 V for RP transistors, 3 V for PL, and 3.3 V for IP.

The results seen in Figs. 3.24 and 3.25 show the same trends predicted in Eq. (3.15). We can see that each kind of transistor is responsible for a different level-shifting equal to its threshold voltages. In SOS, we can thus obtain voltage shifts of 0.7 V, 0.3 V, or 0 V just by

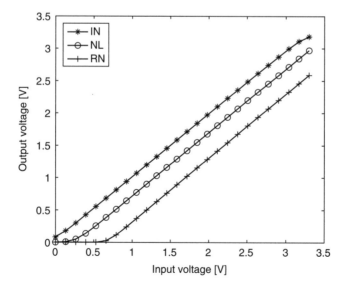

FIGURE 3.24 SOS NMOS source followers simulated response to a sweep of the input voltage V_{in} and with a supply voltage $V_{DD} = 3.3$ V. All transistor sizes were $[W, L] = [2 \mu m, 2 \mu m]$. The bias voltages for M2 gate transistors were 0.7 V for RN transistors, 0.3 V for NL, and 0 V for IN.

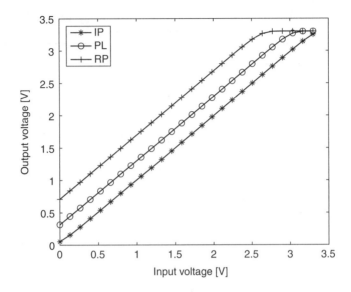

FIGURE 3.25 SOS PMOS source follower–simulated response to a sweep of the input voltage V_{in} and with a supply voltage $V_{DD} = 3.3$ V. All transistor sizes were $[W, L] = [2 \ \mu m, 2 \ \mu m]$. The bias voltages for M2 gate transistors were 2.6 V for RP transistors, 3 V for PL, and 3.3 V for IP.

using the appropriate transistor and without having to resort to very large transistors to reduce the effective threshold, as commonly done in bulk CMOS.

We also simulated the noise performance of the source-follower amplifiers. The result of noise simulations of the source follower in Fig. 3.10 are seen in Figs. 3.26 and 3.27 for NMOS and PMOS amplifiers, respectively.

The data in Figs. 3.26 and 3.27 show a region at low frequency where flicker noise is prevailing up to frequencies of approximately 1 MHz. That is the corner frequency with thermal noise, in this case with a value of 1 fV for NMOS and 10 fV for PMOS amplifiers. The second inflection in the curve at high frequency is the roll-off of the amplifier bandwidth. RN transistors followers show increased flicker noise when compared to lower-threshold NL and IN. In the case of PMOS, the type of transistor does not change the noise performance, at least in these simulations.

We have computed a noise model for the RN NMOS SOS source-follower amplifier reported here. The total noise $v_{e,Total}^2$ for an amplifier like the one reported in Fig. 3.23(A) is expressed by Eq. (3.16), where $v_{e,MX}^2$ are the contribution of transistor X. Notice that the input transistors contribute to noise directly, whereas the NMOS loads are

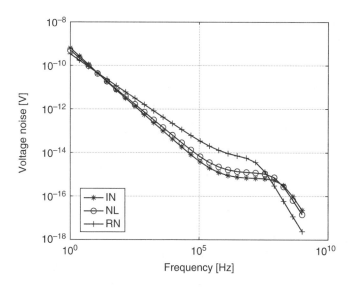

FIGURE 3.26   SOS NMOS source follower–simulated noise response. All transistor sizes were $[W, L] = [2 \, \mu m, 2 \, \mu m]$. RN transistors followers show increased flicker noise when compared to lower-threshold NL and IN.

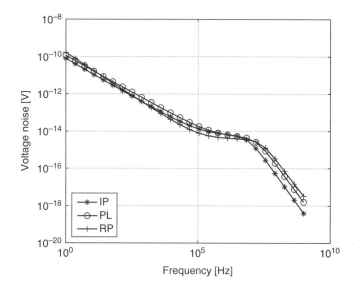

FIGURE 3.27   SOS PMOS source follower–simulated noise response. All transistor sizes were $[W, L] = [2 \, \mu m, 2 \, \mu m]$. The type of PMOS transistor does not change the noise performance.

reduced by the ratio of their transconductance and the input transconductance.

$$v_{e,Total}^2 = v_{e,M1}^2 + \left(\frac{g_{m2}}{g_{m1}}\right)^2 v_{e,M2}^2 \tag{3.16}$$

The theoretical noise model can be divided into flicker noise and thermal noise components. The total noise of the amplifier is the sum of the two components. Equations (3.17) and (3.18) report the model equation for flicker and thermal components, respectively.

$$v_{flicker}^2 = \frac{2K_{M1}}{f^{AF}W_1L_1C_{ox}}\frac{K_{M2}\mu_{M2}L_{M1}^2}{K_{M1}\mu_{M1}L_{M2}^2}df \tag{3.17}$$

$$v_{thermal}^2 = \frac{8kT}{3g_{m1}}\left(1 + \sqrt{\frac{\mu_{M2}(W2/L2)}{\mu_{n1}(W1/L1)}}\right)df \tag{3.18}$$

Equations (3.17) and (3.18) reduce to Eqs. (3.19) and (3.20) when the transconductance for the two transistors M1 and M2 is identical. This value is twice the value of one transistor.

$$v_{flicker}^2 = 2\frac{2K_{M1}}{f^{AF}W_1L_1C_{ox}}df \tag{3.19}$$

$$v_{thermal}^2 = 2\frac{8kT}{3g_{m1}}df \tag{3.20}$$

In Eqs. (3.17) and (3.18), K_n, K_p are the flicker coefficients for both NMOS and PMOS transistors, AF is the flicker exponent parameter, C_{ox} is the gate oxide capacitance, and μ_p and μ_n are the mobility of hole and electrons.

By computing the sum of Eqs. (3.17) and (3.18) for the RP source-follower amplifier case, we obtain the input-referred voltage noise power spectral density in Fig. 3.28. K_n and K_p have an extracted value of 5×10^{-24}, and AF a value of 0.8.

3.6 SOS Differential Amplifiers

A differential amplifier is one of the most important components for the design of linear circuits in CMOS. It is also the input stage of most operational amplifier architectures, and its performance is the main consideration in the design of such amplifiers. The differential input stage has the advantage of providing an output that is relatively independent to the DC level of the inputs, and thus is less affected by bias voltages shifts. The output of the differential amplifier is an amplified version of the difference between the two input voltages. A schematic of an SOS differential amplifiers is given in Fig. 3.29.

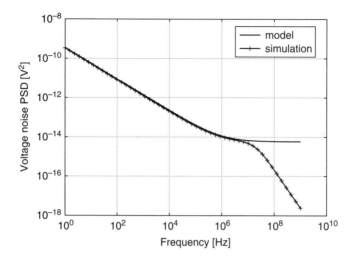

FIGURE 3.28   SOS RN source follower voltage noise power spectral density-simulated response and computed model. The amplifier schematic is reported in Fig. 3.23(*A*).

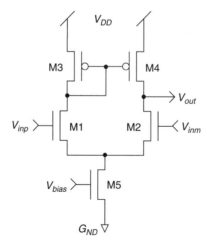

FIGURE 3.29   SOS differential amplifiers DC simulation. Transistors M1 and M2 are the input transistors, M3 and M4 are the current mirror load transistors, and M5 is the tail current source transistor. In general, M1, M2, M3, and M4 are pairwise identical in size.

This schematic is identical to a standard bulk CMOS implementation, and it is composed by a differential pair of transistors M1 and M2 of identical size, topology, and geometry. PMOS transistors M3 and M4 are a current mirror load, offering a voltage output between the input and load transistors. Transistor M5 is the tail current source of the amplifier and defines the current flowing in the two branches of the inputs.

The amplifier in Fig. 3.29 is a very common input stage and is frequently used in CMOS circuit design and operational amplifiers. Inputs have a common mode voltage, which is generally the middle of the supply voltage. This choice maximizes the output voltage swing. An imbalance between the voltage at the two inputs will force the tail current to flow more on one branch than the other. The output voltage can swing between the supply voltage and ground. Notice that the output voltage level is usually identical to the voltage at the gate of load transistors M3 and M4.

The minimum common mode voltage is given by Eq. (3.21), or the sum of the gate voltage of M1, M2 and the saturation gate voltage of M5 (defined at the desired tail current).

$$V_{CM,\min} = V_{GS1} + V_{DS5,sat} \tag{3.21}$$

The maximum common-mode voltage is given by Eq. (3.22), to keep M1, M2, M3, and M4 all in saturation. Notice that this value approximates V_{DD} in SOS circuits, due to the lack of body effect for transistors M1 and M2.

$$V_{CM,\max} = V_{DD} - V_{GS3} + V_{th1} \tag{3.22}$$

The maximum output voltage is given by Eq. (3.23), which is the maximum voltage that keeps load transistors M3 and M4 in saturation.

$$V_{out,\max} = V_{DD} - V_{DS4,sat} \tag{3.23}$$

The minimum output voltage is given by Eq. (3.24), to keep the input transistors M1, M2 in saturation.

$$V_{out,\min} = V_{DS5,sat} + V_{th1} \tag{3.24}$$

Finally, the minimum supply voltage allowed for the differential amplifier is given by Eq. (3.25). Here, V_{GS3} is the gate voltage of M3 load transistors. This voltage needs to be low enough to allow for the bias current to flow and change the output voltage.

$$V_{DD,\min} = V_{GS3} + V_{DS1,sat} + V_{DS5,sat} \tag{3.25}$$

The small-signal operation of the difference amplifier can be described by Eq. (3.26), where the small signal differential gain A_{vd} is obtained as a function of the transconductance of input transistors $M1$, g_{m1} and the parallel of the output resistance of r_{o2} and r_{o4} (because these are usually equal, the parallel is just $r_{o2}/2$)

$$A_{vd} = \frac{v_{out}}{v_{in}} = g_{m1}\frac{r_{o2}}{2} \qquad (3.26)$$

In general, a differential pair can be designed with different goals, such as large gain, large swing, low noise, low power, and many more. Here we report a design that represents a typical compromise between these goals. We have conducted simulations of three kind of SOS differential amplifiers, using the three kind of available transistors—intrinsic, regular, and low threshold. All differential amplifiers use the topology in Fig. 3.29, and use an NMOS transistor as the input. We have simulated three amplifiers—one with RN NMOS input, one with NL NMOS input, and one with IN NMOS input. The RN NMOS input is the reference differential amplifier that can be designed in any bulk CMOS process and only uses 0.7 V threshold transistors. To obtain a fair comparison of the advantages and disadvantages, we have kept a few condition constant for all three kinds of amplifier. The maximum tail current for the simulations was set to 50 μA for all amplifiers. This current was reached as the input terminal was at 3.3 V. The current source transistors type (tail) does not make a difference in the circuit, only the tail current amplitude does. We used a supply of 3.3 V and an input DC voltage of 1.6 V (middle of rails) for all amplifier inputs. We have used this value to be able to maximize the output swing with large signal inputs. The input transistor sizes are identical for all amplifier types and were $[W/L] = 50$ μm/2 μm, whereas the tail current source for the RN amplifier was 100-μm/ 2-μm and was biased at 0.8 V. For the NL and IN amplifiers, we used an IN transistor of 30 μm/2 μm biased at 0 V. All PMOS mirror loads were 4 μm/4 μm, and are of the RP, PL, and IP types for amplifiers labeled RN, NL, and IN, respectively.

Figure 3.30 shows a DC simulation of the output voltage of all three amplifiers when input to M2 was tied to 1.6 V and the input to transistor M1 (input voltage in Fig. 3.30) was swept from 0 V to the supply voltage of 3.3 V. The region around 1.6 V input is the highest gain region for each amplifier, followed by saturation at approximately the supply voltage when the input was larger than 1.6 V. The region where the input voltage was 0 V to approximately 1.5 V shows that the amplifier is behaving like a voltage follower (refer to Fig. 3.24). Similar to the voltage follower response, the linear response of $V_{out} = V_{in} - V_{th}$. Because the voltage threshold is different for the three input transistor types (for RN is 0.7 V, for NL is 0.3 V, and for IN is 0 V), the response of the three amplifier is different in this region. The follower region is due to the fact that the current induced on transistor M1 is much

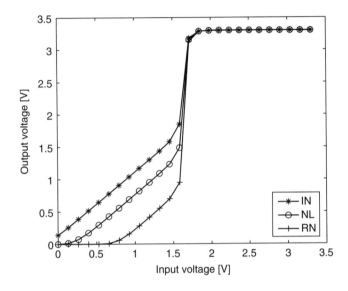

FIGURE 3.30   SOS differential amplifiers DC simulation. The input to M2 was tied to 1.6 V and the input to transistor M1 was swept from 0 V to the supply voltage of 3.3 V. The amplifier operates as a voltage follower when the input is less than 1.5 V, then shows a region of high gain around 1.6 V and a saturation to the supply voltage afterward.

smaller than the saturation current of M5 (set to 50 μA for all amplifiers). When M5 current saturates, the gain increases significantly, as can be seen in Fig. 3.30 with the input voltage around 1.6 V.

Figure 3.31 shows an AC simulation of the voltage gain of all three amplifiers when an AC input is given at transistor M1, whereas the DC level of both M1 and M2 was set to 1.6 V. The simulations were conducted with a 20-fF load capacitor at the output. The results show that the RN differential amplifier implementation has the highest voltage gain, whereas NL and IN amplifiers respectively reach about a half and a third of the RN gain. Referring to Eq. (3.26), amplifiers NL and IN offer progressively lower transconductance values, as can be seen in Figs. 1.5 and 1.6 in Chap. 1. Even if the transconductance of NL and IN is lower than RN, the drain current of NL is higher than RN and the one of IN higher than NL, and this reflects in higher bandwidth for NL when compared to RN and IN when compared to NL. The bandwidth of the RN differential amplifier is 20 MHz, 50 MHz for NL and 80 MHz for IN. This is one of the main advantages of using intrinsic- and low-threshold transistors in a differential amplifier: they can increase the bandwidth at the expense of voltage gain. Also, they can be used to operate at low supply voltages.

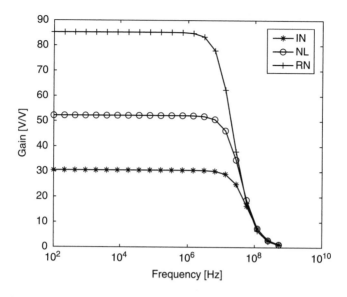

FIGURE 3.31 SOS differential amplifiers AC simulation. The simulations were conducted with a 20-fF load capacitor at the output. The RN differential amplifier has the highest voltage gain, whereas NL and IN amplifiers reach about a half and a third of the maximum RN gain, respectively. The bandwidth of the RN differential amplifier is 20 MHz, 50 MHz for NL, and 80 MHz for IN.

Figure 3.32 shows a transient simulation of the three types of differential amplifiers with a large-signal 10-KHz input sine wave. The input amplitude was 1.6 V with a DC level of 1.6 V and was applied to input transistors M1, whereas transistor M2 was tied to 1.6 V. The transient response reveals aspects of both the AC and DC simulations but also shows the distortion of the input when the signal is large. As can be seen in Fig. 3.32, the input is amplified by the AC gain in Fig. 3.31 when its value is around 1.6 V. When the input is below 1.6 V, the amplifier behaves as a voltage follower, with a threshold shift corresponding to the type of transistor used (for RN is 0.7 V, for NL is 0.3 V, and for IN is 0 V). At inputs higher than 1.6 V, the output of the amplifier saturates to the supply voltage, as shown in the DC simulation and Fig. 3.30.

In conclusion, the design of SOS differential amplifier is very similar to standard-bulk CMOS when a regular threshold transistor (RN) are used. In this case, the main advantage is the absence of body effect, a feature that increases the output swing of the amplifier with respect to bulk CMOS implementations. The RN-type differential amplifier can benefit from the use of a IN transistor types as current source M5,

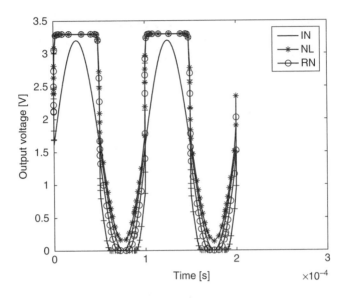

FIGURE 3.32 SOS differential amplifiers large-signal transient response with a 10KHz large-signal input sine wave. The input amplitude was 1.6 V with a DC level of 1.6 V and was applied to input transistors M1, while transistor M2 was tied to 1.6 V.

a choice that decreases the size of that transistor and allows to remove a bias from the design, since and IN M5 can be biased at 0 V.

NL and IN implementations of the differential pair amplifier should be used only when large bandwidths are desired or when low-voltage power supplies are used.

We have simulated the noise performance of the SOS differential amplifiers and report the results in Fig. 3.33. We have computed a noise model for the RN-type SOS differential amplifier reported here. The total noise $v_{e,Total}^2$ for an amplifier like the one reported in Fig. 3.29 can be expressed by Eq. (3.27), where $v_{e,MX}^2$ are the contributions of transistor X. Notice that the input transistors contribute to noise directly, whereas the PMOS loads are reduced by the ratio of their transconductance and the input transconductance.

$$v_{e,Total}^2 = v_{e,M1}^2 + v_{e,M2}^2 + \left(\frac{g_{m3}}{g_{m1}}\right)^2 \left(v_{e,M3}^2 + v_{e,M4}^2\right) \qquad (3.27)$$

The theoretical noise model can be divided into flicker-noise and thermal-noise components. The total noise of the amplifier is the sum

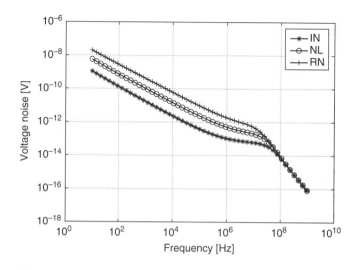

FIGURE 3.33   SOS differential amplifiers voltage noise simulated response.

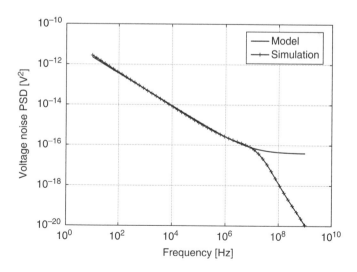

FIGURE 3.34   SOS RN differential amplifiers voltage noise power spectral density–simulated response and computed model.

of the two components. Equations (3.28) and (3.29) report the model equation for flicker and thermal components, respectively.

$$v_{flicker}^2 = \frac{2K_{n1}}{f^{AF} W_1 L_1 C_{ox}} \frac{K_{p3}\mu_{p3}L_{n1}^2}{K_{n1}\mu_{n1}L_{p3}^2} df \qquad (3.28)$$

$$v_{thermal}^2 = \frac{8kT}{3g_{m1}} \left(1 + \sqrt{\frac{\mu_{p3}(W3/L3)}{\mu_{n1}(W1/L1)}}\right) df \qquad (3.29)$$

In Eqs. (3.28) and (3.29), K_n, K_p are the flicker coefficients for both NMOS and PMOS transistors, AF is the flicker exponent parameter, C_{ox} is the gate oxide capacitance, and μ_p and μ_n are the mobility of hole and electrons.

By computing the sum of Eqs. (3.28) and (3.29) for the RN differential amplifier case, we obtain the input-referred voltage noise power spectral density in Fig. 3.34. K_n and K_p have an extracted value of 5×10^{-24} and AF a value of 0.8.

CHAPTER 4

Design of SOS Operational Amplifiers, Comparators, and Voltage References

4.1 Operational Amplifiers in SOS

In this section, we present an example of a fabricated operational amplifier in SOS technology. The amplifier design was targeting intermediate operational-amplifier performance, with an emphasis on compact size, low-power consumption, and good noise performance (Gray et al., 2001; Baker, 2005).

A schematic of the SOS operational amplifier is given in Fig. 4.1. The design uses a three-stage implementation with internal capacitive compensation. The first stage is a conventional differential pair with an active mirror load (for more details, refer to Sec. 3.6). The second stage is a source follower (refer to Sec. 3.5.3) coupled to a current gain output stage. The source follower and output stage of the operational amplifier in Fig. 4.1 is shown in detail in Fig. 4.2. The source-follower–input-transistor M1 is biased by a current source M6. The bias current of this stage limits the amount of current that the operational amplifier can source. Because operational amplifiers have to drive large capacitive or resistive loads, it is important that they offer large-output current-driving capabilities. For this reason, a class AB output stage is added to transistors M2 and M3. The class AB amplifier is obtained by level shifting the output of source followers M1 and M6 with two diode-connected transistors M4 and M5. These two transistors provide the biasing for output stages M1 and M2. When the input to M1 is only a DC bias voltage, the source-follower current

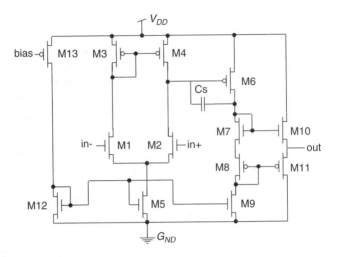

FIGURE 4.1 An SOS operational amplifier schematic. The design uses a three-stage implementation with internal capacitive compensation. The design targets low power, low noise, and reduced size for on-chip analog signal processing.

is mirrored to a fixed value by M1 and M2. If a more positive input is applied to M1, M2 turns on and M3 turns off. The opposite occurs when the input becomes more negative. Thus, the output stage in Fig. 4.2 lowers the power consumption during input swings and keeps it constant during DC input.

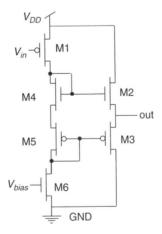

FIGURE 4.2 Schematic of the output amplifier of the SOS operational amplifier in Fig. 4.1. The amplifier is a class AB source follower with current gain stage.

Device Name	Transistor Type	Size [W,L]
M1, M2	RN	15 μm, 5 μm
M3, M4	RP	70 μm, 5 μm
M5	RN	30 μm, 5 μm
M6	RP	70 μm, 5 μm
M7	RN	15 μm, 2 μm
M8	RP	50 μm, 2 μm
M9, M12	RN	15 μm, 5 μm
M10	RN	300 μm, 2 μm
M11	RP	500 μm, 2 μm
M13	RP	2 μm, 2 μm
C_s	MIM capacitor	5 pF

TABLE 4.1   Device sizes of a fabricated SOS operation amplifier of Fig. 4.1.

One of the limitation of the class AB output stage in Fig. 4.2 is the reduced output swing. NMOS transistor M2 is a poor pull-up device, as it shuts off when V_{out} is higher than $V_{DD} - V_{thn}$. Similarly, M3 is a poor pull-down device, as it turns off at its voltage threshold when V_{out} is lower than V_{thp}. For this reason, this stage is not recommended for low-supply voltages because it subtracts approximately $2V_{th}$ from the available output voltage range (Baker, 2005).

This design employs only regular threshold devices because the design power supply was 3.3 V. The amplifier consumed a maximum current of 20 μA during operation.

We have simulated the response and characteristics of the SOS operational amplifier in Fig. 4.1 with the device types and parameters reported in Table 4.1. Figure 4.3 reports the results from an AC magnitude simulation of the operational amplifier. The open-loop gain of the amplifier was 72 dB (4000) and a gain bandwidth product of 3 MHz.

Figure 4.4 reports the results from an AC phase simulation of the operational amplifier. The phase margin is 40° at 3 MHz, where the gain of the amplifier reduces to 1.

Figure 4.5 reports the results from a transient simulation of the operational amplifier connected as a follower (in- and out-shorted). For this simulation the input voltage was a 10-KHz sine wave with a 100 mV amplitude centered at a 1.6 V DC.

Figure 4.6 reports the results from a transient simulation of the operational amplifier connected as a follower (in- and out-shorted) as

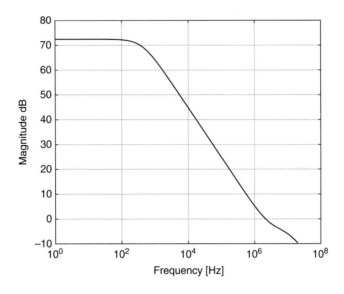

FIGURE 4.3 AC magnitude simulation response of the operational amplifier in Fig. 4.1.

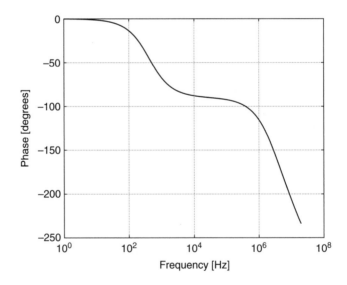

FIGURE 4.4 AC phase simulation response of the operational amplifier in Fig. 4.1.

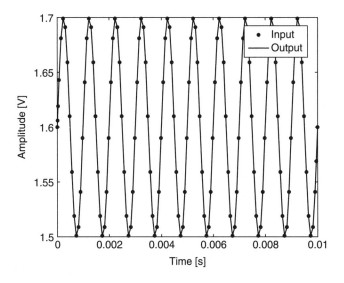

FIGURE 4.5 Transient simulation of the operational amplifier in Fig. 4.1 connected as a follower (in- and out-shorted). The input was a 10-kHz sine wave of 100 mV amplitude and 1.6 V DC.

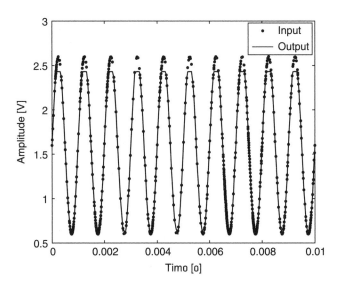

FIGURE 4.6 Transient simulation of the operational amplifier in Fig. 4.1 connected as a follower (in- and out-shorted). The input was a 10-kHz sine wave of 1 V amplitude and 1.6 V DC.

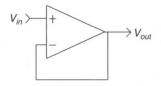

FIGURE 4.7 Schematic of the operational amplifier of Fig. 4.1 connected as a follower (in- and out-shorted).

shown in Fig. 4.7. For this simulation the input voltage was a 10-KHz sine wave with a 1-V amplitude centered at a 1.6-V DC. Notice that the amplifier saturates at approximately 2.4 V and 0.6 V. The range of this operational amplifier is reduced by the class AB output stage.

We have measured and computed a model of the input-referred voltage power spectral density of the operational amplifier in Fig. 4.1. The amplifier was connected as a follower (in- and out- shorted and gain of 1), as shown in Fig. 4.7. The result are reported in Fig. 4.8, where both the measurements and theoretical model are shown between 10 Hz and 20 kHz. The theoretical noise model of the operational amplifier voltage noise is identical to the one presented for the differential pair amplifier in Sec. 3.6. The differential pair of this stage

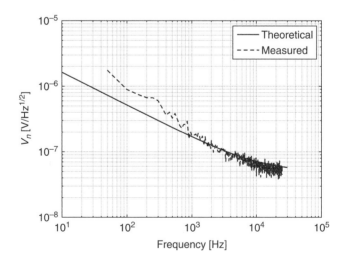

FIGURE 4.8 Noise voltage power spectral density of the operational amplifier in Fig. 4.1. Both measured and theoretical computed model are presented.

DC gain	72 dB
Gain bandwidth product	3 MHz
Input common mode range	0.5 V_{pp}
Output swing	0.7 V_{pp}
Slew rate	3.3 V/μs
Bias current at 3.3 V	20 μA

TABLE 4.2 Performance of the operational amplifier used in the design.

is the significant noise contribution because it is the input and first-gain stage. The following gain stages effectively do not contribute to the added noise because the signal is now amplified by the gain of the first stage.

Table 4.2 reports a summary of the SOS operational amplifier design.

4.2 Comparator Circuits in SOS

A voltage comparator is a circuit that compares an instantaneous value of an input signal V_{in} with a reference voltage V_{ref} and produces a logic output depending on whether the input is larger or smaller than the reference level. A comparator design is essential for analog-to-digital converters (ADC) because its conversion speed is limited by the decision-making response time of the comparator. In addition, the accuracy of a comparator determines the precision of the analog-to-digital conversion.

A comparator is essentially a high-gain operational amplifier (op-amp) designed for open-loop operation. Unlike an op-amp, a comparator does not require frequency compensation in order to maximize the gain, it also provides a digital output only. Any operational amplifiers in Sec. 4.1 can be used as comparator when operated in open loop (Gray et al., 2001; Baker, 2005).

The transfer curve of a differential comparator is shown Fig. 4.9. In this figure, the negative input of the comparator in the figure connects to the input voltage V_{in}. When the input is greater than V_{ref}, the output is low (V_{OL}); otherwise it is high (V_{OH}). In Fig. 4.9, we represented two typical nonideal effects of a comparator—an offset voltage and a finite gain. Ideally, a comparator would have an infinite gain and the slope at V_{in} equal V_{ref} would be infinite. In reality, the differential gain has

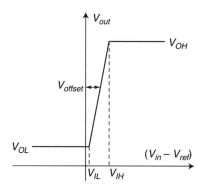

FIGURE 4.9 Transfer function of a voltage comparator. When the input V_{in} is higher (lower) than V_{ref}, the output voltage is V_{OH} (V_{OL}). A finite voltage gain prevents a digital decision between V_{IL} and V_{IH}. An offset voltage can also shift the transfer curve of the comparator.

a finite value (A_v), therefore, there is bound to be a small range of $V_{in}-V_{ref}$, between V_{IL} and V_{IH}, where the output is not clearly digitally determined. Another nonideal effect of the differential comparator is the input-referred DC offset voltage. In the absence of the offset voltage, the comparator DC transfer curve will be symmetrical around $V_{in} - V_{ref} = 0$. However, because of a finite value of V_{offset}, the DC transfer curve of a practical differential comparator is shifted.

Notice that the gain A_v of a comparator is related to its resolution in bits. As an example consider a 8-bit comparator on a 1-V input range: it needs to resolve $1/2^8 \times 1/\sqrt{12}$, or approximately 1.1 mV. If we suppose a 1-V supply, the comparator needs to have a gain of 1 V/1.1 mV equal to 887.

Speed is another important parameter of a comparator. In most applications, it is required that, after an input level change, the comparator must switch between two output levels within the shortest specified amount of time. This response time limits the performance of this comparator and, in turn, the speed of the ADC. The response time is a function of the input voltage difference, the gain, and the current consumption of the circuits. The speed is directly proportional to the conversion rate of the comparator. If the comparator is used in a 1-Ms/s 8-bit ADC that needs 8 comparisons per sample, then the comparator needs to be at least capable of 8 MHz operation (usually at least twice this value).

A typical high-performance comparator block diagram is reported in Fig. 4.10. The comparators can be designed to operate in three stages: the first being a gain amplification stage, as a simple differential

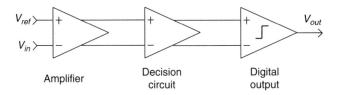

FIGURE 4.10   Typical block diagram of a high-performance comparator. An input stage amplifies the signal and feeds it to a decision stage. The digital output stage converts a differential signal into a buffered digital output.

amplifier. The second stage provides a further differential output amplification and, at the same time, a latchlike functionality. This component is truly a mixed-signal block that converts the differences in the input in a differential pseudo-digital output. The final stage of the comparator is a digital buffer that converts differential signals into a digital output.

In this section we present and discuss the design of SOS comparators. We report the design of a high-performance comparator design and an ultralow power comparator design.

4.2.1 SOS High-Performance Comparators

In this section, we present the design in SOS of a high-performance comparator, with resolutions equal or higher than 8-bits (Baker, 2005). The comparator was implemented by three stages, as represented in Fig. 4.10. A schematic of the three-stage SOS comparator is given in Fig. 4.11.

The comparator consistors of three stages: an input preamplifier (Stage 1 in Fig. 4.11), a positive feedback decision circuit (Stage 2), and an output buffer (Stage 3).

The comparator in Fig. 4.11 uses an active load configuration with fast decision circuit. At the frontend there is a preamplification stage made with a differential amplifier with active diode-connected loads (transistors M13 and M14). The input NFETs M11 and M12 of the differential pair are the intrinsic (IN) transistor type and the depletion type. The IN type of transistor in SOS lacks a diffusion step in fabrication. These devices provide better matching, given lower channel doping levels. In addition, intrinsic devices do not show signs of accentuated kink effects. This allows for better matching, as we eliminate one cause of mismatch in the source/drain doping step. In addition, IN-type transistors provide more gain at zero-threshold voltage. Intrinsic devices in the differential amplifier of the comparator take advantage of the SOS process without introducing drawbacks, as the floating

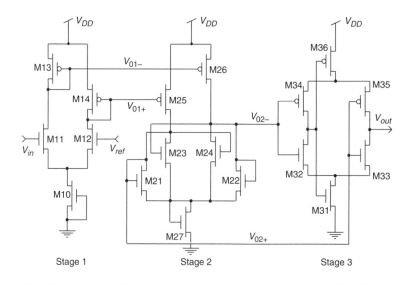

FIGURE 4.11   High-precision SOS Comparator. The comparator is composed of three stages: stage 1 = amplification stage, stage 2 = the decision stage, and stage 3 = the digital output stage. The input stage is a differential pair with output current mirrors. The second stage is a fast-triggered latch. The third stage is a static digital latch.

body effect. The input differential pair is also the most sensitive part of the ADC, as it provides the initial gain for the conversion. The load PMOS transistors M13 and M14 are also IP intrinsic. This provides better matching in the currents of the two branches of the differential amplifier and the zero-threshold allows for operation close to the rail voltage. In addition, the active load provides high differential gain and a differential output for the next stage. M10 is an IN NMOS type with the gate connected to the ground and properly sized to bias the first stage. Notice that load transistors M13 and M14 are mirroring the input stage current for second-stage inputs M25 and M26.

The gain of the differential amplifier is given by Eq. (4.1) as a function of the input NMOS transconductance g_{M1} and the load conductance g_{DS3} of the PMOS mirror devices M3 (or M4) of the first stage in Fig. 4.11.

$$A_{v1} = \frac{g_{M1}}{2g_{DS3}} \tag{4.1}$$

The frequency response of this first stage is given by Eq. (4.2) and the parameters in Eq. (4.3), as a function of the MOSFET device small-signal output resistances r_{DS} and overlap gate to drain capacitances

g_{GD}. The output capacitance C_L is the major capacitive component and is generally due to the gate capacitance of the input of stage 2.

$$t_r = R_l C_l \qquad (4.2)$$

$$R_l = r_{DS1} || r_{DS3}$$

$$C_l = C_{GD1} + C_{GD3} + C_L \qquad (4.3)$$

Note that the frequency response is higher due to the lack of substrate parasitics. The propagation delay of the differential pair can be give by: $T_p \sim 0.69\ t_r$ (Geiger et al., 1990). SOS Spectre simulations showed a switching time on the order of 1 ns.

Stage 2 is the decision circuit: a mixed-mode circuit that commutes the output digital signal (*out*) as fast as possible. This is the core of the comparator as it is capable of discerning millivolt-level signals. It uses two cross-coupled feedback inverting amplifiers, implemented with transistors M25, M23 and M24, M26, respectively. If the voltage of the gate of transistor M25 (V_{01-}) lowers below the gate voltage of M26 (V_{01+}), the gate of M24 rises due to cross coupling. This in turn pulls down the output drain of M26 (V_{02-}). At the same time, the drain of M25 (V_{02+}) pulls up, creating a difference between the two outputs V_{02}.

The decision circuit employs diode connected transistors M22 and M21 in parallel with the NMOS M23 and M24. These transistors act to steal current from the output of the decision stage V_{02}. This increases the speed of the comparator, as transistors M23, M24 are allowed to turn off and swing the outputs V_{02} can fall below the threshold of M23, M24. NMOS M27 is used to shift the output of the decision circuit up by V_{th} and provide balanced noise margin for the following digital stages.

Stage 3 uses cross-coupled inverters in a latch configuration. M34, M32 and M35, M33 are the two inverters. M35, M33 provide the digital output voltage of the comparator V_{out}. Notice that the other output of M34, M32 is the input of yet-another inverter M36, M31, which is used to steer the power to the other inverters. If V_{02-} is higher than V_{02+}, the gate of M36 and M31 lowers, disconnecting V_{DD} and allowing V_{out} to fall to ground. Similarly, V_{out} is pulled to V_{DD} when V_{02-} is lower than V_{02+}. This stage is similar to a digital latch and provides a digital output.

Two or more digital inverters can be added to the circuit in Fig. 4.11 to sharpen the transition and obtain some additional voltage gain.

The decision circuit and output stages of the comparator are feedback digital circuits that are not affected by floating body effects, as the additional kink current only helps to obtain a faster decision. They are therefore built out of regular threshold devices similarly to bulk CMOS digital circuits.

Transistor name	Transistor Type	Size [W, L]
M11, M12	IN	10 μm, 2 μm
M13, M14	IP	12 μm, 1 μm
M10	IN	4 μm, 0.8 μm
M25, M26	IP	12 μm, 1 μm
M23, M24	RN	1.2 μm, 0.8 μm
M27	RN	1.2 μm, 0.8 μm
M31, M32, M33	RN	1.2 μm, 0.8 μm
M34, M35, M36	RP	1.2 μm, 0.5 μm

TABLE 4.3 Device sizes of a fabricated SOS high-performance comparator as in Fig. 4.11.

We have simulated an SOS comparator as shown in Fig. 4.11 with the devices sizes summarized in Table 4.3. The comparator output gain resulted $A_V = 1653$. This value of gain is much higher than the required minimum for an 8-bit ADC and can be easily improved for 10-bit performance. The voltage gain per stage resulting from the simulations was 1.38 for the differential pair, 20.5 for the fist stage of the decision circuit, 16.5 for the second, and 3.5 for the output inverters.

4.2.2 SOS Low-Power Comparator

In this section, we present the design in SOS of an ultralow-power comparator designed to consume the absolute minimum while providing 8-bit equivalent performance. The comparator was implemented by a switched capacitor clocked circuit, and it is thus different from the high-performance design of Fig. 4.10. A schematic of the three-stage SOS comparator is given in Fig. 4.12.

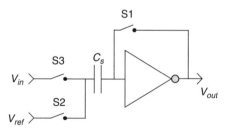

FIGURE 4.12 Low-power SOS comparator. This comparator uses a switched capacitor circuit to both amplify the difference between the input V_{in} and a reference voltage V_{ref} and provide a digital output V_{out}.

This comparator uses a single gain stage to amplify the input differences and provide a pseudo-digital output. One or multiple inverters are generally used after the output V_{out} of the circuit in Fig. 4.12. The operation of the comparator is divided into two stages and is as follows:

1. Reset Phase: Switches S1 and S2 are closed, while S3 is left open. The reference voltage V_{ref} is stored on the capacitor C_s, together with any correlated noise. The inverter input is also initialized to its logic threshold.

2. Compare Phase: Switch S2 first opens, then S3 is closed, and S1 is opened. The input V_{in} is connected, and the comparator output V_{out} changes state when this voltage exceeds V_{ref}.

Figure 4.13 shows a schematic of the inverter used in the low-power SOS comparator in Fig. 4.12. Two PMOS and two NMOS are connected in series. M4 is separated by a capacitor with the input, which is switching between the input signal V_{in} and the reference voltage V_{ref}. Three transistors (M1, M2, and M3) operate in the subthreshold region in

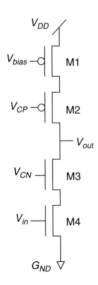

FIGURE 4.13   Cascoded inverter used as amplifier for the low-power SOS comparator. Two PMOS and two NMOS are connected in series. M4 is separated by a capacitor with the input, which is switching between the input signal V_{in} and the reference voltage V_{ref}. Three transistors (M1, M2, and M3) operate in the subthreshold region in order to obtain maximum gain from a single-stage amplifier.

Device Name	Transistor Type	Size [W, L]
M1	RP	10 μm, 2 μm
M2	IP	10 μm, 2 μm
M3	IN	10 μm, 2 μm
M4	RN	10 μm, 2 μm
C_s	MIM capacitor	1 pF

TABLE 4.4   Device sizes of a fabricated SOS low-power comparator as in Fig. 4.12.

order to obtain the maximum gain from a single stage amplifier. Two intrinsic transistors were used for M2 and M3, to eliminate the need of additional biases for the cascoded inverter. A correlated double-sampling (CDS) technique was used in the comparator to eliminate the correlated noise, such as the mismatch in devices. A cascoded inverter implementation offers two advantages over the basic common-source amplifier: first, higher gain due to the larger output resistance; second, higher operational speed due to the reduced Miller capacitance.

We have simulated an SOS comparator as reported in Fig. 4.12 with the devices sizes summarized in Table 4.4. SOS simulations indicate that the comparator consumes 200 nW using a 1.5-V power supply while operating at a maximum speed of 10 M comparisons per second (10 MHz). The cascoded inverter reported a maximum gain of 60 dB from AC simulations. The biases for the cascoded inverter in Fig. 4.13 were set to 0.6 V for V_{bias}, 0 V for V_{cp}, and $V_{DD} = 1.5$ V for V_{cn}, respectively.

4.3   SOS Bandgap References

In Sec. 3.4 we introduced simple circuits to provide stable current references independent of the supply voltage (Baker, 2005). In this section, we present the design of an SOS voltage reference. A voltage reference is a circuit used to generate a fixed voltage independent of the exact power supply voltage and independent of temperature variation within a specified range. It is usually referred to as a PVT reference, where PVT stands for process, voltage, temperature invariance. The SOS bandgap reference presented here is a circuit that combines both a PTAT (proportional to temperature) and a CTAT (complementary to temperature) current sources to provide temperature stability. CTAT and PTAT components of the reference currents are canceled to provide temperature independence. The design presented here targets

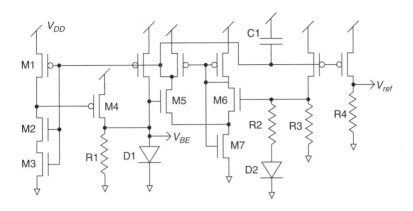

FIGURE 4.14 Schematic caption of the circuit.

very low voltages (1–1.5 V or less) and is aimed at obtaining a stable reference of a few hundreds of millivolts. The design takes advantage of the low-voltage and low-power features of the SOS process and devices to provide a voltage reference for ultralow power energy scavenging and distributed sensors applications.

Figure 4.14 represents a schematic caption of the bandgap reference. In this voltage reference, M4, D1, D2, and R2 form a PTAT circuit, where the terminal V_{be} is proportional to temperature. The voltage reference is generated by two circuit branches containing two diodes, D1 and D2. Diode D2 is an array of K-parallel D1 diodes. The PTAT current through D1 can be expressed by Eq. (4.4), where V_T is the thermal voltage of the diode ($V_T = kT/q$).

$$I_{D1, PTAT} = \frac{V_T \cdot \ln(K)}{R2} \tag{4.4}$$

By adding two resistors, R1 and R3, to steal away current from the PTAT circuit, a CTAT circuit is formed. The CTAT current through D1 can be expressed by Eq. (4.5), where V_{D1} is the voltage across diode D1, and we suppose that R1 = R3 = L · R2.

$$I_{D1, CTAT} = \frac{V_{D1}}{L \cdot R2} \tag{4.5}$$

A differential pair amplifier is formed by transistors M5, M6, and M7 plus the top two PMOS loads. This five-transistor transconductance stage ensures that the voltages at the gates of M5 and M6 are equal using the feedback loop from the drain of M6 (differential pair output) to the gate of M7. This differential amplifier also forces the current through all the top PMOS to be identical (Banba et al., 1999). This current is mirrored by all the top PMOS transistors, all equivalent in size to M1. The reference output voltage depends on the diode built-in voltage V_D and the thermal voltage V_T. This difference in diode sizes affects the V_{ref} output and helps to compensate the negative temperature coefficient of V_D. The V_{ref} output can be expressed by Eq. (4.6), where R4 = N · R2. Notice that the sum of the PTAT and CTAT currents through D1 is the same as the current through R4, which determines V_{ref}.

$$V_{ref} = R4 \cdot (I_{D1,PTAT} + I_{D1,CTAT}) = V_T \cdot N \cdot \ln(K) + \frac{N}{L} \cdot V_{D1} \quad (4.6)$$

The temperature behavior of the bandgap reference can be determined by differentiating Eq. (4.6) by temperature, as expressed in Eq. (4.7):

$$\frac{\partial V_{ref}}{\partial T} = \frac{\partial V_T}{\partial T} \cdot N \cdot \ln(K) + \frac{N}{L} \cdot \frac{\partial V_{D1}}{\partial T} \quad (4.7)$$

Usually, $\partial V_T / \partial T$ is around 0.085 mV/°C, while $\partial V_{D1} / \partial T$ is around −1.6 mV/°C. To eliminate the dependance of V_{ref} on temperature, we can set $L = 1.6/(\ln(K) \cdot 0.085)$.

Transistors M1, M2, and M3 form a start-up stage to prevent instabilities when the circuit is powered-on. Compensation capacitor C1 is used to stabilize the gate terminal of the PMOS transistors of the current mirror.

We have fabricated an SOS voltage reference as reported in Fig. 4.14 with the devices size summarized in Table 4.5. For low-voltage operation, we take advantage of the multithreshold MOS devices available in the SOS process (Peregrine, 2008a). Low-threshold PMOS (PL) transistors are used as current sources for the entire bandgap circuit. Notice also that the input NMOS transistors of the five-transistor transconductance amplifier are also the low-threshold NL-type. The NL input transistors allow the transconductor to operate at low voltages and provide high gain. The diodes are native devices of the PG type. The resistor are SOS native high-resistivity silicon strips of the SN type and the capacitor is a MIM type. Use of the SOS technology and its six-types of MOS devices extends the operation of the BGR to very-low power supplies and, thus, is a fundamental component for a low-power system.

Based on devices sizes from Table 4.5, we have designed the reference voltage to be $V_{ref} = 550$ mV with a minimum power supply of less than 1 V. The values of N, K, and L are given in Table 4.6. A bare die was bonded onto a DIP16 chip carrier and 10 pF capacitors were connected to V_{ref} and V_{be} nodes to eliminate high-frequency parasitic effects and node ringing. The packaged circuit and other components were assembled on a printed circuit board (PCB). The measured

Device Name	Transistor Type	Size [W, L]
M1	PL	10 μm, 2 μm
M2, M3	RN	2 μm, 55 μm
M4	PL	10 μm, 1 μm
M5, M6	NL	6 μm, 2 μm
M7	NL	2 μm, 30 μm
D1	PG	10 μm, 1.6 μm
D2	PG	(K = 8) 10 μm, 1.6 μm
R1, R3	SN	500 KΩ
R2	SN	53 KΩ
R4	SN	345 KΩ
C1	MIM capacitor	175 fF

TABLE 4.5 Device sizes of a fabricated SOS voltage reference as in Fig. 4.14.

average current consumption of the bandgap circuit was 15 μA in the range of power supply voltages of 1–1.5 V.

We measured the performance of the bandgap circuit by evaluating the temperature measurement capabilities and the stability of the output V_{ref} with a power supply of 1.2 V. In order to measure the supply voltage dependence of V_{ref} and V_{be}, the supply voltage was swept both from 0 V to 1.5 V; the results are reported in Fig. 4.15. Notice that the bandgap circuit needs at least 0.9 V to operate correctly and for both the voltages V_{ref} and V_{be} to stabilize.

Figure 4.16 shows the dependence of signals V_{be} and V_{ref} on temperature where the supply voltage is kept 1.2 V. V_{be}, which is equal to kT/q, voltage of the bandgap reference is linearly proportional to

Parameter	Value
K	8
L	9.05
N	2.15
V_T	25 mV
V_{ref}	550 mV

TABLE 4.6 Parameters for the fabricated SOS voltage reference of Fig. 4.14.

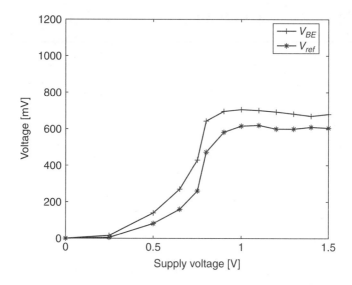

FIGURE 4.15   Measured output reference voltage V_{ref} versus supply voltage obtained from the fabricated SOS voltage reference of Fig. 4.14 with parameters from Tables 4.5 and 4.6.

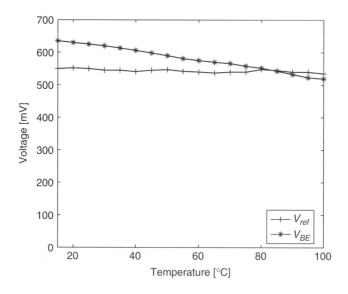

FIGURE 4.16   Measured output reference voltage V_{ref} and temperature signal V_{BE} versus temperature obtained from the fabricated SOS voltage reference of Fig. 4.14 with parameters from Tables 4.5 and 4.6.

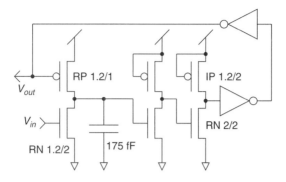

FIGURE 4.17 Schematic of the VCO used to convert analog signals into a digital clock with varying frequencies.

room temperature as expected. The voltage V_{ref} was designed to be approximately 550 mV. Notice that in Fig. 4.16 the V_{ref} signal is constant with temperature in the 15–100°C range. The V_{ref} voltage ripple is approximately 6%, a satisfactory result given the low reference voltage.

The V_{be} voltage of the bandgap reference can be used to measure the temperature of the circuit. We have designed a voltage-controlled oscillator (VCO) circuit to operate as a digital data converter and communication circuit. The VCO schematic is presented in Fig. 4.17.

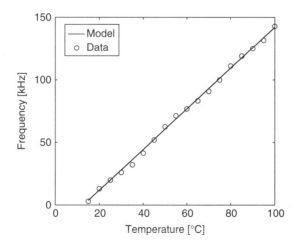

FIGURE 4.18 Measured output frequency of the VCO in Fig. 4.17 as a function of frequency. The VCO is connected to the V_{BE} of the fabricated SOS voltage reference of Fig. 4.14 with parameters from Tables 4.5 and 4.6.

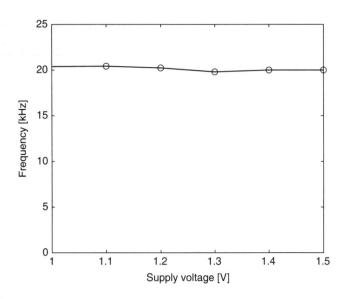

FIGURE 4.19 Measured output frequency of the VCO in Fig. 4.17 as function of the supply voltage. The VCO is connected to the V_{BE} of the fabricated SOS voltage reference of Fig. 4.14 with parameters from Tables 4.5 and 4.6.

The circuit is a self-reset asynchronous oscillator that generates a square wave signal whose frequency depends on the input voltage. Here the V_{BE} signal is converted from the voltage reference into the *temp* square-wave output. The core of the oscillator is a capacitor-feedback integrator based on linear discharge. The input transistor converts an input voltage into a nonlinear current that drains the capacitor of its reset-state charge. When the capacitor is discharged, a feedback loop composed of four inverters provides a delayed reset signal to restart the integration. We use intrinsic transistors to implement two of the inverters in the feedback loop to provide a delay before communicating the reset signal to the integrator.

Figure 4.18 represents the frequency of the square wave of signal *temp* as a function of temperature for a V_{DD} of 1.2 V. Notice the linearity of the output frequency as it increases with room temperature. A linear model of the frequency of the *temp* signal versus temperature resulted in a conversion factor of 1.6 kHz/°C.

Figure 4.19 represents the frequency of the square wave of signal *temp* as a function of the power supply voltage from 1–1.5 V. The output frequency remains stable at 20.5 KHz with a temperature of 25°C, showing the supply voltage independence of the PTAT output V_{BE}.

CHAPTER 5

Digital Circuit Design in SOS

5.1 Introduction

SOS features six types of different MOSFETs devices. These devices can be used to operate digital circuits at different supply voltages while keeping large operational frequencies.

5.2 SOS Inverter Characteristics

A typical SOS inverter schematic is given in Fig. 5.1. An inverter is composed of an NMOS device M1 and a PMOS device M2. These devices can be of any combination of regular, low, and intrinsic threshold SOS MOSFET devices. The choice of the transistor type is discussed in the rest of this section.

In this section, we report on the simulation results of several inverter circuits. We used the following types of transistors: NMOS devices were regular threshold (0.7 V RN), low threshold (0.3 V NL), and intrinsic (0 V IN). PMOS devices were regular threshold (0.7 V RP), low threshold (0.3 V PL), and intrinsic (0 V IP).

The first inverter circuit we characterize is a regular threshold transistor inverter, where the NMOS is an RN type and the PMOS an RP type. Figure 5.2 shows the DC simulations results of a inverter with both NMOS and PMOS with size [W, L] of [2 μm, 2 μm]. The transistors sizes in this stage were chosen to show the operation of the inverter as an analog push–pull amplifier. The large length of the device eliminates short-channel effects from the simulations. We used a power supply of 3.3 V, which is the nominal SOS operating voltage for the FC process. The load was an identically sized inverter with a capacitance of approximately 40 fF. The DC simulation shows a logic threshold of 1.6 V.

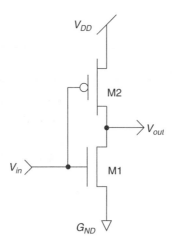

FIGURE 5.1   An SOS inverter schematic. NMOS M1 and PMOS M2 can be of any combination of regular-, low-, and intrinsic-threshold SOS MOSFET devices.

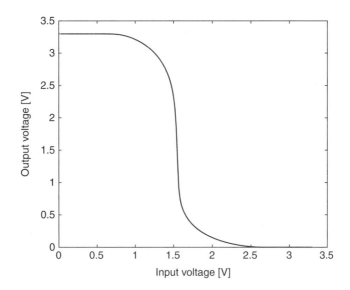

FIGURE 5.2   DC simulation of a regular threshold SOS inverter. RN NMOS and RP PMOS sizes are [W, L] of [2 μm, 2 μm], and simulations used a power supply of 3.3 V. The load was an identically sized inverter with a capacitance of approximately 40 fF.

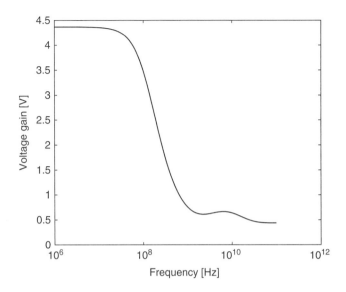

FIGURE 5.3 AC simulation of a regular threshold SOS inverter. RN NMOS and RP PMOS sizes are [W, L] of [2 μm, 2 μm], and simulations used a power supply of 3.3 V. The load was an identically sized inverter with a capacitance of approximately 40 fF.

Figure 5.3 reports the AC simulations results of the same inverter circuit mentioned above. The maximum voltage gain is 4.3 V/V at 1.6 V DC input, and the bandwidth is approximately 500 MHz. The bandwidth of the inverter was here considered the frequency at which the inverter voltage gain becomes one.

Figure 5.4 shows transient simulations results of of the same inverter circuit mentioned above. The input frequency was set to 500 MHz.

Subsequently, we simulated three inverters types: regular threshold, low threshold, and intrinsic. Each inverter uses both PMOS and NMOS with the same threshold type. Each MOSFET in the three types of inverters used both NMOS and PMOS with sizes [W, L] of [5 μm, 0.5 μm]. Transistor sizes in this stage were chosen to show a typical digital gate configuration.

The first inverter circuit we characterize is a regular- threshold transistor inverter, where the NMOS is an RN type and the PMOS an RP type. We used a power supply of 3.3 V, which is the nominal SOS operating voltage for the FC process. The load was an identically sized inverter with a capacitance of approximately 25 fF. Figure 5.5 reports

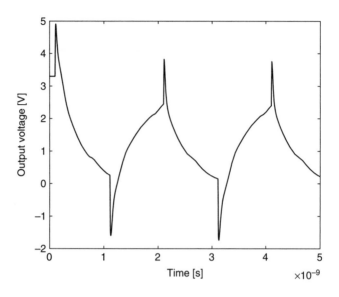

FIGURE 5.4 Transient simulation of a regular threshold SOS inverter. RN NMOS and RP PMOS sizes are [W, L] of [2 μm, 2 μm], and simulations used a power supply of 3.3 V. The load was an identically sized inverter with a capacitance of approximately 40 fF.

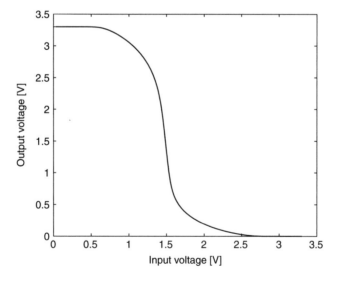

FIGURE 5.5 DC simulation of a regular threshold SOS inverter. RN NMOS and RP PMOS sizes are [W, L] of [5 μm, 0.5 μm], and simulations used a power supply of 3.3 V. The load was an identically sized inverter with a capacitance of approximately 25 fF.

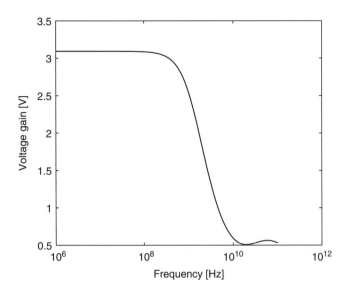

FIGURE 5.6   AC simulation of a regular threshold SOS inverter. RN NMOS and RP PMOS sizes are [W, L] of [5 μm, 0.5 μm], and simulations used a power supply of 3.3 V. The load was an identically sized inverter with a capacitance of approximately 25 fF.

DC simulations results for the regular threshold inverter. The DC simulation shows a logic threshold of 1.6 V.

Figure 5.6 shows AC simulations results of the same inverter circuit mentioned above. The maximum voltage gain is 3.1 V/V at 1.6 V DC input, and the bandwidth is approximately 5 GHz. The bandwidth of the inverter was here considered the frequency at which the inverter voltage gain becomes 1.

Figure 5.7 shows transient simulations results of of the same inverter circuit mentioned above. The input frequency was set to 1.6 GHz.

The second inverter circuit we characterize is a low-threshold transistor inverter, where the NMOS is an NL type and the PMOS an PL type. We used a power supply of 1.5 V. The load was an identically sized inverter with a capacitance of approximately 25 fF. Figure 5.8 depicts DC simulation results for the regular threshold inverter. The DC simulation shows a logic threshold of 0.65 V.

Figure 5.9 shows AC simulation results of the same inverter circuit mentioned above. The maximum voltage gain is 7.5 V/V at 0.65 V DC input, and the bandwidth is approximately 3 GHz. The bandwidth

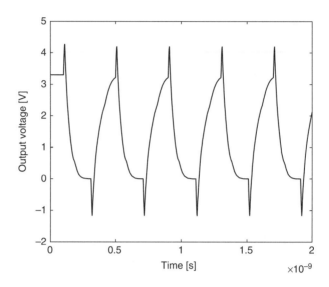

FIGURE 5.7 Transient simulation of a regular threshold SOS inverter. RN NMOS and RP PMOS sizes are [*W, L*] of [5 μm, 0.5 μm], and simulations used a power supply of 3.3 V. The load was an identically sized inverter with a capacitance of approximately 25 fF.

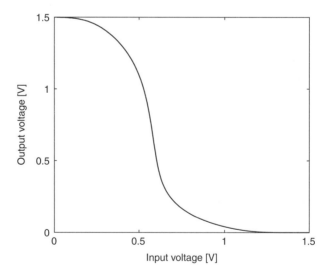

FIGURE 5.8 DC simulation of a low threshold SOS inverter. NL NMOS and PL PMOS sizes are [*W, L*] of [5 μm, 0.5 μm], and simulations used a power supply of 1.5 V. The load was an identically sized inverter with a capacitance of approximately 25 fF.

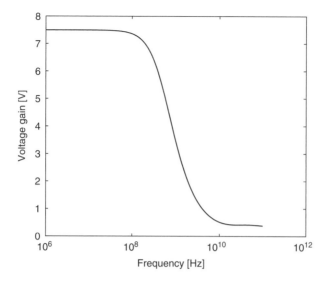

FIGURE 5.9   AC simulation of a low-threshold SOS inverter. NL NMOS and PL PMOS sizes are [W, L] of [5 μm, 0.5 μm], and simulations used a power supply of 1.5 V. The load was an identically sized inverter with a capacitance of approximately 25 fF.

of the inverter here was considered to be the frequency at which the inverter voltage gain becomes 1.

Figure 5.10 shows transient simulations results of of the same inverter circuit mentioned above. The input frequency was set to 1.6 GHz.

The third inverter circuit we characterize is an intrinsic transistor inverter, where the NMOS is an IN type and the PMOS an IP type. We used a power supply of 0.5 V. The load was an identically sized inverter with a capacitance of approximately 25 fF. Figure 5.11 depicts DC simulations results for the regular threshold inverter. The DC simulation shows a logic threshold of 50 mV.

Figure 5.12 shows AC simulations results of the same inverter circuit mentioned above. The maximum voltage gain is 2.2 V/V at 50 mV DC input, and the bandwidth is approximately 2.5 GHz. the bandwidth of the inverter was here considered the frequency at which the inverter voltage gain becomes one.

Figure 5.13 shows transient simulations results of of the same inverter circuit mentioned above. The input frequency was set to 1.6 GHz.

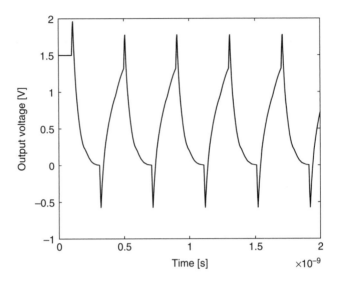

FIGURE 5.10 Transient simulation of a low-threshold SOS inverter. NL NMOS and PL PMOS sizes are [W, L] of [5 μm, 0.5 μm], and simulations used a power supply of 1.5 V. The load was an identically sized inverter with a capacitance of approximately 25 fF.

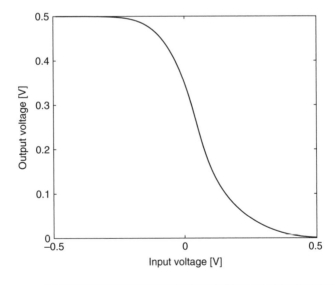

FIGURE 5.11 DC simulation of an intrinsic threshold SOS inverter. IN NMOS and IP PMOS sizes are [W, L] of [5 μm, 0.5 μm], and simulations used a power supply of 0.5 V. The load was an identically sized inverter with a capacitance of approximately 25 fF.

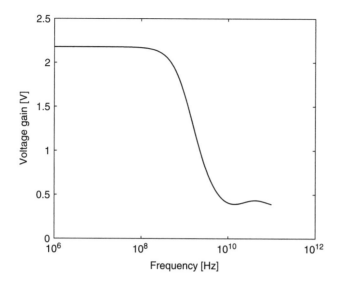

FIGURE 5.12   AC simulation of an intrinsic threshold SOS inverter. IN NMOS and IP PMOS sizes are [W, L] of [5 μm, 0.5 μm], and simulations used a power supply of 0.5 V. The load was an identically sized inverter with a capacitance of approximately 25 fF.

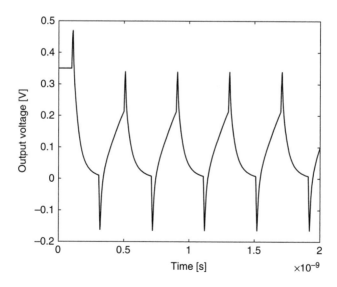

FIGURE 5.13   Transient simulation of an intrinsic threshold SOS inverter. IN NMOS and IP PMOS sizes are [W, L] of [5 μm, 0.5 μm], and simulations used a power supply of 0.5 V. The load was an identically sized inverter with a capacitance of approximately 25 fF.

These simulations show how the different inverters can be used with different power supplies to obtain similar characteristics. The gains and bandwidths of the three kind of inverters just shown are very similar, but the power supplies are 3.3 V, 1.5 V, and 0.5 V, respectively, for regular-threshold, low-threshold, and intrinsic types. The designer can choose between these types of inverters and power supplies. Needless to say, low-threshold and intrinsic transistors can operate also at the nominal 3.3 V supply but will consume larger power and obtain much faster switching speeds.

CHAPTER 6

Design of SOS Data Converters

6.1 Introduction

Engineering typically requires us to interface with physical ambient signals, such as light, sound, temperature, etc. As an interface between sensing environment and digital processing module, digital-to-analog and analog-to-digital converters (DAC, ADC) are crucial in bridging signal sensing and digital processing in applications. In this chapter we describe the design and development of high-performance DACs and ADCs in SOS technology. After the first five chapters of this book, an example of a fabricated design addresses punctiliously the advantages of SOS technology and critical steps of the design. Both DACs and ADCs are an ideal example because they contain both analog and digital circuits and are the interface between sensors and processing circuits.

Recent developments in CMOS technology, as well as the recent popularity of SOI technologies, require a reevaluation of traditional design of both DACs and ADCs. The newest technology, in fact, offers several possible improvements even for simple textbook implementations. As an example, SOI technologies add advanced new devices, as lower-threshold transistors. These devices can be used to simplify the design of the ADC, to increase the performance and to lower power consumption. An additional benefit of SOI technology is an increase in speed due to the reduction of capacitance of the digital control logic. All these advantages can be attained without reducing feature sizes. This is one of the major benefits of SOI and SOS technologies.

In this chapter, we present two types of DAC topologies: capacitive and resistive ladder. Also, we present the design and characterization of two types of ADC topologies: successive-approximation converters and asynchronous sigma–delta converters.

6.2 SOS DACs

DACs are important circuit component that provide an interface between digital processing circuits and analog sensors and other electrical physical quantities. In this section, we present two types of DAC topologies: capacitive and resistive ladder.

6.2.1 SOS Capacitive DAC Converters

Capacitive DACs are the most frequent DACs in CMOS processes. The precision of CMOS processes in the manufacturing of large arrays of identical capacitors is the reason for the popularity of these kind of converters. Precision up to 12–13 bits can be obtained with capacitive DACs without the use of correction algorithms or correction circuits. For this reason, this kind of DAC is used for low-power high-end analog converters and high-precision bias voltage generators with digital control (Baker et al., 1998). The design of capacitive DACs in SOS can benefit from a reduced parasitic capacitance and an increased precision. Alternatively, the size of the capacitor can be reduced while keeping the precision the same. This allows to lower the power consumption used when charging and discharging the capacitors.

A schematic of a capacitive DAC, also named "charge-sharing" DAC, is given in Fig. 6.1. It is composed of a parallel array of binary weighted capacitors. The output of the capacitance array is connected to an operational-amplifier buffer (or a voltage follower). The capacitor C is the "unit capacitance" of the array. Its minimum value has to be much larger than the parasitic capacitances of the wiring and amplifier inputs, typically 10–100 fF or more. After an initial reset and discharge of the capacitors to ground, the digital inputs D_0 to D_{n-1} switch each capacitor of the binary array to either V_{ref} or ground (G_{ND}). The output V_{out} is then a function of the capacitive voltage division between V_{ref} and ground. The "DAC cells" are composed of simple switches that connect the bottom plates of the capacitors either to V_{ref} or ground. The output driver stage of the DAC cells have to be sized according to the capacitor size they are to drive. This requires each DAC cell to have a different exponentially larger driver to be able to charge the capacitors at the same time. Successive approximation register (SAR) cells can be either sized with an exponentially higher current (balanced design) output buffer or can be all made the same and use the largest driver (worst-case design).

The output voltage V_{out} can be expressed mathematically as a function of the digital inputs D_0 to D_{n-1} by Eq. (6.1).

$$V_{out} = \sum_{k=0}^{n-1} D_k 2^{k-n} V_{ref} \qquad (6.1)$$

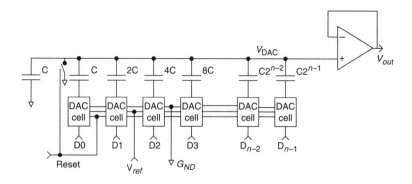

FIGURE 6.1 A capacitive charge-scaling DAC. The output V_{out} is a function of the capacitive voltage division between V_{ref} and the ground, programmed by the digital inputs D_0–D_{n-1}.

The voltage V_{DAC} in Fig. 6.1 is the analog output voltage from the converter. An operation amplifier is used to decouple this node from the rest of the circuitry and to provide voltage buffering. The use of an amplifier and its input capacitance are the major limitation to this kind of converter. The input capacitance is a parasitic capacitance that limits the performance of the converter in two ways: first, it imposes a minimum on the unit capacitance, thus limiting the minimum power consumption. Second, it limits the frequency of operation of the converter by imposing a minimum total capacitance. The total capacitance of this DAC is 2^N unit capacitances C, or 256 in case of a 8-bit DAC and 1024 for a 10-bit DAC.

This DAC is used in charge redistribution analog-to-digital converters, such as the successive-approximation ADCs (refer to Sec. 6.3.1).

6.2.2 SOS Capacitive DAC Accuracy

The limitations on the conversion precision in a SOS capacitive DAC is directly proportional to capacitor matching. We now show how to calculate the integral nonlinearity (INL) and differential nonlinearity (DNL) figures of a DAC (Baker et al., 1998). Figure 6.2 is a graphic representation of the DNL and INL. The DNL is the actual increment in height of the transition n minus the ideal transition, as can be seen in Eq. (6.2).

$$\text{DNL}_n = (V_{out,n} - V_{out,n-1}) - V_{ref}/N \qquad (6.2)$$

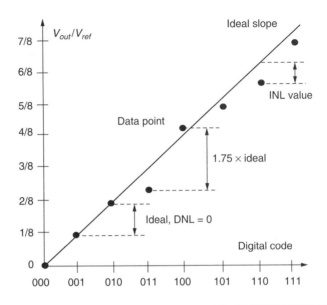

FIGURE 6.2 Definition of differential non linearity (DNL) and integral non linearity (INL). The DNL is the actual increment in height of the transition n minus the ideal transition. INL is the output value of code n minus the output value of the reference line at that point.

INL is the output value of code n minus the output value of the reference line at that point, as can be seen in Eq. (6.3).

$$\text{INL}_n = V_{out,n} - V_{ideal,n} \qquad (6.3)$$

The maximum INL can be derived by Eq. (6.4), where C is the unit capacitance used in the capacitive array.

$$|\text{INL}_{max}| = 2^{N-1}(C + |\Delta C_{max,\text{INL}}|) - 2^{N-1}C = 2^{N-1}|\Delta C_{max,\text{INL}}| \qquad (6.4)$$

This corresponds to the error of the converter as the MSB switches value from 0 to 1. The maximum ΔC that will result in an INL of less than $1/2$ LSB is given by Eq. (6.5).

$$|\Delta C_{max,\text{INL}}| = \frac{0.5C}{2^{N-1}} = \frac{C}{2^N} \qquad (6.5)$$

The maximum DNL can be derived by computing Eq. (6.6). Once again, this maximum value tends to occurs as the converter is in the middle of the scale (MSB error). This corresponds to the jump from code 0111...1111 to 1000...0000, as expressed in Eq. (6.6).

$$DNL_{max} =$$

$$\left[2^{N-1}(C + |\Delta C|_{max,DNL}) - \sum_{k=1}^{N-1} 2^{k-1}(C - |\Delta C|)_{max,DNL} \right] - C \quad (6.6)$$

Equation (6.6) can be turned into Eq. (6.7) and then Eq. (6.8).

$$DNL_{max} =$$

$$2^{N-1}(C + |\Delta C|_{max,DNL}) - (2^{N-1} - 1)(C - |\Delta C|)_{max,DNL} - C \quad (6.7)$$

$$DNL_{max} = 2^N - 1 \left| \Delta C_{max,DNL} \right| \quad (6.8)$$

With the maximum, ΔC, which results in a DNL less than $1/2$ LSB, is expressed by Eq. (6.9).

$$\left| \Delta C_{max,DNL} \right| = \frac{0.5C}{2^N - 1} = \frac{C}{2^{N+1} - 2} \quad (6.9)$$

This is the maximum ΔC that will result in a DNL of less than $1/2$ LSB.

6.2.2.1 The Split-Array Capacitive DAC

The design of capacitive DACs has the advantage of being simple to design and analyze, while at the same time providing excellent accuracy. On the other hand, as the number of bits requirement increases, so does the area of the capacitive array. The area is, in fact, exponential with the number of bits. As an example, if a minimum capacitance of 0.1 pF is used, a $N = 16$-bit DAC will require a total capacitance $C_{tot} = 2^N \cdot 0.1 \text{ pF} = 2^{16} \cdot 0.1 \text{ pF} = 6.55 \text{ nF}$. With a capacitance density of 5 fF/μm^2, typical in CMOS processes, the area required to implement this capacitance is at least 1.3 million μm^2 or 1.3 mm^2. This capacitance requires a fairly large silicon area and thus can make a capacitive DACs expensive, power hungry and slow.

In order to overcome this limitation, the split-capacitor array can be used instead of the full-capacitor array. In a bulk CMOS process, the floating plate of each capacitor in the array couples with the substrate, thus forming a parasitic capacitance that introduces a systematic error in the capacitor divider (Baker et al., 1998). The SOS process is immune to this problem because its substrate is an insulator. This split-capacitor configuration is thus a topology that can only be directly translated in the SOS process.

An 8-bit example split-array DAC is reported in Fig. 6.3. The difference between this and the charge redistribution DAC of Fig. 6.1 is that

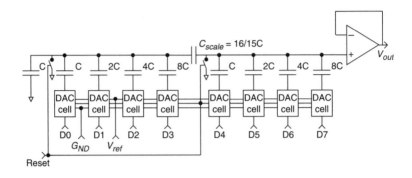

FIGURE 6.3   A capacitive charge-scaling DAC with split array topology. The output V_{out} is a function of the capacitive voltage division between V_{ref} and ground, programmed by the digital inputs D_0–D_{n-1}.

the split array is composed of two arrays (most and least significant bit—MSB and LSB) coupled by a scaling capacitor C_{scale}.

The two split-capacitor arrays are initially reset to ground by the Reset signal. In order to output a converted analog voltage, the DAC cells connect the bottom terminal of the capacitors to either V_{ref} or ground, depending on the binary inputs D_0–D_{n-1}. The scaling capacitor C_{scale} value is given by Eq. (6.10), where C_{MSB} and C_{LSB} are, respectively, the sum of all MSB and LSB unit capacitors in the arrays.

$$C_{scale} = C \cdot \frac{\sum C_{LSB}}{\sum C_{MSB}} \tag{6.10}$$

In Fig. 6.3, the scaling capacitor C_{scale} value is 16/15C, as 16 is the sum of all LSB unit capacitors and 15 is the sum of all MSB unit capacitors.

As an example, suppose we present the digital input string D_7 to D_0 "1000-0000". After an initial reset, the LSB array forms a total capacitance of 16C because all capacitors bottom plate are tied to ground. The MSB array has three capacitors tied to ground and one (the MSB) tied to V_{ref}. This simple capacitive circuit can be analyzed and the output voltage will then be $V_{out} = V_{ref} \, 8/((16/15 \, || \, 16) + 7 + 8) = 1/2 \, V_{ref}$, which is the expected result.

The split-array DAC uses approximately $2^{N/2+1}$ unit capacitances C, or a total of 32C for a 8-bit DAC. This is a saving of $2^{N/2-1}$ when compared to a full binary array.

6.2.2.2 The C–2C Ladder Capacitive DAC

In order to further minimize the capacitor array area, a "C–2C" capacitor ladder can be used. As shown in Fig. 6.4, the capacitor chain is

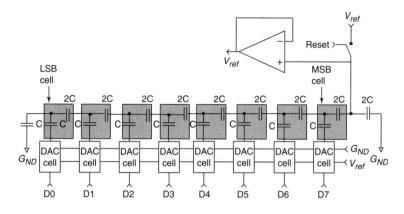

FIGURE 6.4   A capacitive charge-scaling DAC with a C–2C ladder topology. The output V_{out} is a function of the capacitive voltage division between V_{ref} and ground.

composed of eight identical capacitor cells and two grounded capacitors at either end. Each capacitor cell includes a floating 2C capacitor and a 1C capacitor (C is the unit capacitance), which is driven by the input digital bits D_0 to D_{n-1}. The 2C capacitor is floating because it is not directly charged or discharged by a voltage source.

Similarly to the split capacitor case, we point out that in a typical bulk CMOS process, the floating plate of the capacitors couples with the substrate–forming parasitic capacitance, which introduces errors in the capacitive dividers. This error can propagate in the cells of a C–2C array. The SOS process is immune to this problem because its substrate is an insulator, thus the C–2C architecture can be easily implemented in the SOS process only. MSB and LSB capacitor cells are connected to ground via a 2C capacitor and a 1C capacitor, respectively. The C–2C capacitor ladder has at least two advantages over a conventional binary scaled capacitor array. The C–2C capacitor ladder is area efficient when compact compared to a conventional full-capacitor array. The C–2C ladder DAC uses $3N + 1$ unit capacitances C. An 8-bit DAC with C–2C capacitor ladder needs only 25 unit capacitances C, while a binary scaled capacitor array needs 256 C and a split array of 32 C. Moreover, because the individual cell capacitors are smaller, charging and discharging of the capacitors consume less power. Finally, because all the SAR cells drive the same capacitive load, all cell drivers are the same and do not need to be optimized for each load.

6.2.2.3 Layout of SOS Capacitive DACs

Capacitive DACs need very precisely matched capacitors arranged in a binary array. The capacitors thus must be laidout and organized to minimize mismatch. When a large number of bits is required, the matching of the smallest capacitances (C, 2C, 4C, etc.) becomes increasingly more difficult to control (Geiger et al., 1990; Baker et al., 1998; Razavi, 2000; Hastings, 2005). To provide the required precision up to 12–13 bits, the capacitors need to be designed as multiple of the smallest unit capacitance C.

In order to minimize mismatch, all the capacitive structures are organized as an array of instances of a single well-characterized device (unit capacitance C). Using the same device to create larger devices reduces mismatch due to fabrication and geometry, and also due to interconnections. All other capacitances were multiples of C in powers of 2. As an example, in Fig. 6.5 we provide an example of the arrangement of the capacitive ladder in a split-capacitor DAC array.

The capacitive ladder is divided into two identical arrays. The arrangement is symmetric in the horizontal plane and minimizes mismatch by providing the same boundary conditions to both subladders. Grounded dummy capacitors are added at the periphery of the capacitor array in order to minimize neighborhood mismatch. With the added dummies, each of the capacitors in the array has the same

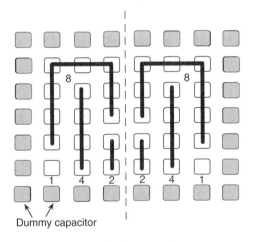

FIGURE 6.5 Layout arrangement of a capacitive array for SOS capacitive DACs. All capacitors are multiples of a small unit capacitance C. Capacitors are arranged in a radial fashion with common centroid to reduce mismatch and fabrication-derived layout differences. Grounded dummy capacitors are added at the periphery to minimize neighborhood mismatch.

identical set of neighbors. The connection between the capacitor follows a cross-coupled scheme typical of analog design in bulk CMOS. We use a spiral arrangement of larger to small capacitors in order to shield smaller units inside larger ones. Smaller capacitances are more sensitive to mismatch, as they carry information about the LSB. A common-centroid arrangement of the capacitor is also used to reduce mismatch (Geiger et al., 1990; Hastings, 2005). Finally, notice that the design of interconnects and top metal layers, on the other hand, assumes higher importance in the design of the ADC. In fact, although the substrate capacitances are greatly minimized, interconnections and line crosstalk are the critical points in the SOS process.

6.2.3 SOS Resistive DACs

Resistive DACs in SOS can be designed to be compact and low power by taking advantage of the built-in resistor in the technology. We have designed a resistive multiplying DAC (mDAC) in SOS. The DACs multiply an input voltage by a number specified by the digital inputs $D_0–D_{n-1}$. Implementation of the mDAC is shown in Fig. 6.6.

The mDAC produces an output voltage V_{out} that is the input voltage V_{in} multiplied or scaled by a number specified by the binary inputs

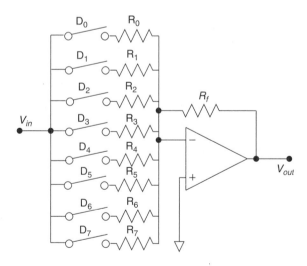

FIGURE 6.6   An SOS multiplying DAC (mDAC). The mDAC produces an output voltage V_{out}, which is the input voltage V_{in} multiplied or scaled by a number specified by the binary inputs $D_0–D_{n-1}$.

D_0 to D_{n-1}. When switches D_i are closed, the output of the mDAC is given by Eq. (6.11).

$$V_{out} = -\sum_{i=0}^{N} D_i \frac{R_f}{R_i} \cdot V_{in} \qquad (6.11)$$

6.3 SOS DACs

In today's market, there is a great variety of contending ADC architectures. The most popular ADC architectures are SAR, pipelined, flash, and sigma-delta ($\Sigma\Delta$) ADCs. Each one of these architectures has a preferential strength. Table 6.1 shows qualitatively how flash, $\Sigma\Delta$, pipelined, and successive approximation (SA) architectures differ with respect to the number of comparators and of comparison cycles per conversion (Baker et al., 1998; Razavi, 2000).

The number of comparators is linearly related to the system's total power consumption and is also proportional to the total system area. The sampling speed is linearly related to the number of comparison cycles per conversion. When selecting one ADC architecture, a trade-off exists between the number of comparators and comparison cycles. For example, a flash ADC performs one conversion each cycle with $2^N - 1$ comparison operations at the same time. On the other end, a successive approximation and $\Sigma\Delta$ ADCs requires one comparator but needs multiple comparisons to converge to a final result. A pipelined ADC is an intermediate option, which performs multiple simultaneous comparisons in different stages and converts at a faster speed than a SA converter. Figure 6.7 shows how different ADC architectures relate to ADC resolution, sampling speed, and power consumption.

In this chapter, we present two types of SOS ADCs topologies: successive-approximation converters (Sec. 6.3.1), and asynchronous sigma-delta converters (Sec. 6.4). These two architectures are chosen because of their low-power operation, their simplicity, and their good sampling speeds. Both are a good trade-off in both power and speed,

ADC Type	SAR ADC	$\Sigma\Delta$ ADC	Pipelined ADC	Flash ADC
Number of comparators	1	1	P	$2^n - 1$
Conversion cycles	n	2^n	$P \times 2^{\frac{n}{p}}$	1

TABLE 6.1 Comparison of ADCs architectures, number of comparators required, and conversion cycles.

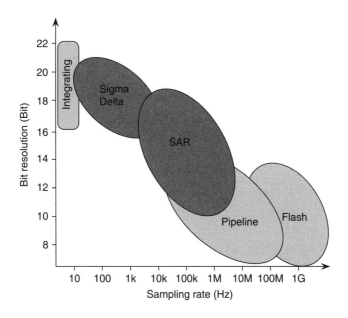

FIGURE 6.7 Comparison of ADC architectures, resolutions, and sampling rates. In this chapter, we present two SOS topologies: a SA and a $\Sigma\Delta$ converter.

but the SA ADC is generally faster, whereas the $\Sigma\Delta$ converters are generally more power efficient. For these reasons, these two topologies have been chosen as sample SOS designs and are presented here.

6.3.1 SOS Successive Approximation Analog-to-Digital Converters

SA ADCs represent a successful compromise between power consumption, small footprint and high resolution for CMOS circuit implementation (Zhou et al., 1997; Walden, 1999; Montezapour and Lee, 2000; Promitzer, 2001; Kuttner, 2002; Scott et al., 2003; Yang and Sarpeshkar, 2005; Gambini and Rabaey, 2007; Verma and Chandrakasan, 2007).

The ADC that we address in this chapter is a mixed-mode analog-digital circuit performing an algorithmic conversion of an analog input voltage into an 8-bit binary code. Conversion is performed by recursively comparing a stored reconstructed voltage to the sampled input until the desired precision is met (Baker et al., 1998). Figure 6.8 is a typical schematic of an SA ADC.

SA ADCs have the advantage of using low power, as they reuse elements in an algorithmic fashion (from this derives their name).

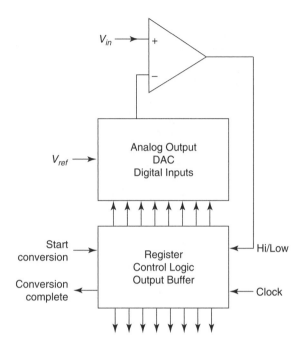

FIGURE 6.8 An SA ADC. An input voltage is compared to fractions of a reference voltage in a sequence of steps. A digital control logic supervises the algorithmic conversion and stores the converted result. The conversion ends after N cycles or when a predefined precision has been met.

The number of analog components being active during conversion is limited to a single high-performance comparator. The rest of the circuitry employs low-power fast-switching digital logic to coordinate the algorithmic conversion. Similarly, SA ADCs occupy a small footprint because, again, of the reutilization of the same circuit components. Also, the use of a single comparator prevents space–performance trade-offs typical in Flash or other kinds of ADCs. The comparator can be optimized for performance without large penalties in the layout area. In addition, SA ADCs are able to report high resolutions currently up to 16–18 bits. These resolutions are achieved with an external calibration circuit. Algorithmic conversions rely on area matching of capacitive devices, thus a high number of bits (12 bits) can be obtained even without the need of error-correcting techniques.

The absence of parasitics in the SOS CMOS fabrication process simplifies the design of ADCs. The circuit design is, in fact, freed from the unaccounted bulk CMOS capacitances that generate digital signal feedback and corrupt the performance of the ADC at high operating frequencies (Peregrine, 2008a). In addition, passive components like capacitors, whose size is the main source of mismatch in algorithmic converters, can be fabricated in SOS with high precision and without the additional uncertainty value of the bulk CMOS passive capacitances to the substrate.

We have designed and discussed here the realization of the first fabricated 8-bit SAR ADC in the FC 0.5-μm SOS CMOS technology offered by (Peregrine, 2008a). This designed was developed for low-power application and was able to perform at 1 mW power consumption. We also present the design and performance evaluation of three advanced SAR ADCs—a high-performance 10-bit and a 8-bit resolution, and one with 8-bit resolution and an ultralow power consumption below 1 μW. The design of these four converters encompasses many possible applications and trade-offs. The 8-bit ultralow power SAR ADC can be used in application where precision is not paramount but that require monitoring physical quantities for extended times while running on batteries. The first 8-bit SAR ADC was designed to provide 1 Ms/s performance and low-power operation. The 10-bit SAR ADC was designed to provide both performance and precision, with 1 Ms/s and a few milliwatts of power. The proposed ADCs were all targeted for low-power applications and average conversion performance needed in data acquisition, systems on a chip, embedded systems, and vision systems employing column parallel architecture.

6.3.2 Operation of a SAR

The SAR is a classic CMOS mixed-mode analog-digital circuit that is capable of performing analog-to-digital conversion as well as digital-to-analog conversion internally. The input voltage is compared to a reconstructed internal analog voltage in a series of steps.

Figure 6.9 illustrates a schematic caption of the ADC system. A split array of capacitive elements is used in the ADC to perform the algorithmic conversion. The capacitor array performs as a DAC. Capacitor sizes are arranged in powers of two and divided into two banks to reduce the total area of the capacitor array, as explained in the next paragraph. Notice that this architecture allows us to sample/hold the input on one of the capacitor arrays, thus saving the extra capacitance that would be needed to sample the input if this was provided on the negative terminal of the comparator. The ADC performs a binary search over all possible quantization steps (a number of successive approximations or steps equal to the number of bits of the comparator) to converge on the final digital output value. This is done by copying the input voltage to be converted into an internal V_{DAC} voltage, successively adjusting the capacitors voltages with steps of approximation. Successive approximations are a series of capacitive charging,

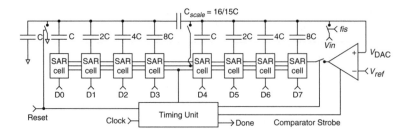

FIGURE 6.9 Split capacitive array and successive approximation register. The capacitor array is split into two smaller arrays to minimize area. SAR cells store the bits of the conversion. A comparator successively tests a fraction of the reference signal V_{ref} with the converted internal signal V_{DAC}. The ADC algorithmic sequence is controlled by a digital timing unit. Notice that this architecture allows us to sample/hold the input on one of the capacitor arrays.

comparing the value V_{DAC} with the input and updating the output register (SAR cells in Fig. 6.11) (Baker et al., 1998; Zhou et al., 1997). At the end of the conversion, the SAR register will contain the binary representation of the input voltage, with one bit per SAR cell.

A conversion begins with a global reset (signal: *reset*) that discharges all the capacitors in the array (*reset* signal in Fig. 6.9). This signal starts the conversion and therefore corresponds to the *start conversion* signal found in most algorithmic ADCs. The capacitor plates are connected to ground during the reset phase, thus eliminating all the charge they previously contained. In addition, the *reset* signal brings all the bits in the SAR cells to a zero value. The input signal V_{in} is then sampled on the capacitive array by strobing the signal *fis*; this is done by keeping the capacitors lower plate connected to ground. Notice that the signal is sampled on the MSB bank for charge distribution of the capacitive network. At this point, the capacitor array stores the input voltage V_{in}, so $V_{DAC} = V_{in}$. The input signal must be smaller than the input reference voltage V_{ref}, whose overall stability affects the conversion directly. The algorithmic conversion begins by adding $V_{ref}/2$ to V_{DAC} and comparing the value to V_{ref} (an example of a 4-bit conversion is given in Fig. 6.10). If the value is larger, the corresponding capacitor is reset and its SAR cell (MSB) in the register is cleared. If the value is smaller, the MSB will be set to a logical one. The second most significant bit is then tested. Now a voltage of $V_{ref}/4$ is added to the capacitor array voltage V_{DAC} and the value is once again compared to V_{ref}.

The algorithm follows, as described above, for all the remaining N bits, or until the desired precision is met. The voltage on the upper plates of the capacitor array slowly converges to the input value

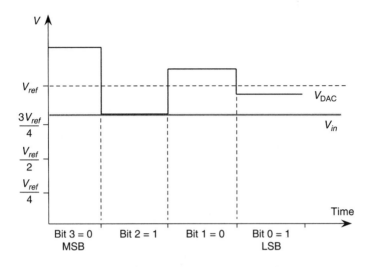

FIGURE 6.10   Example of a $N = 4$-bit conversion. The ADC successively compares a digitally controlled fraction of the reference signal V_{ref} (V_{DAC}) against the input analog voltage V_{in}. After N conversion steps, the signal V_{DAC} converges to the input V_{in} with a precision of less than 1/2LSB.

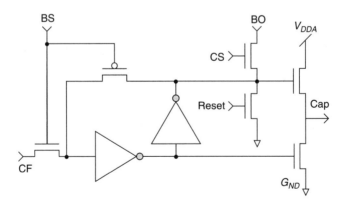

FIGURE 6.11   SAR cell. Two inverters act as a static RAM cell, controlled by the sample signal BS and the Reset. An output inverter toggles the capacitor Cap voltage to either the reference V_{DDA} or ground.

V_i until the desired precision is reached and the V_{DAC} voltage will converge to V_{ref} [Eq. (6.12)].

$$V_{DAC} = V_{in} + \sum_{k=0}^{N-1} D_k 2^{k-N} V_{ref} \tag{6.12}$$

Equation (6.12) gives a value of the V_{DAC}, (see Fig. 6.9) voltage at the internal input node of the capacitor array. This value is compared with the reference voltage V_{ref} at each iteration k. D_k is the digital output value of the kth SAR cells (Baker et al., 1998).

6.3.3 Capacitive Ladder and Charge Scaling

The basic structure of the ADC is the capacitive ladder in Fig. 6.9. The capacitive ladder implements a charge-scaling DAC, which algorithmically minimizes the error between its output voltage and the analog input to be converted, as described in the previous section. This is a very popular architecture in CMOS technology because it employs a switched capacitors network, which is easy to model, to control digitally and provide good linearity with the devices available in the process. $N + 1$-weighted capacitors, where N is the number of bits of the ADC (here, 8), are switched on or off to generate an output voltage V_{out}. V_{out} is a function of the voltage division between capacitors. The total capacitance of a SAR would generally be approximately $2^N C$, where C is the minimum capacitance used for the least significant bit. The use of a full SAR would require a very large chip area because of the size of the largest capacitor. In fact, the lowest capacitance C is lower bounded for precision, and this bound reflects on the largest capacitance, which then uses a large silicon area. In addition, the penalty to charge a large capacitance is a high-power consumption. The disadvantages of the use of a full SAR suggests the use of a more space- and power-aware architecture. The *split capacitive array* (SCA) architecture was chosen. SCA provides a good trade-off between capacitor size and accuracy, especially when taking advantage of the ideality of the SOS substrate, which greatly reduces crosstalk and spurious bulk capacitances that would limit the precision of the ADC. Figure 6.9 is a pictorial representation of the SCA architecture. SCA uses the properties of parallel and series passive elements to create an equivalent capacitance of smaller size but is effectively seen as a logarithmically increasing capacitive ladder.

Figure 6.9 represents the two sides of the capacitor array, connected by a scaling capacitor C_{scale}. The scaling capacitor makes the upper MSB bank of capacitors seem to have a larger capacitance value than their effective size, allowing for equivalent savings in terms of area and power consumption. The scaling capacitor value is given by Eq. (6.10). For an 8-bit precision ADC, the C_{scale} computed value is 16/15C (Baker et al., 1998).

Figure 6.11 represents a successive approximation register cell (Zhou et al., 1997). The cell is responsible for charging and discharging the capacitors in the array and for storing the digital code for each bit of the ADC. The comparator flag (CF) signal allows the comparator to set the value of the bit after comparing the capacitor array voltage to the input voltage. The bit select (BS) line activates only the active bit for each conversion step. BS is set by the timing logic circuitry. The CS signal selects the bits for readout on the bit out line BO. The Reset signal stores a zero value at the beginning of the conversion and also discharges the capacitor. The cap line is connected to the capacitor in the array belonging to the SAR cell. This line sets and resets the capacitor value at the beginning of the conversion or after a comparator decision.

The voltage V_{DDA} in Fig. 6.11 can be either an analog power supply, or it can be a voltage reference whose value is smaller than the analog power supply. In either case, the precision of the ADC is influenced primarily by the value of this supply voltage, therefore this supply needs to be properly buffered, filtered, and stabilized to provide noise immunity. Finally, notice that the SAR cell does not introduce charge injection on the capacitor arrays because the capacitor driver is an inverter operating as a bimodal switch.

6.3.4 SAR ADC Design in SOS and Optimization

The design of a successive approximation analog to digital converter in SOS is in many ways simplified. Taking advantage of the novelties of the process requires a careful study of the analog components and the use of the appropriate devices. When we take advantage of the benefits of the SOS process, the design is more easily handled.

Linear capacitors in SOS were designed using a metal–insulator–metal (MIM) type device provided by the Peregrine FC process. This kind of capacitor has large densities, and is equal to MOS capacitors while having the advantage of being linear devices.

Analog SOS circuit advantages are higher current drive, short interconnections, and reduced capacitance. Because of the lack of substrate crosstalk and coupling, noise and latch-up are minimized. Passive components are also freed from the plague imposed by parasitic elements, typically capacitances and parasitic diodes to the substrate.

On the other hand, analog design in SOS can be affected by floating body effects of active devices. In particular, the kink effect in the drain current to voltage relation (as in Chap. 2 and in Culurciello et al., 2002). In case of need, artificial body contacts can be placed around critical devices to remove floating body effects. Floating body effects have been recorded in regular threshold devices, both PMOS and NMOS, whereas the effect is not present or is negligible in lower threshold devices, as reported in Chap. 2. This suggest the use of low-threshold devices as an additional measure to avoid floating body effects.

Another issue is thermal conductivity at room temperature (0.4 W/cm-K), which is about three times worse than in silicon (1.4 W/cm-K). Temperature can, in fact, influence the behavior of critical analog circuits as the comparator. This is not a problem for the successive approximation ADC because of the relatively low speed of conversion and the low component density around the critical analog components. In addition the SOS SAR ADCs were designed to perform with low-power operation of a few milliwatts at most.

The performance of a successive approximation ADC greatly depends on the capacitive ladder, the comparator, and the switching logic.

6.3.4.1 Design of Capacitor Array in SOS

Figure 6.5 represents the arrangement of the capacitive ladder, and, in particular, of a SCA and layout arrangement for a SAR ADC implemented in SOS.

As an example, in one of our designs the reference capacitor size was $C = 100.6$ fF. The scaling capacitor C_1 size is 107.3 fF approximated for layout restrictions to 107.2 fF. The error resulting from this approximation can be calculated as follows. The error in C_1 is 0.1%. The maximum error due to the scaling capacitor occurs when the code switches from 00001111 to 00010000 (code 31 to 32). This corresponds to the LSB of the upper bank of capacitors. Because of the scaling, the error seen from the upper bank of capacitors is 0.456% from a parallel of a 16/15 C capacitor with 0.01 C error and a 15 C capacitor.

The sequence of bits to the corresponding capacitor to charge is the following: 0 1 2 3 - 4 5 7 to 8 1 4 2 - 2 4 1 8. The second series of numbers correspond to the capacitor arrangement. Because the activated capacitor is in sequence (1 2 4 8), the addressing is interleaved to minimize digital crosstalk in the capacitive ladder.

In the 0.5 μm FC SOS process, resistor RN matching is 0.4% for a 4-μm width device and 0.2% for a 27-μm width device. Capacitor matching is 1% for a 5×5 μm device, 0.5% for a 10×10 μm device, and 0.1% for a 25×25 μm device.

Note that the SOS design not only removes parasitic capacitance to the substrate, but also greatly minimizes the parasitic effect of fringe capacitances. A typical value of capacitance per unit in the SOS process is 0.57 fF/μm^2, whereas fringe capacitance is limited to about 28 aF/μm. In a standard 0.5 μm CMOS process the fringe capacitance is on the order of 60 aF/μm. These values show how SOS capacitor matching is superior to standard CMOS because of the lower fringe capacitance and also because of the extremely low impact of fringes as compared to the desired capacitors. The availability of better matching allows for a reduction in capacitor size and, ultimately, in power consumption of the ADC. The size of the capacitor must be chosen according to the matching statistics of the process. The smallest unit capacitance have to match to at least 50% precision with each other, equivalent to a half LSB. If the fabrication company provides statistical data, the size of the minimum capacitance can be calculated according to the data. As an example, consider a process with the matching properties derived from Eq. (6.13), where α and β are parameters, A is the capacitor area and σ_c is the standard deviation of capacitance in percent.

$$\sigma_c^2 = \frac{\alpha}{A} + \frac{\beta}{A^{3/2}} \, [\%\mu m^2] \qquad (6.13)$$

From Eq. (6.13) we can compute the minimum size for a desired matching (the value is $\sigma_c = 50\%$ for the ADC). The resulting value of capacitance should be used to minimize power consumption and silicon area consumption.

6.3.4.2 Input Switches, Logic, and SAR: SOS Design

Sampling the input voltage on the capacitive ladder requires a MOS switch-to-transfer charge. The charge injection of the switch must be lower than the LSB of the ADC. As an example, in one of our designs we used a pass gate with transistor size of (width, length) 3×0.5 μm. The charge injection of the switch is given by $C_{gate}/2$ where it is supposed that at the opening time of the switch, the charge accumulated in the channel spreads evenly between source and drain. Data resulted from simulations of this pass gate indicated that the switch introduces a charge injection smaller than 5 mV on a 1.6-pF capacitor (this value is the largest capacitance of the SAR ladder in our low-power 8-bit SAR ADC example) during sampling of the input voltage. The charge injection can be eliminated using a dummy switch that drains the charge away as the switch commutes.

Noise on the voltage reference V_{ref} line can affect the performance of the ADC significantly. The use of an internal voltage reference is highly recommended and was not included in this first version of the successive approximation ADC. An internal reference reduces the noise coupled to the interconnection and wires of an external source. In addition, particular care has to be taken during the layout of the V_{ref} distribution line. Noise immunity in SOS is intrinsically higher due to the lack of substrate crosstalk. Additional immunity can be guaranteed by separating the V_{ref} line from digital logic lines by a few multiples of the minimum distance between metal lines and by avoiding overlapping metal lines.

Another concern in the design is how to size individual cells of the SAR. Cells output drivers should be sized in exponential progression, as each cell has to drive an exponentially increasing capacitor. This complicates the design, as each cell must comply to specification and needs to be able to charge the capacitance in the allotted time to 1 LSB of precision. In addition, the cell drivers would not match, as the charge injection would be related to driver size. Because the use of a SA capacitive ladder reduces the differences between sizes of LSB and MSB, however, we decided to use a unique cell design for the entire SAR. This in turn puts a lower bound on the minimum cycle time, or the time it takes to charge the biggest capacitance of the split array (8C).

6.3.4.3 Comparator

The comparator is a critical component of the ADC because not only does it have to provide an accuracy of comparison better than $1/2$ LSB, but it also has to compare with very high speed, if high sampling rates are desired. In critical analog components like the comparator of the successive approximation here presented, the novelty of the SOS process and its devices allow for design simplification and improved performance. A description of the comparator has been given in Sec. 4.2.1.

6.3.5 A High-Performance 8-Bit SAR ADC

We present measured data on a fabricated and tested an 8-bit SAR ADC manufactured in the SOS process. The SAR ADC was the first of its kind to be designed, fabricated, and tested on the SOS process (Walden, 1999; Scott et al., 2003; Fu and Culurciello, 2006; Yang and Sarpeshkar, 2005; Gambini and Rabaey, 2007; Verma and Chandrakasan, 2007). It was designed to both take advantage and serve as an example design of SOS mixed-mode circuits. The target was a power consumption of 1 mW and a sample rate of 1 Ms/s. These are midrange characteristics that make this converter a great compromise between power, efficiency, resolution, and performance.

The SAR ADC schematic diagram is seen in Fig. 6.9 and described in Sec. 6.3.2. The fabricated SOS ADC circuit occupies an area of $450 \times 315 \ \mu m^2$ and $916 \times 790 \ \mu m^2$ with output pads. Figure 6.12 is a micrograph of the fabricated chip. The minimum size capacitor for the array is 100.6 fF and the total input capacitance at the conversion input is 3.2 pF, excluding the pad capacitance. The ADC operates with a power supply of 3.3 V nominally. The reference voltage V_{ref} was set to 2.7 V throughout testing. The operational voltage can be decreased to 2.9 V with unaccounted loss of precision.

A description of the comparator used in this design has been given in Sec. 4.2.1.

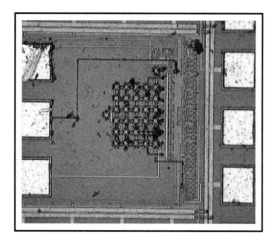

FIGURE 6.12   A micrograph of the fabricated 8-bit SOS ADC die. Visible is the capacitor array (30 + 1 capacitors), the digital control logic on the right side and some of the output pads.

The power consumption of the chip is divided into analog and digital power supplies. The analog power supply provides currents for the charging and discharging of the capacitors, the DAC and the comparator. The digital power supply provides current to the digital logic, clocking logic, and output buffers. The analog power supply uses 0.77 mW over the operational range, 1 kHz to 32 MHz. Digital power consumption accounts for 0.79 mW at 1-kHz operation. The digital power consumption increases with the operating frequency because of the high capacitive load of the output buffers and pads. A numerical fit of the digital power consumption P_d is given by Eq. (6.14).

$$P_d = 0.78 + 1.2 \cdot 10^{-7} (f[Hz]) \quad [mW] \qquad (6.14)$$

This value, when related to the general formula $P_d = f_{clock} C_L V_{DD}^2$ at the described parameters, gives $C_L = 11$ pF. This capacitance is an effective model of the dynamic power dissipation of the circuit at high frequency. Static power dissipation for the ADC is 1.55 mW. Figure 6.16 shows a plot of both digital and analog power consumptions versus ADC clock operating frequency with a 3.3-V supply.

The maximum conversion input range was measured as 2.1 V (0.2–2.3 V) using a 3.3-V supply with V_{ref} set to 2.7 V. The lower bound of the conversion range is due to the first stage of the comparator, which stops operating for very low input voltages, or when the current source of the differential pair exits the saturation region. The upper bound is due again to the comparator's first stage. For high-input voltages,

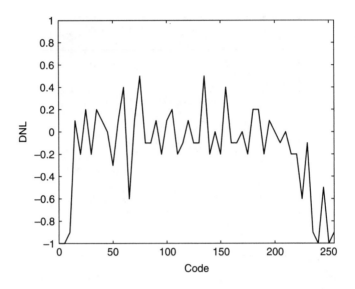

FIGURE 6.13 ADC DNL at 1 KHz for the 8-bit SAR ADC fabricated in the SOS process.

the internal reconstruction voltage V_{DAC} becomes higher than the rail voltage, thus the comparator is unable to resolve.

The 8-bit ADC performed with an average of 0.18 LSB differential nonlinearity (DNL) and a mean integral nonlinearity (INL) of 0.87 LSB. Figures 6.13 and 6.14, respectively show plots of DNL and INL as a function of the output digital code. These data were collected at a clock frequency of 1 kHz, corresponding to a sampling frequency of 38 S/s. At this rate, the ADC has 8 bits of resolution and 8 bits of accuracy.

The frequency range of operation of the ADC is 500 Hz–32 MHz, which corresponds to a sampling frequency of 19 S/s–1.23 MS/s, as 26 clock ticks are necessary to complete a conversion.

The number of bits was 7.92 bits for operational frequency of up to 3 MHz (115 kS/s), whereas decreased to 7 bits at a clock frequency of 10 MHz (384 kS/s). Figure 6.15 represents the number of bits as a function of the operational frequency of the ADC clock. At faster speeds, precision dropped significantly due to insufficient settling time for charging and discharging of the capacitor array as well as comparator settling time.

Total harmonic distortion (THD) was measured by sampling a 2-V peak-to-peak 1-kHz sine wave with a 1.38-MHz clock (53 KS/s). A plot of the measured FFT spectrum for the sampled 1 kHz waveform is given in Fig. 6.17. THD was measured to be 29.42 dB. The spurious

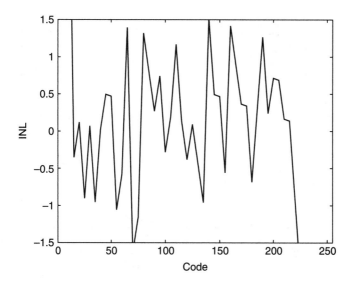

FIGURE 6.14   ADC INL at 1 KHz for the 8-bit SAR ADC fabricated in the SOS process.

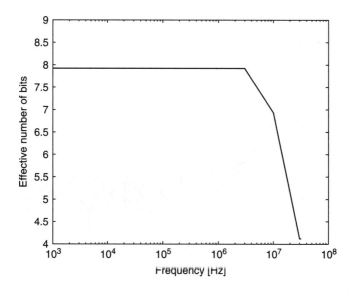

FIGURE 6.15   Effective number of bits for an input voltage $V_{in} = 2.01$ V for the 8-bit SAR ADC fabricated in the SOS process.

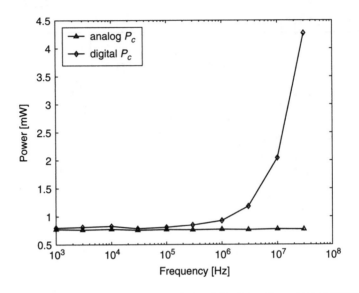

FIGURE 6.16 Power consumption (P_c) versus operational clock frequency for the 8-bit SAR ADC fabricated in the SOS process.

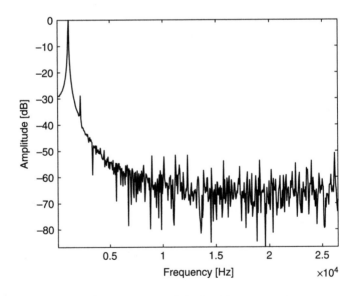

FIGURE 6.17 Measured FFT spectrum for a KHz sine waveform ($2\ V_{pp}$) sampled at 53 KS/s for the 8-bit SAR ADC fabricated in the SOS process.

free dynamic range (SFDR) for the same input was measured to be 29.09 dB.

A common figure of merit (FOM) used to compare ADC designs is given in Eq. (6.15) (Scott et al., 2003).

$$\text{FOM} = \frac{P_d}{2^{ENOB} \cdot f_s} \tag{6.15}$$

The FOM value for this SOS ADC is 5 pJ/conversion.

6.3.6 A High-Precision 10-Bit SOS SAR ADC

We designed, fabricated, and tested a high-precision 10-bit SAR ADC in SOS. The designed targeted high bit counts performance analog-to-digital conversion while maintaining low power operation at the same time. The target for this design was a power consumption of 5 mW and a sample rate of 1 Ms/s. These characteristics make this converter competitive with lower feature-size bulk CMOS ADCs in power efficiency, resolution, and performance.

The design uses an SCA to preserve the silicon area and to reduce power consumption at the same time. In this 10-bit ADC, the unit capacitance C was sized to be 215 fF. A schematic of the 10-bit ADC is similar to the one reported in Fig. 6.9, but with an additional capacitor for each side of the split array. The left array contained the capacitances C, C, $2C$, $4C$, $8C$, and $16C$, whereas the right array contained C, $2C$, $4C$, $8C$, and $16C$. The scaling capacitor C_{scale} has a value of $32/31\ C$ in this 10-bits ADC. In this 10-bit ADC, the SA has a total capacitance of approximately 64 unit capacitors C (13.7 pF in total), whereas a binary-scaled array would need 1024 unit capacitors, or 220 pF. The silicon area savings are thus on the excess of 16. We mention again that the SCA is not easily implementable in a bulk CMOS process because of the parasitic capacitances to the substrate. This implementation and the following results are only possible with the SOS process and its features.

The 10-bit ADC uses a fast precision comparator like the one described in Sec. 4.2.1.

The die micrograph is shown in Fig. 6.18. Die size is $480 \times 340\ \mu m^2$ and $890 \times 630\ \mu m^2$ with pads. The converter operates at 3 V power supply. The reference voltage was set to from 1.5 to 2.5 V during testing.

Figure 6.19 shows the analog and digital power consumption at different sampling speeds when $V_{ref} = 1.5$ V. The analog power consumption is 2.3 mW. Table 6.2 shows the total power consumption with three external reference voltages at a conversion rate of 500 kS/s. When the V_{ref} is 2.5 V, the total power is 1.8 mW.

The ADC samples are from 1.8 KS/s to 1 MS/s. Figures 6.20 and 6.21 show the INL and DNL collected at a sampling speed of 500 kS/s

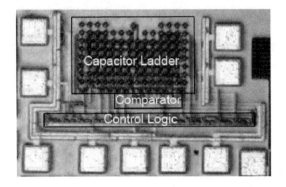

FIGURE 6.18   Micrograph of the high-performance 10-bit SAR ADC fabricated in the SOS technology. Die size is 480 × 340 μm^2 and 890 × 630 μm^2 with pads.

and 1.5-V reference voltage. The ADC performed with an average of 0.38 LSB INL and an average DNL of 0.23 LSB. The maximum INL is 0.79 LSB, and the maximum DNL is 0.65 LSB. The input range was measured as 1.44 V. A plot of Faot Fourier Transform (FFT) spectrum for a sampled 1.67-kHz sine waveform is given in Fig. 6.22. The THD for this converter is 62 dB. The SFDR is 65 dB.

The FOM value for this SOS ADC is 1 pJ/conversion.

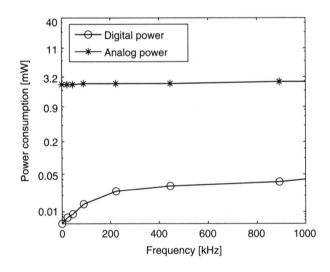

FIGURE 6.19   Analog and digital circuits power consumption versus sampling frequency of the 10-bit SAR ADC fabricated in SOS.

Reference voltage (V)	1.5	2	2.5
Total power consumption (mW)	2.4	2	1.8

TABLE 6.2   Total power consumption with different
reference voltages.

6.3.7 A Low-Power 8-bit SOS SAR ADC

We designed, fabricated, and tested a low-power 8-bit SAR ADC in
SOS. The designed targeted low-power applications. The target for
this design was a power consumption of 5 μW and a sample rate of
300 Ks/s. These characteristics make this converter competitive with
lower feature-size bulk CMOS ADCs in power efficiency, resolution,
and performance.

The design uses an SCA to preserve silicon area and, at the same
time, reduce power consumption. In this 8-bit ADC, the unit capacitance C was sized to be 215 fF, and the total capacitance amounts to
6.8 pF. A schematic of the 8-bit ADC is reported in Fig. 6.9. The design
uses a switched capacitor comparator to save power. The comparator
design is reported in Sec. 4.2.2.

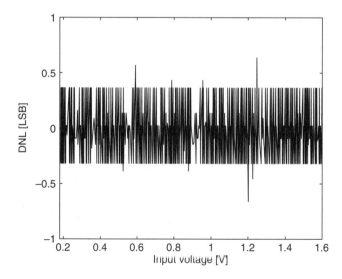

FIGURE 6.20   DNL versus input voltage at the sampling speed of 500 kS/s
for the 10-bit SAR ADC fabricated in the SOS process. The ADC performed
with an average DNL of 0.23 LSB and a maximum value of 0.65 LSB.

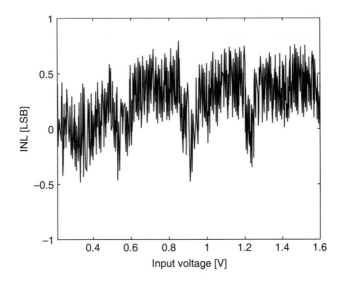

FIGURE 6.21 INL versus input voltage at the sampling speed of 500 kS/s for the 10-bit SAR ADC fabricated in the SOS process. The ADC performed with an average INL of 0.38 LSB and a maximum value of 0.79 LSB.

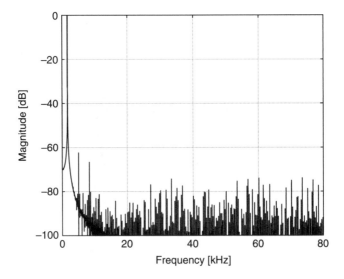

FIGURE 6.22 Measured FFT spectrum for a 1.67-kHz sin waveform sampled at 200 kS/s for the 10-bit SAR ADC fabricated in the SOS process. The THD for this converter is 62 dB. The SFDR is 65 dB.

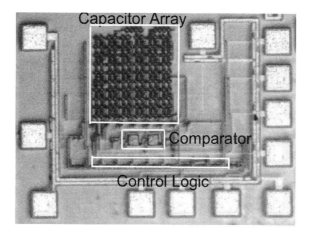

FIGURE 6.23 Micrograph of the 8-bit SAR ADC fabricated in the SOS technology. The die size is 620 × 780 μm² and 770 × 1120 μm² with pads.

The die micrograph is shown in Fig. 6.23. The die size is 620 × 780 μm² and 770 × 1120 μm² with pads. The converter operates at a 1.5-V power supply. The reference voltage was set to 1 throughout testing.

Figure 6.24 shows the analog and digital power consumption at different frequencies. The power consumption of ADCs includes the dynamic and static part. The power for analog components provides current to charging and discharging of the capacitors and operating of the comparator. The digital power supplies the control logic and the output buffers the driving pads and off-chip load. It increases linearly with the frequency. Digital power becomes negligible when ADC is used in a system-on-a-chip. The power consumption of analog power part is 3 μW. The ADC operates from a 22 kHz to 4.5 MHz clock rate, which corresponds to a sampling frequency from 2 kS/s to 409 kS/s.

Figures 6.25 and 6.26 show the INL and DNL. The data are collected at a clock frequency of 22 kHz, corresponding to a sampling frequency of 2 kS/s. The ADC performed with an average of 0.12 LSB DNL and an average of 0.38 LSB INL. The max INL is 0.82 LSB. Notice an increase of the DNL and INL in the middle of the input range, due to the difference between the computed scaling capacitor (C_{scale} in Fig. 6.9) and actual capacitance available in the SOS process.

The effective number of bits (ENOB) was 7.62 bits for a operational frequency of up to 3.4 MHz (309 kS/s), whereas it decreased to 7 bits at a clock frequency of 4.5 MHz (409 kS/s). Figure 6.27 represents the ENOB as function of operational clock frequency. THD was measured at 100 kS/s. A plot of the FFT spectrum for a sampled 1-kHz sine

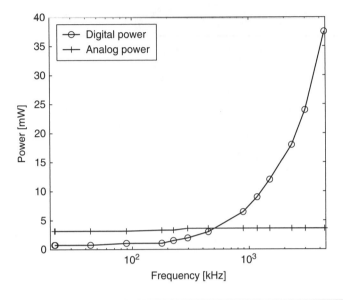

FIGURE 6.24 Analog and digital circuits power consumption versus operational clock frequency for the 8-bit SAR ADC fabricated in the SOS process. The digital supply is 1.5 V.

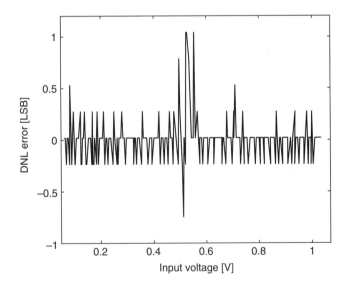

FIGURE 6.25 INL versus input voltage at 2 kS/s for the 8-bit SAR ADC fabricated in the SOS process. The ADC performed with an average INL of 0.38 LSB and maximum value of 0.82 LSB.

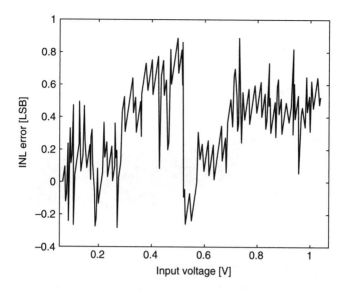

FIGURE 6.26   DNL versus input voltage at 2 kS/s for the 8-bit SAR ADC fabricated in the SOS process. The ADC performed with an average DNL of 0.12 LSB.

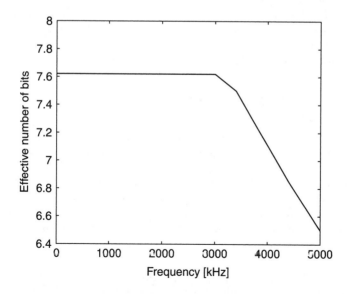

FIGURE 6.27   Effective number of bits for $V_{in} = 720$ mV

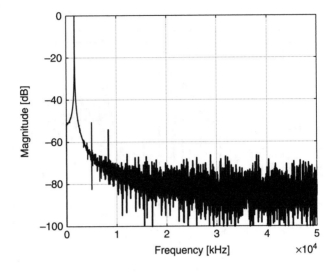

FIGURE 6.28   Measured output waveform FFT spectrum for an input 1.67-kHz sine waveform sampled at 100 kS/s for the 8-bit SAR ADC fabricated in the SOS process. THD was measured to be 52 dB. The SFDR for the same input was measured to be 50 dB.

waveform is given in Fig. 6.28. THD was measured to be 52 dB. The SFDR for the same input was measured to be 50 dB. The input range was measured as 0.97 V using a 1.5-V supply with reference voltage V_{ref} set to 1 V.

The FOM value for this SOS ADC is 20 fJ/conversion.

6.3.8   A C-2C Ladder Ultralow Power SOS SAR ADC

The capacitive array of a SAR ADC can be minimized by using a C-2C DAC as the one presented in Fig. 6.4. We designed an ultralow power SAR ADC in SOS by taking advantage of the C-2C ladder DAC structure (Fu et al., 2006). Our design is the first C-2C ADC in CMOS technology with satisfactory performance. SOS lack of substrate parasitics allows to manufacture floating capacitors and thus significantly reduce the area of a capacitor array. The C-2C SAR ADC uses 25 unit capacitances C, whereas a split array configuration uses 32 and a full binary array uses 256 C.

The ADC schematic is presented in Fig. 6.29. The capacitor ladder is composed of eight identical cells (C-2C) and two grounded C capacitors at either end. Each capacitor cell includes a floating 2C capacitor and a C capacitor, all driven by an SAR. The MSB and LSB capacitor

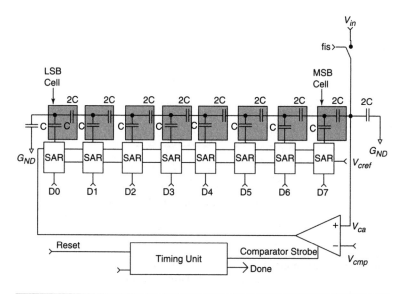

FIGURE 6.29   A schematic of the SAR ADC with C-2C capacitive ladder topology implemented in the SOS fabrication process.

cells are connected to ground via, respectively, a 2C capacitor and a single C capacitor. The design uses a switched capacitor comparator to save power. The comparator design is discussed in Sec. 4.2.2.

The output of the capacitor chain V_{ca} is the voltage between the floating 2C capacitance in the MSB cell and the grounded 2C capacitor at the same end. It is expressed in Eq. (6.16).

$$V_{ca} = V_{in} + \sum_{k=1}^{N} \frac{D_k 2^{N-k}}{2^N \times 3} V_{cref} \tag{6.16}$$

where N is the total number of bits in the ADC (8 in this case). D_k is the digital bit stored in the kth successive approximation register. The capacitor reference voltage V_{cref} is the voltage of the capacitor bottom plate when the successive approximation register is set to 1. The capacitor reference voltage V_{cref} was set to 1.6 V with a 1.1-V power supply and 2.3-V with a 1.5-V supply.

The die micrograph is shown in Fig. 6.30. The die size is 300 × 700 μm² and 640 × 1070 μm² with pads. The converter operates at 1.1 V to a 1.5-V power supply. The reference voltage was set to 1 for a 1.5-V power supply and 0.8 V for 1.1 V.

Figure 6.31 shows power consumption against operational frequency at 1.1 V and 1.5 V power supply. P_a is the analog part of power consumption, P_d is the digital part of power consumption.

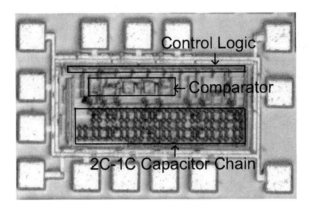

FIGURE 6.30 Micrograph of the 8-bit C-2C ladder SAR ADC fabricated in the SOS process. The die size is 300 × 700 μm² and 640 × 1070 μm² with pads.

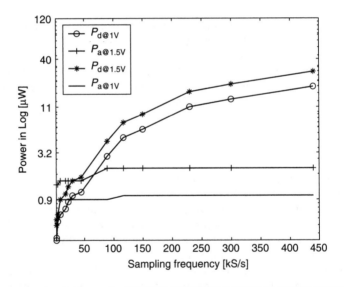

FIGURE 6.31 Power consumption against operational frequency at 1.1 and 1.5 V power supply for the C-2C 8-bit SAR ADC fabricated in the SOS process.

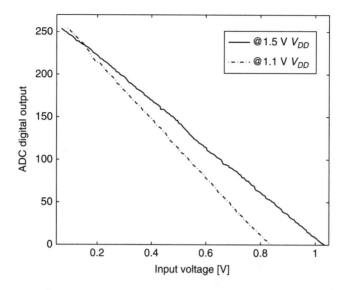

FIGURE 6.32 Transfer function curve between analog input and digital output at 1.1 V and 1.5 V power supply for the C-2C 8-bit SAR ADC fabricated in the SOS process.

Figure 6.32 shows a transfer function curve between analog input and digital output at 1.1 and 1.5 V power supply.

The FOM value for this SOS ADC is 8 fJ/conversion.

6.3.9 SOS SAR ADC Summary

We presented four successive approximation ADCs fabricated in the SOS technology. All ADC designs targeted low-power sensor interfaces and untethered sensing applications. The performance of the SAR ADCs is summarized in Table 6.3.

6.4 SOS Asynchronous $\Sigma\Delta$ Analog to Digital Converters

Sigma-delta ($\Sigma - \Delta$) converters are a special kind of ADC typically referred as oversampling converters. These converters sample the input signal at a much higher rate than the Nyquist rate $f_s = 2 \cdot f_{in}$, where f_{in} is the input signal frequency. They are called oversampling converters because they typically require a high number of cycles per conversion. This kind of converter has the advantage of very high resolutions and

ADC Type	8-Bit Split Array ADC	8-Bit Split Array ADC	8-Bit C-2C Ladder ADC	10-Bit Split Array ADC
Power supply	3.3 V	1.5 V	1.1 V–1.5 V	3 V
Reference voltage	2.7 V	1 V	0.8 V–1 V	1.5 V–2.5 V
Power consumption	1.5 mW	3 µW@1.5 V	900 nW@1.1 V / 1.35 µW@1.5 V	1.8 mW@3 V
Operational clock rate	500 Hz–32 MHz	22 kHz–4.5 MHz	22 kHz–4 MHz	21.6 kHz–12 MHz
Sampling rate	19 S/s–1.23 MS/s	2 kS/s–409 kS/s	2 kS/s–400 kS/s	1.8 kS/s–1 MS/s
Resolution	8	8 bit	8 bit	10 bit
Active area	450 × 315 µm^2	615 × 780 µm^2	300 × 700 µm^2	480 × 340 µm^2
Input range (Reference voltage)	2.1 V	0.96 V (1 V)	0.84 (1 V)	1.44 V (1.5 V)
Unit capacitance	100.6 fF	215 fF	215 fF	215 fF
DNL	0.18 LSB	0.12 LSB	0.70 LSB	0.382 LSB
INL	0.87 LSB	0.38 LSB	1.02 LSB	0.23 LSB
THD	31.48 dB@1 kHz	52 dB@2 kS/s	48 dB@2 kS/s	62 dB@500 kS/s
SFDR	31.65 dB@1 kHz	50 dB@2 kS/s	44 dB@2 kS/s	65 dB@500 kS/s
FOM	5 pJ/conv	20 fJ/conv	8 fJ/conv	1 pJ/conv

TABLE 6.3 Summary of the performance of the SOS SAR ADCs.

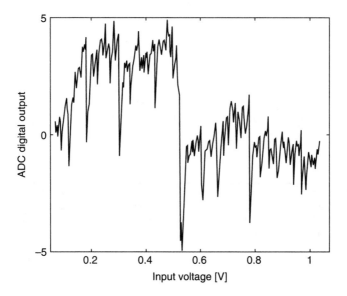

FIGURE 6.33 INL versus input voltage at 2 kS/s for the C-2C 8-bit SAR ADC fabricated in the SOS process.

low-power operation at the expense of sampling speed. Because a large number of cycles is needed to obtain a sample, these converters either require fast digital reconstruction circuits or low-input bandwidth signals.

In this section, we present a special kind of oversampling converter: an asynchronous $\Sigma\Delta$ converter. This converter is composed of an integrator and one or two voltage comparators. The input of this converter is nominally a current that is integrated on a capacitor. The input can also be a voltage passed through a resistor or a switched-capacitor resistor.

Figure 6.34 is a block diagram representation of a current-mode asynchronous converter. It is composed of an analog front end with an integrator and two comparators and a digital counter, latch, and shift register.

As shown graphically in Fig. 6.35, the input current I_{in} is integrated over a capacitor until the integrator output reaches either of the compare voltages V_{plus} or V_{minus}. At the end of the integration, the change in the comparator's state generates a pulse. The converter starts a counter at each reset and stops it when a pulse is generated by the comparators. The counter value is then latched and serially communicated to a computer-based data-acquisition system by means of a shift register. Because the output data are oversampled and synchronized by means

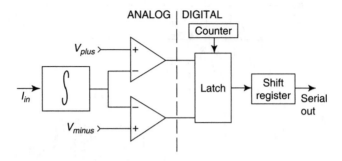

FIGURE 6.34 Asynchronous $\Sigma\Delta$ ADC topology. An input current is integrated and converted to a voltage that is compared against two thresholds. Once one of the threshold is passed, the free-running counter latches a countervalue.

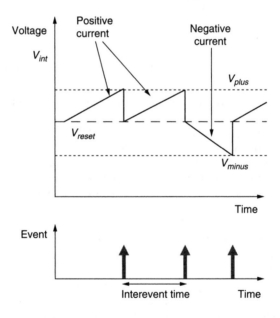

FIGURE 6.35 Asynchronous $\Sigma\Delta$ ADC mode of operation. Positive current generate positive events, whereas negative currents generate negative events. Each event latches a value of a free-running digital counter. The digital difference between these values is the converted representation of the analog input current.

of an external clock, the time between two integration pulses can be measured accurately. The difference between two latched values thus yields the integrator's reset frequency. This frequency can be used along with the system transfer function in Eq. (9.13) to calculate the input current.

$$I_{in} = C_{int} \Delta V f_{pulse} \qquad (6.17)$$

In Eq. (9.13), I_{in} is the input current, C_{int} the integration capacitance used in the integrator, and ΔV is the voltage swing of the integrator before one threshold is passed (the difference between V_{reset} of the integrator and the comparator voltages: $\Delta V = |V_{reset} - V_{minus}|) = |V_{reset} - V_{plus}|$).

The asynchronous circuit here described is functionally identical to a first order $\Sigma \Delta$ modulator (Baker et al., 1998; McIlrath, 2001). The only design parameter is the value of the integration capacitance C_{int} in Eq. (9.13). This value sets the minimum signal level for a specific minimum sampling frequency. As an example, if the minimum sampling rate is 10 kHz (f_{pulse}), and the $\Delta V = 0.1$ V with a $C_{int} = 100$ fF, then the minimum current that can be input is 1 pA. This calculation also applies to voltage input: using a 1 MΩ resistor, the minimum input voltage is then 1 μV.

The maximum signal that can be converted is a function of the f_{pulse} and the internal counter. As shown in Fig. 6.35, a positive input signal will generate two event pulses. Their interevent time can be measured by means of a running counter f_{cnt}. The maximum signal can be calculated with Eq. (6.18).

$$I_{in-min} = C_{int} \Delta V f_{cnt} \qquad (6.18)$$

As an example, if a maximum counter frequency of 100 MHz is used, $\Delta V = 0.1$ V with a $C_{int} = 100$ fF, then the maximum input is 1 μA. This value can be increased to 100 μA if $\Delta V = 1$ V and $C_{int} = 1$ pF. This also corresponds to a voltage value of 100 V when a 1-MΩ resistor is used to convert the voltage into an input current to the integrator.

Notice that a $C_{int} = 100$-fF integration capacitance is close to the minimum value that guarantees linearity in the conversion. C_{int} can be reduced, but its values should be much higher (at least 10 times) than the integrator input capacitance. This capacitance is usually of the order of 10 fF when an operation amplifier is used.

6.4.1 Result in SOS Technology

We have designed a current-mode asynchronous sigma-delta converter both in a 0.5-μm bulk-CMOS process and in the SOS process. The results of the bulk CMOS design (Laiwalla et al., 2005) can be

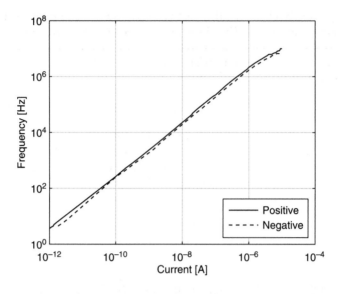

FIGURE 6.36 Measured output pulse frequency of the SOS current-mode asynchronous $\Sigma\Delta$ converter as a function of positive and negative input currents, as integrated on a 600-fF capacitor. The dynamic range of the patch-clamp amplifier was measured to be 7 decades.

compared to the ones designed in the SOS process (Laiwalla et al., 2006a). The SOS design was 30% more power efficient, 40% lower noise, and five times faster than the bulk CMOS counterpart. This advantage has also been claimed for digital circuits (Kuo and Su, 1998; Park et al., 1999; Buchholtz et al., 2000).

We tested the SOS current-mode asynchronous $\Sigma\Delta$ converter by sourcing a range of input currents from a few microamperes to a picoampere while recording the frequency of the output pulse. The amplifier was powered at 3.3 V with an Agilent 3631A DC power supply. The voltage noise on the power supply and the bias voltages was measured as less than 2 mV RMS. The input currents were sourced by applying a voltage across a megaohm resistor. A Keithley 2400 Source Meter was used to source and measure the input current. The frequency of the output pulse was measured using a Tektronic TDS2014 Four-Channel Digital Storage Oscilloscope. We obtained a linear transfer function across the entire range of tested currents in the range (3 pA–10 μA), as shown in Fig. 6.36. The output pulse frequency was in the range (3 Hz–10 MHz) and was observed to increase in discrete steps when very high currents were sourced as quantization noise due to low oversampling became dominant. We operated the device with clock frequencies up to 50 MHz and were thus able to extend the upper limit of the current measurements. The use of fast clocks and SOS process make this device one of the largest dynamic-range current measuring system reported (Laiwalla et al., 2006a).

6.5 Summary

Advancements in process technology, such as the available SOS process, create opportunities for improving and optimizing existing architectures, especially toward enhanced mixed-mode analog/digital circuits.

CHAPTER 7

Photosensitive Circuits

7.1 Introduction

In this chapter we describe the design, layout, and characteristics of photodetectors in both bulk silicon and SOS processes. The focus of the chapter is the use of the photodetectors for the design of SOS active pixel sensors and imaging arrays.

Although many pixel designs and optimizations have been presented in the literature during the last decade, a conventional three-transistor photodiode-type active pixel sensor (PD-APS), implemented in standard CMOS bulk technology, became the most popular imaging circuit design due to its low cost, compactness, simplicity of operation, and low-power dissipation (Fossum, 1995, 1997). Design of state of the art pixels for imaging applications usually involves the minimization of power dissipation and readout noise, maximization of photo collection efficiency and linear light-to-voltage conversion region. Wide dynamic range imaging is also desired to try to match the exceptional capabilities of the human visual system of 10 orders of magnitude in dynamic range (Schrey et al., 1999; Yang et al., 1999; Schanz et al., 2000; McIlrath, 2001; Culurciello et al., 2003). Light sensitivity, the resulting photocurrent, and integration time all influence the dynamic range of an image and is discussed in this chapter.

Recent progress in SOI technologies, as well as noise analysis and novel circuits, changed significantly the perspective of designers of imaging devices (Andreou et al., 1995; Zhou et al., 1997; Schrey et al., 1999; Yang et al., 1999; Pain et al., 2000; Schanz et al., 2000; McIlrath, 2001; Culurciello et al., 2003). SOI reduction of the parasitic effects due to the bulk allows for improved noise immunity and higher signal-to-noise ratios. The lack of wells, typical isolation tools in bulk silicon design, also significantly changes the design of photodiodes, relocating the collection region of the diode from the buried horizontal surface of the well design to a superficial vertical placement in SOI.

7.1.1 Advantages of SOS Photodetectors and Image Sensors

Photosensitive devices and circuits can be implemented in the SOS technology as a direct translation of bulk CMOS circuits. On the other hand, an SOS implementation provides several advantages when compared to a bulk CMOS.

The first advantage is the possibility of backside illumination due to the transparent substrate property of the SOS technology. Sapphire has uniform light transmission property from UV (200 nm) to infrared (5500 nm), as can be seen in Fig. 7.1 (Andreou et al., 2001). The advantage of backside illumination is an increase in effective light-sensitive area as the silicon layer is not blocked by other fabrication layers. In addition, the transparency of the SOS substrate to wavelengths from infra-red to ultra-violet opens opportunities for applications in high-speed free space interconnects and 3D integration (Andreou et al., 2001).

A second advantage of SOS is the availability of six different types of MOSFET transistors, with multiple threshold voltages. This allows flexible and more efficient pixel design, widening the dynamic range of the sensor and significantly reducing power dissipation of the imager. Also, because each transistor resides in an isolated island, wells are not needed. This allows pixels to use multiple devices without incurring area penalties. As an example, a PMOS can be used as reset transistor in an APS pixel. This replacement leads to increased output swing of the sensor, because no voltage drop exists when PMOS is used for reset.

Another advantage of SOS is crosstalk elimination between neighbor pixels. In SOS each pixel is fully isolated from the others, preventing from photogenerated electron-holes diffusion from one pixel

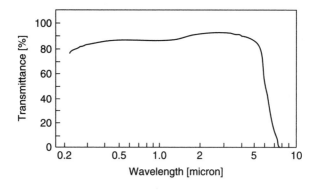

FIGURE 7.1   Optical transmittance of sapphire (Andreou et al., 2001).

to the other. This prevents the blooming effect from occuring at high light intensities.

Finally, the low-power advantages of the SOS process allows us to decrease the power consumption of image sensors. A reduction of power usage by 30% can be achieved without the need of scaling the supply voltages. Large dynamic ranges and good image quality can still be obtained.

SOS implementation also has a few disadvantages. The shallow silicon layer reduces light sensitivity at higher wavelengths. Also, when using backside illumination, all peripheral circuits cannot be shielded from the incident light and thus can incur in nonlinear behavior.

We discuss in this chapter a framework for designing photosensitive elements that will be used in Chap. 8 to implement an SOS image sensor array.

7.2 Photodetector Design: General Theory

In the process of designing a photosensitive circuit, a precise characterization of the photosensitive devices and the efficiency in light conversion is the first important step. Figure 7.2 is an schematic diagram of typical photodetector devices. In Fig. 7.2(A) a photoconductor had width W and length L is illustrated. Incident photons are indicated by an arrow hitting the surface of the detector. Each photon has and energy $E_p = h\nu$ and can generate an electron-hole pair in

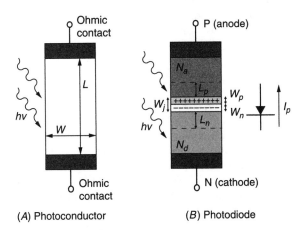

(A) Photoconductor   (B) Photodiode

FIGURE 7.2   Photodetectors schematic diagram: (A) a photoconductor made of semiconducting material, (B) a photodiode or reverse-biased PN junction photodetector.

the photodetector with probability η. Here, h is Planck's constant (6.626×10^{-34} Js), v (or f) is the photon frequency (c/λ), and η is the quantum efficiency (QE) of the detector, defined as the ratio of the number of output electrons to input photons. QE is a property of the detector material. Notice that the photoconductor in Fig. 7.2(A) can be implemented by a phototransistor, where the transistor body is the collection area. In Fig. 7.2(B), a reverse-biased photodiode PN junction is presented. W_j is the depletion region width of the reverse-biased diode, and W_p and W_n are the extension of the depletion regions, respectively, on the P and N sides. N_d and N_a are, respectively, the doping of the semiconductor regions P and N. I_p is the photocurrent generated by the incident photons in the device.

The most important quantity to characterize in a photodetector is the output photocurrent provided by a given input light intensity. This quantity, which for a photodiode captures the efficiency of light conversion, is known as *responsivity*. Specifically, responsivity is the ratio of the output photocurrent I_p to the input optical power P_{opt} (given by the number of photons N_p by the photon energy E_p). Modeling the responsivity using experimentally measurable parameters, the responsivity is related to the quantum efficiency as in Eq. (7.1) (Sze, 1981, 1990).

$$\Re = \frac{I_p}{P_{opt}} \left[\frac{A}{W}\right] = \frac{\eta q}{hf} = \eta \frac{\lambda(\mu m)}{1.24} \left[\frac{A}{W}\right] \qquad (7.1)$$

where q is the charge of an electron, f is the frequency of light incident on the photodiode, and λ is the wavelength of the light. Details about relevant figures of merit for photodetectors will be subsequently discussed and calculated for a precise determination of the expected output photocurrent.

7.2.1 Quantum Efficiency

The photocurrent is directly proportional to the responsivity of the device, which is a function of the quantum efficiency. The quantum efficiency η of the photodetector measures the quantity of electron-hole pairs collected at the contacts, that have been generated by the incident photons. We can model the quantum efficiency using only available physical device characteristics or utilizing only experimentally measurable parameters. Modeling the quantum efficiency with device parameters requires an understanding of the physical properties of the photosensitive material. Using this depiction, the quantum efficiency is given by Eq. (7.2) (Sze, 1981, 1990).

$$\eta = (1 - r)\zeta[1 - e^{-\alpha L_d}] \qquad (7.2)$$

where r is the optical power reflected at the detector surface (thus, $1 - r$ is the effectively transmitted light power), ζ is the fraction of electron-hole pairs, which contribute to the photocurrent, α is the absorption coefficient of the detector material in cm^{-1}, and L is the width of the absorption region of the detector. Note that this equation is referred to as the *internal quantum efficiency* and is a property of the material and varies with the wavelength of the incident light (Gupta, 2000).

A second model for the quantum efficiency of a photodetector takes into account the photogenerated current I_p produced by the absorption of photons with energy hf, which has a specific optical power per unit area, P_{opt}.

$$\eta = \frac{I_p/q}{P_{opt}/hf} \tag{7.3}$$

The expression in Eq. (7.3) measures the *external quantum efficiency* of a photodetector because all of the parameters used can be determined experimentally (Gupta, 2000). Depending upon the available parameters for a specific fabrication process, either model can be used.

The absorption coefficient α of the photodetector material is a property of the material itself and can be experimentally quantified. The absorption depends on the temperature of the system, especially for indirect bandgap semiconductors like silicon. As the temperature increases, the absorption coefficient shifts toward longer wavelengths. In addition, the absorption coefficient is strongly influenced by the radiation's wavelength. Moreover, for a given semiconductor, a noticeable photocurrent can only be generated over a specific range of wavelengths related to its bandgap configuration.

Figure 7.3 (Moini, 1994) shows a plot of the absorption coefficient α for silicon as a function of wavelength. The values of α in the visible wavelengths range of 0.4 to 0.7 μm are in the range 10^5 cm^{-1} to 10^1 cm^{-1}. These parameters shows us how the thickness of the detector material is important for the conversion of photons with different energy. As can be seen in Fig. 7.3, photons at 400 nm (violet-blue) only need a 0.1-μm-thick material to be absorbed, whereas red photons at 700 nm need a 10-μm-thick material. Over the visible range of wavelengths, the coefficient α can be approximated with a linear fit in the logarithmic plot of Fig. 7.3, giving the resulting symbolic expression of Eq. (7.4) (λ is expressed in centimeters). This simplification produce errors in the lower wavelength, but, overall, is a good fit of the parameter α over the visible range of wavelength.

$$\alpha = 10^{-62500\lambda + 7.875} \left[\frac{1}{\mu m} \right] \tag{7.4}$$

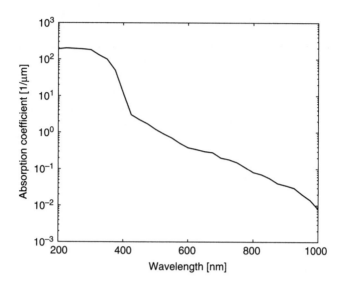

FIGURE 7.3 Absorption coefficient for silicon versus wavelength (redrawn from Sze, 1981). Notice that as longer wavelengths need thick silicon detectors in order to be absorbed, whereas shallow detectors are sufficienct to detect short wavelengths.

Not all of the incident light reaching the detector is converted into photocurrent. Only photons with energy larger than the bandgap energy of the semiconductor generate electron-hole pairs. The electron-hole pair generation rate G_L (photons/s/m) as a function of the material depth can be determined by means of the absorption coefficient and the incident light intensity using Eq. (7.5) (Gupta, 2000).

$$G_L(x) = \Phi_0 \alpha e^{-\alpha x} \tag{7.5}$$

G_L of Eq. (7.5) is a function of the depth of the device and thus needs to be integrated to obtain the total number of pair generated. $\Phi_0 = (1 - R) \, P_{opt}/Ahf$ is the incident photon flux (the total number of photons that impinge the photodetector per unit time and per unit area A). P_{opt} is the incident optical power converging on the junction, and α is the absorption coefficient of silicon.

7.2.2 Photodiode Photocurrent Models

Here, we report two possible ways to calculate the photocurrent in a photodiode. These models differ in that one is a device dependent model and one is experimentally determined. The experimental model uses measurable parameters such as quantum efficiency and input optical signal power to determine the photocurrent.

$$I_p = \Re P_{opt} = \frac{q\eta}{hf} P_{opt} = q\eta\Phi \tag{7.6}$$

This form of the photocurrent equation is mostly used when little is known about the physical device parameters, but experiments can be conducted to determine the input optical signal power and the quantum efficiency as previously discussed.

The photocurrent is generated both in the depletion region (W_j) and the region around the detector where carriers can diffuse to anode and cathode. The latter is a region defined by the carrier diffusion lengths L_n and L_p. The photocurrent through the photodiode is given in Eq. (7.7).

$$I_P = q\,AG_L(W_j + L_n + L_p) \tag{7.7}$$

Referring to Fig. 7.2, L_n is the electron diffusion length, L_p is the hole diffusion length, and W_j is the depletion width. Here, L_n and L_p represent the average distance and electron or hole will travel before recombining. A is the physical area of the diode ($A = W \times L$). Equation (7.7) shows that the electron-hole pair generation rate for the volume of the photodiode is the sum of the electron-hole generated in the physical area of the photodiode, and the pairs generated in an area equal to the carriers diffusion length around the diode. The generation rate times the charge of electron yields the photocurrent. L_n and L_P are calculated using Equation set (7.8) (Singh, 1994; Rabaey, 1996).

$$\begin{aligned} L_n &= \sqrt{D_n\tau_n} \\ L_p &= \sqrt{D_p\tau_p} \end{aligned} \tag{7.8}$$

D_n and D_p are the diffusion coefficients and τ_p and τ_n are the lifetimes for electrons and holes, respectively.

To calculate the width of the depletion region W_j, an assessment of the potential at the junction is required. ϕ_0 [Eq. (7.9)] is the built-in equilibrium potential (Rabaey, 1996) that comes from the difference between the charge on the P- and N-sides of the junction when the junction is unbiased.

$$\phi_O = \phi_T \ln\left[\frac{N_A N_D}{n_i^2}\right] \tag{7.9}$$

For a given photodiode, the values of N_A, the acceptor (hole) concentration and N_D, the donor (electron) concentration, are a function of the doping levels for a specific fabrication process. The intrinsic concentration n_i is the electron and hole concentration for a semiconductor in thermal equilibrium. ϕ_T is the thermal voltage, defined as $\phi_T = kT/q$.

$$C_j = \frac{dQ_j}{dV_D} = A_D \sqrt{\left[\frac{\epsilon_{si}q}{2}\frac{N_A N_D}{(N_A + N_D)}\right]\frac{1}{(\phi_0 - V_D)}} \tag{7.10}$$

In Eq. (7.10), ϕ_0 is the junction potential as defined in Eq. (7.9) (Sze, 1981; Singh, 1994; Rabaey, 1996) for an abrupt junction diode. After calculating ϕ_T and ϕ_0, the width of the depletion region, W_j can be calculated (Rabaey, 1996) with Eq. (7.11).

$$W_j = \sqrt{\left[\frac{2\epsilon_{si}}{q}\frac{(N_A + N_D)}{N_A N_D}\right](\phi_0 - V_D)} \tag{7.11}$$

Assuming that an abrupt junction indicates that the density of doping atoms changes drastically from N_a on the P-side to N_d on the N-side of the junction. This is a reasonable approximation if the actual distance over which the junction changes is much smaller than the width of the depletion region itself. Most modern processes feature only abrupt junction diodes.

7.3 Photodiode Design in a Bulk CMOS Process

We now consider a bulk CMOS photodetector obtained with N-diffusion on a P-substrate in a 0.5-μm bulk process (American Microsystems Inc. 1999) with nominal supply of up to 5 V. The layout of such bulk CMOS PN photodiode is given in Fig. 7.4. In the figure, X_j is the junction depth between the N and P regions, L_p and L_p are the diffusion lengths of carriers that can contribute to the photocurrent, and W_j is the depletion region width. The doping of the P-substrate is estimated to be 4.8×10^{15} cm^{-3} and the doping of the N-diffusion layer is 1.3×10^{16} cm^{-3} (MOSIS, 1999). The quantum efficiency, responsivity, and photocurrent will be assessed. The model used to calculate the photocurrent and the quantum efficiency of the PN junction depends upon the physical properties of the photosensitive material. This model is used because of the lack of process specific experimental data available for this photodiode.

7.3.1 Generation Rate, Diffusion Coefficients, and Electron Mobility

The first process-specific semiconductor variables to be calculated are the diffusion coefficients and the electron-hole generation rate per unit area, G_L. We use a P_{opt} of 1 W/m^2, typical in indoor ambient

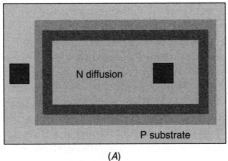

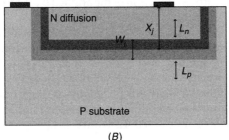

FIGURE 7.4 A bulk CMOS PN photodiode top view (*A*), and side view section (*B*). The vertical PN junction arrangement in bulk CMOS allows a large depletion region to be formed parallel to the surface. X_j is the junction depth between the N and P regions, L_n and L_p are the diffusion lengths of carriers that can contribute to the photocurrent, and W_j is the depletion region width. The darker squares are the device contacts.

lighting. G_L can be calculated for silicon using Eq. (7.5). The average visible wavelength is 0.55 μm and the average absorption coefficient is 10^6 m^{-1}; therefore, $G_L = 2.76 \times 10^{24}$ m^{-3} s^{-1} by means of the linear interpolation in Eq. (7.4).

The final material specific characteristic that needs to be determined is the diffusion coefficients for electrons D_n and for holes D_p. These coefficients are a function of the doping concentration and the temperature. For this specific junction we obtain a value $D_n = 34.95$ cm^2/s and $D_p = 18.94$ cm^2/s (Moini, 1994).

7.3.2 Characterization of the Depletion Region

The width of the depletion region, including the average diffusion lengths for holes and electrons, is to be determined. Minority carrier lifetime values depend upon temperature, doping levels, and fabrication process. Typical values range from 10^{-8}s to 10^{-5}s (Sze, 1981). To obtain a lower estimate of the average electron and hole diffusion lengths, a value of 10^{-8}s will be assumed for minority carrier lifetimes. Moreover, taking the diffusion coefficients, as determined in Sec. 7.2.2, the length of the diffusion for electrons and holes can be calculated using Eq. (7.8). In this case, $L_n = 4.59$ μm and $L_p = 4.32$ μm.

The width of the depletion region will be determined using Eq. (7.11), assuming an abrupt junction. In this case, solving Eq. (7.9), the built-in voltage, ϕ_0 gives 0.68 V. Notice that the width of the depletion region is a function of the applied reverse bias voltage V_d. Because this value will be changing as a function of the photocurrent, a graph of the depletion width as a function of the applied bias can be plotted and examined to give a range of possible widths. From Eq. (7.11), the range of depletion widths is 1.46 to 0.50 μm. A lower bound on the photocurrent can be calculated using the smallest depletion width. In this case, the sum of depletion region and diffusion lengths equals 10.36 μm.

Notice that because the bulk CMOS photodiode has a deep depletion region and the carrier diffusion lengths are large (around 10 μm), the detector is quite efficient in the conversion of photons on all the visible and infrared range (400–1200 nm).

7.3.3 Quantum Efficiency and Responsivity

To compute a first-order approximation of the quantum efficiency for a bulk process, we assume that ζ equals 1. This is a reasonable approximation due to the recent advances in the production of high-quality materials. In addition, we assume that applying an nonreflective coating to the surface of the detector can reduce r. Therefore, η can be expressed by Eq. (7.12).

$$\eta \approx [1 - \exp(-\alpha L)] \qquad (7.12)$$

Note that the following calculations do not take into account specific process data other than the doping concentrations. Therefore, it can be concluded that $\eta = 0.99$ for $\alpha = 3 \cdot 10^3$, the average absorption coefficient for the visible band corresponding to an average wavelength of 0.55 μm for the incident light and using the total depletion width of 10.36 μm as calculated above. Knowing the average wavelength of visible light, and using the value of η stated above, Eq. (7.12), characterizing the responsivity of the PN junction, is solved. $\Re = 0.44$ A/W for the examined 0.5-μm bulk process.

The photocurrent can be calculated in two ways. The first method uses the definition of responsivity as given in Eq. (7.6) and assumes that the input optical power per unit area for ambient light is 1 W/m². Therefore, the resulting photocurrent per unit area $I_p / A = 0.44$ A /m².

The other model to calculate the photocurrent takes into account the physical device characteristics. Using Eq. (7.7) and the parameters determined above, $I_p / A = 0.47 \text{ A}/\text{m}^2$. The two different theoretical models match reasonably well.

To calculate the actual photocurrent, the area of the diffusion region is considered the photosensitive area of the diode due to its horizontal structure. Examples are a $1 \times 1 \ \mu\text{m}^2$ photodiode results in photocurrent $I_p = 0.47 \text{ pA}$; a $10 \times 10 \ \mu\text{m}^2$ reports $I_p = 47 \text{ pA}$; and a $100 \times 100 \ \mu\text{m}^2$ reports $I_p = 4.7 \text{ nA}$.

7.4 SOS Photodetectors

In this section, we introduce some photodetectors that can be realized with SOS technology. Due to thin silicon film and the lack of wells, the design of photodetectors elements in SOS is quite different from that in a standard bulk process. Figure 7.5 shows the

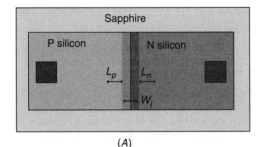

(A)

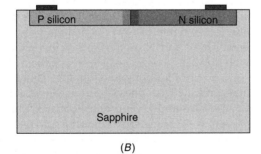

(B)

FIGURE 7.5 A lateral PN junction photo-detector in SOS top view (A), and side view (B). The device is implemented in a thin silicon film of 100 nm and with a superficial junction. The highlighted areas are the photons collection regions, where depletion layers are formed. L_n and L_p are the diffusion lengths of carriers that can contribute to the photocurrent and W_j is the depletion region width. The darker squares are the device contacts.

main difference in detector arrangement and layout, as compared to Fig. 7.4. Bulk CMOS uses a n^+/p-substrate (psub), N-well/p-substrate or p^+/N-well photodiodes, which have a collection region situated at the buried horizontal surface. On the other hand, the collection region of an SOS photodiode is a thin vertical volume between the P and N regions (Andreou et al., 2001; Apsel et al., 2003; Culurciello, 2007; Culurciello and Weerakoon, 2007; Weerakoon and Culurciello, 2007b). The collection regions are the highlighted areas in Fig. 7.5. The collection region of lateral photodiodes (SOS) is much smaller than the one in bulk CMOS. First, its vertical placement significantly reduces its size, and second, the shallow silicon layer in SOS (100 nm) further limits the collection area. A shallow silicon thickness also limits the collection of photons above 400 nm (see Fig. 7.3).

PN junction photodiodes are not the only photodetector that can be obtained in the SOS process. We will introduce a PN junction photodiode, a PIN photodiode, and a phototransistor. All photodiode devices have the limitation in collection area portrayed in Fig. 7.5, but phototransistors can obtain a larger collection region due to the size of their bodies. In the next few sections, we present the layouts, responsivity characterization, and collected data from all these devices.

7.4.1 SOS PN Photodiodes

A PN photodiode is a native structure in the SOS process and it is called the NG diode. The layout of the SOS PN photodiode is illustrated in Figs. 7.5B (side view) and 7.6 (top view). This type of photodiode is a P+/P−/N+ gated photodiode. The device is implemented in a silicon film thickness of 100 nm on a sapphire substrate of thickness of approximately 250 μm. This device has a floating polysilicon gate across its gate junction (see Fig. 7.6) in order to ensure that the oxide above the N and P region interface is of high quality and to minimize leakage currents. The P and N regions are of the RN and RP type with diode layers superimposed. The gate area reduces the doping of the diode (P−) to that of PMOS channel doping. This photodetector has the advantage of simplicity, linearity in the conversion of photons to current, reduced size and layout, and good responsivity. A PN photodiode is a very well characterized device and is used, with small variations, in most recent CMOS and CCD image sensors.

To explain the photodetector operation, the P+/P−/N+ junction can be divided into two junctions: P+/P− and P−/N+. When a small reverse voltage is applied in the range of 0 and 3.3 V, P− is not fully depleted and the width of the depletion region is defined by the P− and N+ doping concentrations and the value of the reverse voltage. The photocurrent is generated by the drift of electron-hole pairs generated in the depletion region (the larger component), and by diffusion

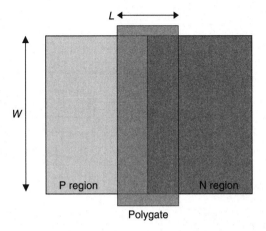

FIGURE 7.6   Layout of a PN photodiode in the SOS process (top view). The P and N regions are abiding under a polysilicon gate. The gate is needed to ensure that the oxide above the N and P region interface is of high quality and reduce leakage currents. P and N regions are of the RN and RP type with diode layers superimposed.

of the generated minority carriers to the depletion region (a smaller component). This is the normal mode of operation of the device.

When a large reverse voltage is applied (above the SOS supply limits of 3.3 V), the depletion region also increases and reaches the P+/P− junction. This extending of the depletion region to the other terminal causes an increase in electric fields (mainly defined by the doping concentrations of P+ and P−) and, therefore, causes a break-through. The minority carriers of one terminal are then accelerated to the other terminal. The total current through the device increases and the dark current also increases. Therefore, it is preferable to operate the photodiode at relatively low reverse voltages to achieve a higher signal-to-noise ratio.

Notice that because most of the depletion region is located under the polysilicon gate, the device quantum efficiency at shorter wavelengths is decreased even further.

7.4.1.1 Measured Characteristics of an SOS PN Photodiode

We tested an array of 1024 (32 × 32) SOS NG photodiodes connected in parallel. The measurements were carried out using a Keithley 617 electrometer by applying different reverse bias voltages and measuring the resulting photocurrent (Fish et al., 2007). The measurement setup lower limitation was 10 pA. The size of each NG diode layout is 6.1 × 6.1 μm. Figure 7.7 shows measured photocurrents and dark

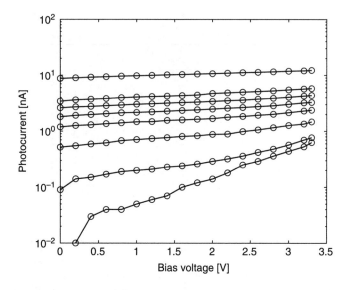

FIGURE 7.7   SOS photodiode data for an array of 32 × 32 NG devices of
6.1 × 6.1 μm² area with the top polylayer left floating. The data are for
front-side illumination with the following set of intensities: [0 (dark), 0.1, 0.4,
0.8, 1.2, 1.6, 2, 5] mW/cm². Notice that the lowest curve is the dark current
data.

currents for the entire array. This data is for a frontside illuminated
device with the following set of intensities: [0 (dark), 0.1, 0.4, 0.8, 1.2,
1.6, 2, 5] mW/cm². Measurements were performed with an incoherent
light source at low illumination levels typical of indoor environments.

The NG devices show dark current levels of 0.1 nA (0.1 pA per
device) at reverse bias voltages under approximately 2 V. The dark
current increases exponentially with the reverse bias, as expected from
the theory. Photocurrents have a linear relationship with the reverse
bias and result in up to 10 nA (10 pA per device) at intensities of
5 mW/cm² and 2-V biases. This data shows that an signal-to-noise
(photocurrent/dark current) of about 100 can be achieved with this
kind of detector.

Figures 7.8 and 7.9, respectively, report measurements on the responsivity and quantum efficiency of the NG photodiode as a function of wavelength. The responsivity of the photodiode was obtained
by measuring light intensities and photocurrents. We used an Horiba
Jobin Yvon Fluoromax-3 as a light source. We varied the source wavelength from 275 to 1000 nm. The light source was measured with
a Newport 818-UV photodiode with known spectral responsivity.

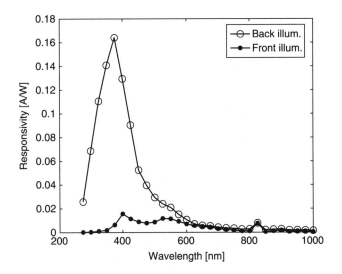

FIGURE 7.8 Responsivity of an SOS NG photodiode with a 6.1 × 6.1 μm²
area with the top polylayer left floating. Data are for both front- and backside
illumination. Notice the difference in responsivity with front- and backside
illumination. The front side spin-on glass significantly reduces the response
near UV.

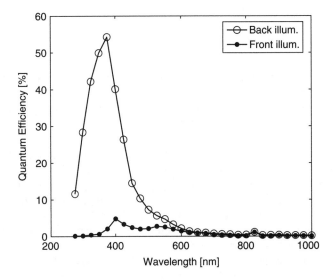

FIGURE 7.9 Quantum efficiency of an SOS NG photodiode with a
6.1 × 6.1-μm² area with the top polylayer left floating. Data are for both
front- and backside are illumination. Notice the difference in responsivity
with front- and backside illumination. The frontside spin-on glass reduces
significantly the response near UV.

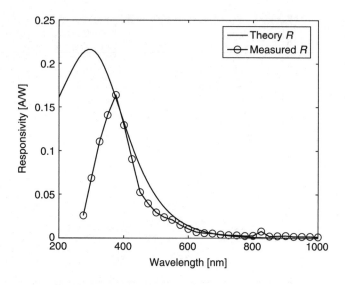

FIGURE 7.10 Theoretical model of the responsivity of an SOS NG photodiode with a 6.1 × 6.1-μm² area with the top polylayer left floating. The theoretical response matches well with back-illuminated devices but not with frontillumination because of the filtering effect of the top SOS layers.

The SOS NG photodiode reported the highest responsivity in the near-UV region. We have illuminated the device from the top side and from the bottom side through the transparent sapphire substrate. Notice the difference in responsivity with front- and backside illumination. The frontside spin-on glass significantly reduces the response near UV. The backside of the SOS dies was not polished. The quantum efficiency was computed from the responsivity of the device using Eq. (7.2).

A theoretical model [Eq. (7.2)] of the responsivity of a back-illuminated SOS NG photodiode is shown in Fig. 7.10. The model is a good match to the data in the wavelengths above 400 nm. At lower wavelengths, the model differs from the data because the backside of the SOS dies was not polished.

On a final note, SOS allows us to implement stacked PN photodiodes. By using PN junctions in series, it is possible to obtain significant dark-current reduction. In these devices, the dark current is reduced because each of the stacked diodes sees only a fraction of the full bias voltage, and lower biases show lower dark currents, as seen in Fig. 7.7.

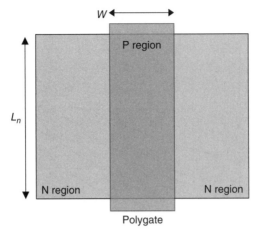

FIGURE 7.11 Layout of a phototransistor in the SOS process (top view). The device is a standard RN-type NMOS with gate left floating and width and length defined as W and L. The collection area is the floating body of the device forming a lateral bipolar transistor (LBT).

7.4.2 SOS Phototransistors

Another photosensitive element that can be implemented in the SOS technology is a lateral bipolar transistor (LBT) (Culurciello, 2007; Culurciello and Weerakoon, 2007; Weerakoon and Culurciello, 2007b) implemented by a standard RN-type MOSFET with the gate left floating. The LBT is obtained from the floating body of a MOSFET. Floating-body MOSFETs have a parasitic bipolar device in parallel, where the drain is the collector, the body is the base and the source is the emitter. For an NMOS, the parasitic bipolar device is an NPN structure. When the gate of a floating-body NMOS is left unbiased unpolarized, the LBT device becomes a photodetector. Hole-electron pairs are generated in the device body and swept to the source and drain terminals. Figure 7.11 shows the top-view layout of a RN photodetectors with a floating gate.

The phototransistor is rarely used in image sensor arrays because its internal gain variations generate large fixed-pattern noise. Its advantages are contained layout, small outlines, and large responsivity due to the internal gain. The disadvantages are gain mismatch and nonlinearity in the response versus light intensity.

7.4.2.1 SOS Phototransistor Measured Characteristics

We tested an array of 1024 (32×32) SOS RN phototransistors connected in parallel. The measurements were carried out using a Keithley 617 electrometer, applying different reverse bias voltages and measuring

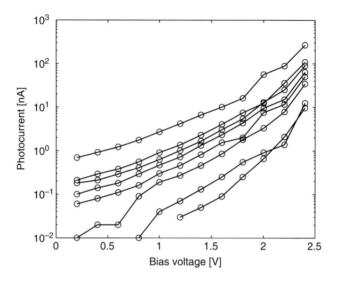

FIGURE 7.12   SOS phototransistor data for an array of 32 × 32 device of
[*W, L*] = [2, 1] μm area with top polylayer left floating. The data are for
frontside illumination with the following set of intensities: [0, 0.1, 0.4, 0.8,
1.2, 1.6, 2, 5] mW/cm². Notice that the lowest curve is the dark current data,
which is lower than 10 pA at low bias voltages.

the resulting photocurrent. The measurement setup was lower limited
to 10 pA. The size of each RN transistor layout is [*W, L*] = [2, 1] μm.
Figure 7.12 shows measured photo- and dark currents for the entire
array. This data is for a frontside illuminated device with the following
set of intensities: [0 (dark), 0.1, 0.4, 0.8, 1.2, 1.6, 2, 5] mW/cm². Measurements were performed with an incoherent light source at low illumination levels typical of indoor environments, which are acceptable
measurements setups for image sensors.

RN NMOS devices show dark current levels of 0.05 nA (0.05 pA/
device) at reverse bias voltages under approximately 1.5 V. The dark
current increases exponentially with the reverse bias, as expected from
the theory. Photocurrents have a linear relationship with the reverse
bias, and result in up to 10 nA at intensities of 5 mW/cm². This data
shows that a signal-to-noise ratio (SNR) of about 200 can be achieved
with this detector.

Figures 7.13 and 7.14, respectively, show measurements on the responsivity and quantum efficiency of the NG photodiode as a function
of wavelength. The responsivity of the phototransistor was obtained
by measuring light intensities and photocurrents. We used an Horiba

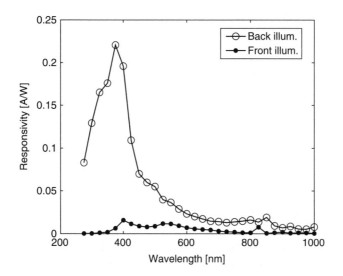

FIGURE 7.13 Responsivity of an SOS RN photo transistor with a
$[W, L] = [2, 1]$ μm with the top polylayer left floating. Data are for both
front- and back-side illumination. Notice the difference in responsivity with
front- and back-side illumination. The front side spin-on glass reduces
significantly the response near UV.

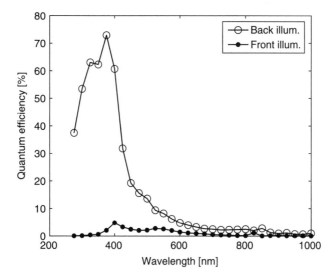

FIGURE 7.14 Quantum efficiency of an SOS RN phototransistor with a
$[W, L] = [2, 1]$ μm with the top polylayer left floating. Data are for both
front- and backside illumination. Notice the difference in responsivity with
front- and backside illumination. The frontside spin-on glass reduces
significantly the response near UV.

Jobin Yvon Fluoromax-3 as a light source. We varied the source wavelength from 275 to 1000 nm. The light source was measured with a Newport 818-UV photodiode with known spectral responsivity.

The SOS NG photodiode reported the highest responsivity in the near-UV region. We have illuminated the device from the top side and from the bottom side, through the transparent sapphire substrate. Notice the difference in responsivity with front- and backside illumination. The frontside spin-on glass significantly reduces the response near UV. The backside of the SOS die was not polished. The quantum efficiency was computed from the responsivity of the device using Eq. (7.2).

7.4.3 SOS PIN Photodiodes

Figure 7.15 represents a typical lateral PIN photodiode implemented in the SOS process (Peregrine, 2008a). The PIN diode differ from typical PN diodes because it contains an intrinsic (undoped) silicon region between the N and P regions. The entire intrinsic region is depleted

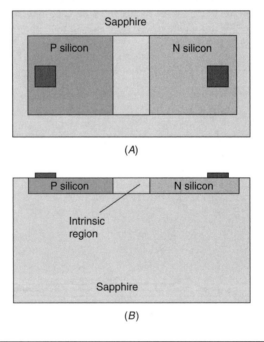

(A)

(B)

FIGURE 7.15 An SOS PIN photodiode layout top view (*A*), and side view (*B*). The diode features an intrinsic silicon region between N and P regions. This intrinsic region increases the overall collection area and can widens the device bandwidth by reducing the diode capacitance.

of carriers and its width can be adjusted to give the PIN diode specific bandwidth, capacitance and collection area. PIN photodiodes are usually preferable to increase the depletion region width of SOS detectors. The intrinsic area can be much larger than the PN diode depletion region, as can be seen in Fig. 7.15 (compared to Fig. 7.5). The PIN photodiode has the advantages of low capacitance, high bandwidth of operation and good responsivity. The disadvantage is a slightly larger layout area than a PN photodiode and higher leakage currents.

As we can see in Fig. 7.16, the PIN photodiode has a wide depletion width, which decreases the junction capacitance and makes the photodiode faster (higher bandwidth) and more efficient in generating photocurrent (larger collection area). In addition, the physical structure of the diode is vertical instead of horizontal, as seen in typical bulk processes. Note that because of the insulation of the SOS substrate, the device performs typically like an ideal diode (Peregrine, 2008a). Virtually no leakage current to the bulk or substrate is present. Notice that a PIN photodiode cannot be fabricated in standard bulk process because the substrate cannot be isolated from the intrinsic region.

A top view of the layout of an SOS PIN photodiode is reported in Fig. 7.15. The N and P regions are both of the intrinsic kind (IN and IP

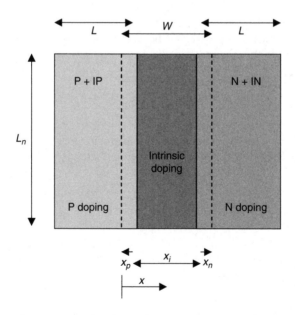

FIGURE 7.16 Layout of a PIN photodiode in the SOS process (top view). Regions P and N are intrinsically doped (doping layers IN, IP). A SD-block layer leaves the region between N and P undoped (intrinsic). This device does not have a polysilicon gate.

doping, respectively). The intrinsic region of the diode is obtained by blocking the source or drain implants (SD-block layer). The intrinsic region width is X_i, whereas the device's total width is L_n.

7.4.3.1 SOS PIN Photodiode Model

In this section, we present an analytical derivation of of the photocurrent in an SOS PIN photodiode as portrayed in Fig. 7.15. In the figure, W is the extension of the depleted region of the diode, L is the extension on the diode terminals, and X_p and X_n are the depletion regions in the P and N regions. The effective depth of the device is given by l_d, which is here set as the thickness of the silicon layer in SOS (100 nm).

Under steady-state conditions, the current through the PIN photodiode can be divided into the drift current density J_{drift}, due to carriers generated in the depletion region, and the much smaller diffusion current density J_{diff} [Eq. (7.13)], due to carriers generated outside the depletion layer.

$$J_{tot} = J_{drift} + J_{diff} \tag{7.13}$$

Supposing that the thickness of the silicon island composing the photodiode are much smaller than the penetration depth α. We can express the carrier concentration generated by the incident photons in a vertical slice of silicon as G. This value is the result of integration over the vertical axis [Eq. (7.14)], because photons will travel a variable length inside the silicon.

$$G = \int_0^{l_d} \Phi_0 \alpha e^{-\alpha y} dy = \Phi_0 (1 - e^{-\alpha l_d}) \tag{7.14}$$

A system of two single-dimension diffusion Eqs. (7.15) is used to calculate the photocurrent density inside the device (Moini, 1994).

$$D_p \frac{\partial^2 p_n(x)}{\partial x^2} - \frac{p_n(x) - p_{n0}}{\tau_p} + G = 0$$

$$\tag{7.15}$$

$$D_n \frac{\partial^2 n_p(x)}{\partial x^2} - \frac{n_p(x) - n_{p0}}{\tau_n} + G = 0$$

where D_p and D_n are the diffusion coefficients of holes and electrons, τ_p and τ_n are the lifetime of excess carriers, and p_{n0} and n_{p0} are the equilibrium concentrations of the minority carriers in the n and p silicon portions of the device.

The above set of equations can be solved by imposing the following set of boundary conditions (referring to Fig. 7.16: $p_{n0}(0) = 0$, $p_{n0}(-L) = p_{n0}$, $n_{p0}(W) = 0$, $n_{p0}(W + L) = n_{p0}$). Having the x axis referenced to the extension of the depletion region in the intrinsic portion of the device simplifies the result. The general solution for the differential equations is given in Eq. (7.16).

$$p_n(x) = p_{n0} + Ae^{\frac{x}{L_p}} + Be^{-\frac{x}{L_p}} + C$$

$$n_p(x) = n_{p0} + De^{\frac{x}{L_n}} + Ee^{-\frac{x}{L_n}} + F$$

(7.16)

With $C = G\tau_p$, $F = G\tau_n$. L_n, L_p are the diffusion lengths of excess carriers. This set of parameters derives directly to fitting the general solution to the differential equation. G defines the generation of minority carriers inside the depleted region of the device due to incident photons.

Parameters τ_p, τ_n, D_p, and D_n were calculated using the empirical formulas for silicon in Eq. (7.17), as a function of the impurity density (Moini, 1994).

$$\tau_p = 1/\left(7.8 \cdot 10^{-13} N_d + 1.8 \cdot 10^{-31} N_d^2\right)$$

$$D_p = \frac{kT}{q}\left(370 + \frac{370}{1 + 1.563 \cdot 10^{-18} N_d}\right)$$

(7.17)

$$\tau_n = 1/\left(3.45 \cdot 10^{-12} N_a + 9.5 \cdot 10^{-32} N_a^2\right)$$

$$D_n = \frac{kT}{q}\left(232 + \frac{1180}{1 + 1.125 \cdot 10^{-17} N_a}\right)$$

The parameters of the above equation can be solved with the boundary condition to obtain the equation set.

$$A = -\frac{e^{L/L_p}\left(Ce^{L/L_p} - C - p_{n0}\right)}{e^{2L/L_p} - 1}$$

$$B = -\frac{Ce^{L/L_p} - C - p_{n0}}{e^{2L/L_p} - 1}$$

$$D = -\frac{e^{W/L_n}(Fe^{L/L_n} - F - n_{p0})}{e^{2L/L_n} - 1}$$

(7.18)

$$E = -\frac{e^{(L+W)/L_n}\left(Fe^{L/L_n} - F - e^{L/L_n}n_{p0}\right)}{e^{2L/L_n} - 1}$$

With all the parameters calculated in terms of the basic property of silicon, we can proceed to obtain the diffusion component of the current density J_{diff}. This is given by Eq. (7.19).

$$J_{diff} = -q\frac{D_p}{L_p}A + q\frac{D_p}{L_p}B + q\frac{D_n}{L_n}De^{\frac{W}{L_n}} - q\frac{D_n}{L_n}E e^{-\frac{W}{L_n}}$$

(7.19)

The drift component of the photocurrent density is given by Eq. (7.20) (Sze, 1981).

$$J_{drift} = q\,\Phi_0(1 - e^{-l_d\alpha}) \qquad (7.20)$$

This value is obtained integrating both on the vertical and horizontal axis of the structure of the photodiode. The drift current density can be also expressed as $J_{drift} = -q\,G$, where G is the value of photocurrent integrated over the vertical axis.

Simplifying the diffusion current density, we can obtain a closed-form solution, which is the final analytical model of the photocurrent density of the SOS PIN photodiode. A simpler derivation of the photocurrent can be obtained by ignoring the very small contribution of the diffusion currents and limiting to the calculation of the drift component.

The quantum efficiency of the device can be estimated using Eq. (7.21) by dividing the current density by the effective light power reaching the device. S is the area of the photodiode (Sze, 1981).

$$\eta = (1 - r)\frac{J_{tot}}{q\,\Phi_0} \qquad (7.21)$$

A plot of the quantum efficiency of a SOS photodiode with area $S = 1\ \mu m^2$ (the area is $L_n \times X_i$) is seen in Fig. 7.17. The large response

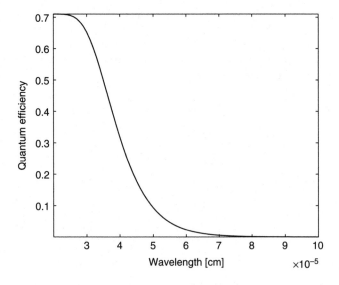

FIGURE 7.17 Quantum efficiency of an SOS PIN photodiode with an area of 1 μm^2, calculated from Eq. (7.21). The quantum efficiency is low at higher wavelengths because of the thin SOS silicon film. The efficiency at lower wavelengths and UV is high.

to lower wavelengths is due to the fact that most of the carriers close to the surface are absorbed by the device and converted into a photocurrent. The shallow silicon layer of the SOS process result in an overall poor performance for higher wavelengths. Notice that the model in Fig. 7.17 is the same as in Fig. 7.10.

The quantum efficiency of an actual device will be masked by the reflection property of the material coating the photodiode itself. Because the light radiation has to travel through a stack of different media, the light incident on the photodiode will be a fraction of the light intensity reaching the top surface of the device. The effective transmission coefficient of the light can be calculated with a simple physics principle by investigating the composition of the top layers of the device.

By understanding the geometry of the PIN diode in SOS, the physical model of quantum efficiency as described above can be improved. This estimate of the quantum efficiency takes into account the path the light goes through to reach the photosensitive area of the PN junction. Figure 7.18 shows the different layers in the SOS process that light has to travel through before reaching the junction (Peregrine, 2003).

Equation 7.22 (Pozar, 1998; Kalayjian, 2000) calculates the transmission coefficient T of incident light energy at the boundaries between two materials where η is the wave impedance for the material and η_0

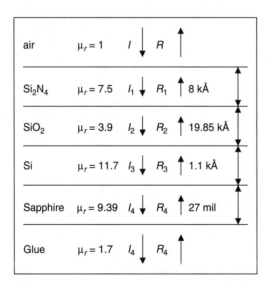

FIGURE 7.18 Light path through SOS wafer and media. I is the incident light, and R is the reflected light. Relative permittivity and layer thicknesses are reported for each layer.

T_0 = air to Si_2O_4 interface = 0.53	$R = I(1 - T_0^2) + R_1 T_0^2 = -0.47$
$T_1 = T_0^2 I + R_1(1 - T_0^2) = 1.16$	$R_1 = I_1(1 - T_1^2) + R_2 T_1^2 = 0.16$
$T_2 = T_1^2 I_1 + R_2(1 - T_1^2) = 0.73$	$R_2 = I_2(1 - T_2^2) + R_3 T_2^2 = -0.27$
$T_3 = T_2^2 I_2 + R_3(1 - T_2^2) = 1.05$	$R_3 = I_3(1 - T_3^2) + R_4 T_3^2 = -0.05$
$T_4 = T_3^2 I_3 + R_4(1 - T_3^2) = 1.39$	$R_4 = I_4(1 - T_4^2) = 0.39$

TABLE 7.1 Transmission and reflection coefficients in a SOS wafer.

is the wave impedance in free space. Because the incident power of the light radiation is the quantity of interest, the theory of propagation of planar waves will be used to solve for the transmission and reflection coefficients at the interfaces.

$$T = \frac{2\eta}{\eta + \eta_0} \quad \eta = \eta_0 \sqrt{\frac{\mu_r}{\epsilon_r}} \quad (7.22)$$

Moreover, the reflection coefficient R is given by $R = T - 1$. Here, μ_r is the relative permeability of the material and ϵ_r is the relative dielectric constant. This approximation assumes a lossless medium this signifies that R, T, μ_r, and ϵ_r are real numbers. Table 7.1 summarizes all equations involved in the calculation of the incident transfer coefficient in a SOS integrated circuit where R and T are the transmission and reflection coefficients as described above and I is the power of the incident light.

The final value of the reflection coefficient is a lower-bound estimate, taking into account only the primary reflections and not all the secondary terms given by the reflection of the reflections. The resulting coefficient of transmission for visible light into the wafer equals 0.71 for the SOS process. This value was used to calculate the quantum efficiency in Fig. 7.17.

Therefore, the internal quantum efficiency can be approximated using Eq. (7.2) by realizing that the variable r is the optical power reflectance, which is equal to R the reflection coefficient. It follows that the term $(1 - r)$ is equal to T, the transmission coefficient as defined in Eq. (7.22).

However, before Eq. (7.2) can be applied, the absorption coefficient and the width of the depletion region need to be determined. Because the purpose is to detect visible light, our photodetector needs to be sensitive to wavelengths between 0.4 and 0.7 μm. Using the data in Fig. 7.3, the absorption coefficient α is between $10^5 cm^{-1}$ and $10^3 cm^{-1}$.

From the Peregrine SOS users manual (Peregrine, 2003) we were able to obtain some of the critical parameters for calculating the width of the depletion region, W_j. The SPICE parameters for the SOS process are as follows: The substrate doping, $n_i = 6 \times 10^{16}$ cm^{-3} and the doping of the N or P diffusion layer is $N_A = 8.86 \cdot 10^{16}$, $N_D = 2.41 \cdot 10^{16}$ cm^{-3}. Note that for a PIN diode the depletion width is composed of the two depletion regions in the N and P diffusion layers in addition to the intrinsic region (refer to Fig. 7.16). The resulting depletion width is given by Eq. (7.23).

$$W = x_n + x_p + x_i = \sqrt{\left[\frac{2\epsilon_{si}}{q} \frac{(N_A + N_D)}{N_A N_D} \right] (\phi_O - V_D)} \qquad (7.23)$$

In Eq. (7.23) ϕ_0 is the built in potential as defined above in Eq. (7.9), which equals 0.77 V for SOS.

$$x_p = x_n = \sqrt{\left[\frac{2\epsilon_{si}}{q} \frac{\phi_0}{N_D} \right]} \qquad (7.24)$$

Notice that the width of the depletion region [Eq. (7.24)] is proportional to the square root of the bias voltage. Because the largest depletion width occurs at the maximum reverse bias voltage, the equation can be solved with $V_d = 3$ V. Using this approximation, the depletion width occurring in the N and P diffusion layers is $x_n = x_p = 1.73$ μm.

Equation (7.2) allows us to calculate the maximum η for the visible spectrum, equal to a value of 0.71. This corresponds to a wavelength of 0.4 μm for the incident light and a corresponding absorption coefficient of 10^5 cm^{-1} (Sze, 1981; Sze, 1990). For the average wavelength of visible light, η equals 0.61. η in maximum if the incident light is ultraviolet, although this maximum is not useful for imaging visible light. This is due to the thickness of the silicon film in SOS. Longer wavelengths cannot be absorbed efficiently in thin silicon films.

7.4.3.2 Measured Characteristics of an SOS PIN Photodiode

The responsivity of a photodetector is given by the current in the device I_p divided by the input light power P_{OPT}. Using Eq. (7.5), assuming to operate at a wavelength of 0.40 μm, and using the value of η calculated above for the SOS process, the average responsivity equals 0.17 A/W.

Knowing the responsivity, we can determine the photocurrent per unit area by applying the definition of responsivity as shown in Eq. (7.6). If $P_{opt} = 10^{-4}$ W/cm^2 for ambient light, the photocurrent per unit area, $I_p/A = 17.18$ μA/cm^2, where A is the photosensitive area.

A plot of the responsivity of a front-illuminated SOS PIN photodetector is given in Fig. 7.19. A fit of the portion of data between 550 and 700 nm of wavelength gives a penetration depth of approximately

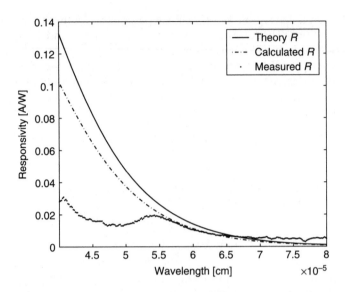

FIGURE 7.19 Comparison of the theoretical maximum and calculated and measured responsivity of a front-illuminated SOS PIN photodiode. The difference in the models and data are due to the top passivation layers in the SOS process. The model matches well only the back-illumnated detectors, as can be seen in Fig. 7.10.

$2.5 \cdot 10^{-6}$ cm. This is the effective surface layer of silicon that we are interested in the photoconversion process. The disagreement between measured data and models in the region of wavelengths 400 to 550 nm is due to the top passivation layers used in the SOS process to protect the circuitry. These spin-on glass layers significantly reduce the response at UV wavelengths and below. There is a great difference in the response of front- and back- illuminated detectors, as can be seen in Figs. 7.8 and 7.9. We have not collected back-illumination data for the SOS PIN photodiode, but we are confident the data will be similar to the one presented for the PN photodiode in Figs. 7.8 and 7.9. Notice instead that if the detectors are back-illuminated, the model can reflect the data with higher accuracy, as can be seen in Fig. 7.10.

In the Peregrine SOS process, the PIN photodiode is a vertical structure, therefore, the PN junction is the small area that connects a P-type to N-type diffusion (refer to Fig. 7.16) as opposed to a bulk photodiode, where the PN junction is immersed in a well and therefore has a photosensitive area is equal to the diffusion area. The total optical current deriving from phototransduction in the SOS process is 0.401 pA/cm^2.

Another important parameter to calculate is the maximum internal junction capacitance of the photodiode, since the time allotted to

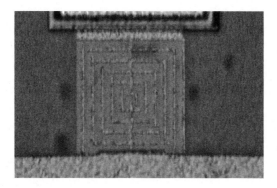

FIGURE 7.20   SOS PIN photodiode with concentric ring construction.

integration is directly proportional to the junction capacitance and the output current. Knowing the width of the depletion region, we can solve for the junction capacitance. For a 10.1-μm-long × 0.11-μm deep photodiode in SOS, $C_j = 6.5 \times 10^{-9}$ F/cm^2 then $C = 7.25 \times 10^{-17}$ F. This capacitance is very small, especially when compared to the one in a bulk process. This makes the PIN SOS photodiode one of the best candidates for high-speed imaging and light sensing (Apsel, 2002).

Because PIN photodiodes have low junction capacitance, they are ideal for high-speed optical communication. We report here on the design of a large PIN photodiode 40 μm × 40 μm with concentric metal rings used as terminals. Figure 7.20 is a micrograph of the layout of the PIN photodiode. The top and bottom metal layers are the contact to the N and P silicon regions, respectively. The metal rings and contacts improve the device bandwidth because they reduce the resistance of terminals in large devices. This is equivalent to a multifinger MOS structure for high-speed and RF circuits. We have designed three devices with different intrinsic regions lengths: 2, 3, and 4.5 μm, respectively.

We now report the measured frequency response of these SOS PIN photodiodes. The measurements were performed with a 785-nm laser diode source impinging the photodiode. We probed the samples with 40-GHz 100-μm pitch ground–signal–ground probes (picoprobe no. 40A-100-C), focused the 785-nm beam onto the sample, modulated the beam with a signal generator (Wiltron 68369B Synthesized Signal Generator 10 MHz-40 GHz), and measured the output of the photodiode directly with a spectrum analyzer (Advantest R3271A 100 Hz-26.5 GHz). The bias across the photodiodes is 2.5 V in all cases. Figure 7.21 shows the resulting frequency responses for the three 40 μm × 40 μm ring PIN photodiodes. All of the responses are normalized to show the measured bandwidth of the devices, denoted

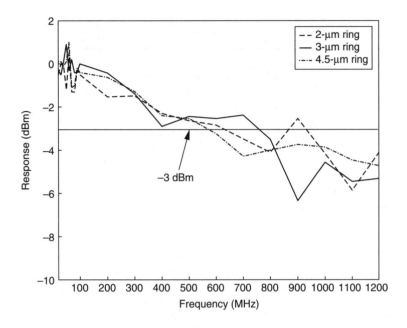

FIGURE 7.21 Frequency response of 40 μm × 40 μm SOS ring photodiodes with varying junction widths. A line designates a 3-dB drop from the DC response. There is no conclusive difference in bandwidth between these three devices.

by a 3-dB drop in response. The bandwidth of the photodiodes was approximately 850 MHz.

For the capacitance of the 40A-100-C GSG GGB picoprobes, we find that the total capacitance to ground of these probes is approximately 2.3 pF. The resistances of the photodetectors with metal ring structures was calculated to be approximately 4 Ω. The 50-Ω input resistance of the spectrum analyzer dominates in the measurement. The bandwidth of 850 MHz can be approximated by a 80-Ω photodiode resistance and a 2.3-pF probe and device capacitance. More accurate photodiode bandwidths can be measured by the addition of an output buffer or amplifier stage with a small gate load on these photodiodes.

Although these speeds seem slow in comparison to GaAs and other direct bandgap detectors, they are actually fast compared to most commercial silicon PIN detectors. Honeywell's highest speed PIN detectors, for instance, have minimum rise times of 5 ns under normal bias conditions. Furthermore, SOS detectors are monolithic in a CMOS process, allowing easy integration of detectors and receiver circuitry for inexpensive packaging of optoelectronic systems at gigabit rates.

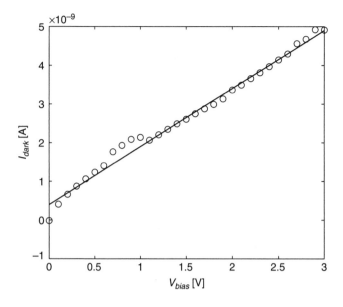

FIGURE 7.22   Dark current versus photodiode bias voltage at 300 K.

Figure 7.22 reports a plot of the dark current for a 92,198-μm^2 PIN photodiode fabricated in an SOS at an ambient temperature of 300 K. This photodiode's size is many times bigger than an APS photodiode, which is in the range of 25 μm^2. The data in Fig. 7.22 has to be properly scaled to be applied to APS design. For a 25-μm^2 device, the dark current at 3-V bias is about 1.62 pA. Notice that this value is approximately 10 times higher than the PN photodiode leakage in Fig. 7.7. For this reason, PIN photodiodes are not good candidates for imaging applications but can be used for digital data, such as optical communications.

7.5 Conclusions

In this chapter, we characterized a variety of photodetectors implemented in the SOS process. We have provided design details and performance evaluations of PN photodiodes, PIN photodiodes, and MOS phototransistors. We have shown that the PN photodiode and the phototransistors structures can deliver high signal-to-noise ratios at the photodetector. Phototransistors also have a large responsivity at the expense of nonlinearity with light intensity. PIN photodiodes provided the largest operation bandwidth at the expense of higher dark currents and lower signal-to-noise ratios. These devices are optimal for optical communication but not ideal for imaging arrays. In summary, we demonstrated that the SOS process is capable of high-quality imaging devices that can be used for assembling large vision systems, just like the ones presented in Chap. 8.

CHAPTER 8

SOS Address-Event Image Sensor

8.1 Introduction

The SOS CMOS process available from Peregrine semiconductors (Peregrine, 2003) is a promising technology for hybrid optoelectronic microsystems, as reported in Andreou et al., 2001. The transparency of the substrate to wavelengths from infrared to ultraviolet (Fig. 7.1 and Chap. 7), opens opportunities for applications in high-speed free-space interconnects and three dimensional integration. One of the most important optical components are image sensor arrays. These devices can be used for a variety of applications, such as imaging (Fossum, 1997; Zheng et al., 2000; Kleinfelder et al., 2001; McIlrath, 2001; Uryu and Asano, 2002; Kucewicz et al., 2004; Choa et al., 2007), three-dimensional optical interconnections (Andreou et al., 2001; Apsel and Andreou, 2001; Apsel et al., 2004; Apsel and Andreou, 2005), adaptive optical wavefront correction (Cohen et al., 1999), ultraviolet imaging (Marwick and Andreou, 2007; Park and Culurciello, 2008a; Park and Culurciello, 2008b). In addition, SOS offers MOS transistors with three different thresholds (refer to Chap. 1), thus enabling the optimization of both analog and digital circuits for low-power image-sensor designs.

The disadvantage of image sensors implemented in this technology is the low quantum efficiency at high wavelengths. This is due to the thin film of silicon in the SOS process on which the photodetectors are implemented. The limitations of the photodetectors and their spectral response is illustrated in Chap. 7.

In this chapter, we present the design of three SOS CMOS image sensor arrays. Two designs are based on an active pixel sensor with analog outputs. One design employs PN photodiodes (in Sec. 7.4.1) as detectors, one uses phototransistors (in Sec. 7.4.2). The third design uses PIN photodiodes 7.4.3 and features an intensity-based digital

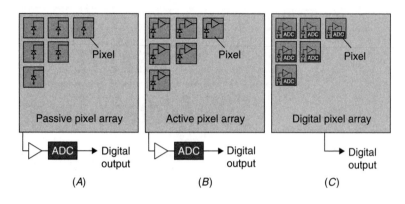

FIGURE 8.1 Types of image sensor arrays: (A) passive pixel sensor array, (B) active pixel sensor array, and (C) digital pixel sensor array.

address-event output interface. The chapter begins with an overview of photodetector circuits and measures of their performance.

8.2 Overview of Photosensitive Circuits

Photosensitive circuits in a large array are called pixels (picture elements). These arrays of photosensitive circuits form imaging arrays or image sensors. Integrated image sensors can be of three different kinds: passive pixel arrays, active pixel arrays, and digital pixel arrays (Moini, 1994; Fossum, 1995; Fossum, 1997). These arrays are illustrated in Fig. 8.1. Passive pixel sensor arrays use passive pixels that convey directly their photosignal to the periphery of the array without the use of an amplifier, as in Fig. 8.1(A). Passive pixel sensor arrays suffer from large noise and mismatch, but pixel sizes are small. Active pixel sensor arrays use an amplifier per pixel to deliver the photosignal to the periphery of the array, as in Fig. 8.1(B). Active pixels require additional area, but have a larger signal-to-noise ratio than passive arrays. A global amplifier and an ADC deliver a digital output signal. The array amplifier and ADC can also be implemented columnwise to reduce the requirements on these components when large arrays or high frame rates are required. Figure 8.1(C) shows a digital pixel array, where each pixel contains an amplifier and an ADC. Digital pixels are large and reduce the percentage of an effective area dedicated to the photodetectors (called the fill factor). Digital pixels deliver the best signal-to-noise ratio performance and can operate at very large speeds with a large number of pixels.

Photodetector circuits (pixels) can be divided into two main categories: "integrating" and "nonintegrating." Integrating pixels collect light for a fixed amount of time and measure the resulting collected electrical signal produced by the photocurrent during the integration time. Thus, integrating pixels sample light intensity at a given rate, therefore, they are continuous-value discrete-time circuits. Nonintegrating photodetectors measure light intensity in real time; thus, they are a continuous-value continuous-time circuit. We will consider here only active integrating pixels.

In this chapter, we present an active pixel sensor as in Fig. 8.1(B) in Sec. 8.4, and we also present a digital pixel sensor array as in Fig. 8.1(C) in Sec. 8.5.

8.2.1 Active Pixels

The most typical pixel circuit in CMOS image sensors is called an "active pixel sensor," or APS, and uses a voltage follower to convey the integrated signal to the periphery of the pixel array. The APS pixel generally uses photodiodes as detectors but can also employ phototransistors, photogates or other photodetectors (Fossum, 1995; Sodini and Howe, 1996; Johns and Martin, 1997). The APS is an efficient alternative compared to passive pixels because each amplifier is only used during the readout; therefore, the power dissipation of each pixel is minimized, but the signal is buffered for speed and precision by the amplifier. Figure 8.2 reports a typical circuit diagram of an APS pixel sensor and its response.

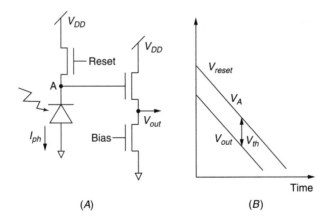

(A)   (B)

FIGURE 8.2 (A) Active pixel sensor (APS) schematic. A photodiode converts photons into a current that is integrated in the capacitance at node A. The follower circuit outputs the integrated voltage on the V_{out} node. The follower active load is biased by the voltage bias. The signal Set resets the integration. Different photodetectors can be used instead of the photodiode. (B) Photointegrated voltage V_{out} versus time.

General operation of the APS pixel is as follows (Fossum, 1997). The combined capacitance of the photodiode and the amplifier gate transistor and its charge at the cathode A in Fig. 8.2 (*A*) is initially reset by means of a switch controlled by the signal *Reset*. Voltage V_A thus resets to value V_{reset}. During photointegration, the photodiode removes the charge at node V_A proportionally to the intensity of incident light. The voltage V_A thus decreases linearly with time for a fixed impinging light and generated photocurrent I_{ph}, as illustrated in Fig. 8.2(*B*). The voltage on the integrating node A is decoupled by a single-stage unity-gain buffer presented with a current source transistor biased with voltage *Bias*. After light integration, the row select is pulled high, and the voltage is presented to the output V_{out}. V_{out} is one voltage threshold (V_{th}) lower than voltage V_A because of the follower transistor. Notice that because SOS provides multiple threshold transistors, the threshold drop at the pixel can be reduce to 0.3 V using low-threshold transistors (Nl, LP) or 0 V using intrinsic transistors (IN, IP).

Notice that all transistors in the bulk CMOS APS are of the same type. This is because a combination of genders would require silicon wells and thus increase the pixel's area (Fossum, 1997). On the other hand, SOI and SOS can use both genders of transistors without having to use additional silicon area. The use of mixed genders is beneficial because although the NMOS can be used for higher amplification, given the higher mobility of electrons, the use of a PMOS as reset switch is desired to increase the initial voltage on capacitance A, reduce the reset noise, and increase the dynamic range of the circuit. Notice that a PMOS is used to reset the pixel; there is also a significant reduction in fixed-pattern noise due to incomplete reset and threshold differences. When a NMOS is used, in fact, these combine to give a large variation in the final reset voltage from pixel to pixel. The use of PMOS instead will guarantee the reset voltage to reach the supply voltage V_{DD}, reducing the effect due to threshold variations and reset timing.

$$V_{reset} - V_{out} = \frac{I_{ph}}{C_{int}} \cdot t_{int} \qquad (8.1)$$

The output signal V_{out} from an APS circuit is given by Eq. 8.1, where I_{ph} is the diode photocurrent, C_{int} the pixel integration capacitance (usually parasitic and not explicit), and t_{int} is the integration time or the inverse of the frame rate.

8.2.2 Photodiodes and Active Pixel Operation

We now describe how the photodetector operates within a pixel circuit. We focus here on a photodiode-based pixel, although the principles are identical for other kinds of photodetectors. In order to accumulate charge on the photodiode, a potential difference across the diode must exist. Applying a high voltage to the gate of the *Reset* transistor (Fig. 8.2) charges the capacitance of node A and thus reverse biases the diode to $V_{DD} - V_{th}$, where V_{th} is the threshold of the reset transistor. The reverse bias raises the potential barrier at the PN junction, which in turn widens the depletion region. When photons reach the depletion region of the photodiode, electron-hole pairs are generated.

As can be seen from Eq. 8.1, the important photodetector parameters are the capacitance C_{int} and the the generated photocurrent I_{ph}. These parameters allow the designer to compute the voltage swing for a specific integration time t_{int}. The capacitance C_{int} is the parallel of the photodiode junction capacitance C_j and the total capacitance at node A in Fig. 8.2. Because the depletion region contains few mobile carriers, it can be thought of as an insulator with dielectric constant ϵ_{si} with the n and p regions acting as the plates of the capacitor. To solve for the junction capacitance per unit area, we can use Eq. (7.10), and the doping levels of the N-type and P-type regions must be known.

The junction capacitance of the diode depends on the initial potential at the cathode, and it varies with the integrating voltage at node A during photocollection. Refer to Fig. 7.6 in Sec. 7.4.1 where the depletion region in a vertical junction photodiode is presented in detail. The junction capacitance is an important parameter because it is directly proportional to the time it takes the pixel to saturate, as can be seen in Eq. (8.1). Notice that the voltage across the diode changes during light integration, as illustrated in Fig. 8.2(*B*). As the depletion region of the diode changes, so does the effective collection area. This effect introduces a modest nonlinearity of V_A versus time in the APS circuit, which is generally not taken into account in a typical imaging application.

8.2.3 Active Pixel Sensors

An array of active pixels is referred to as an active pixel sensor, or APS. This is the most typical CMOS imaging front end (Fossum, 1997). Figure 8.3 shows a schematic of a typical signal path in a CMOS image sensor array, also illustrated in Fig. 8.2(*B*).

The APS arrays are organized in imaging array of pixels, column circuits, and global circuits. Pixels are generally organized in arrays of $N \times M$ elements. Each pixel is addressed by means of digital shift registers that periodically scan the array and sequentially output each pixel voltage. Column circuits usually contain pixel biases, sampling capacitance, and additional circuits to reduce fixed-pattern noise. The global circuits are usually amplifiers to deliver the pixel outputs outside the integrated circuit.

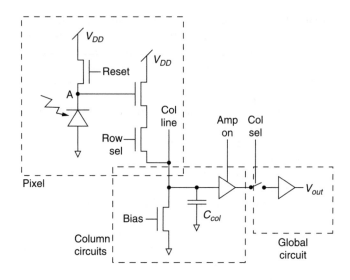

FIGURE 8.3 The signal path of an active pixels sensor also illustrated in Fig. 8.2(*B*). Individual pixels deliver the integrated photosignal to column circuitry and then to a global output amplifier. Digital shift registers (not drawn in this figure) scan the array and provide the column and row signal to deliver each pixel to the global output amplifier. The output is a time-multiplexed analog voltage.

Referring to Fig. 8.3, the pixel circuit is identical to the one in Fig. 8.2, the only difference is that the biasing current source of the voltage follower is migrated to the column circuitry, whereas a switch (column select-col sel) is used to select all the pixels in a column. The column shift register selects the active column. Column circuits have a first voltage buffer. Only one column at a time is connected to the global amplifier by means of the row shift registers. The global amplifier buffers the signal to drive the large off-chip capacitances. The signal is then delivered to an internal or external analog to a digital converter.

8.3 Noise Sources in Phototransducing Circuits

In a photosensitive circuit, there is a variety of noise sources that corrupt the electrical signal generated by incident photons. Fixed pattern variations, dark current, thermal and shot noise, and reset and sampling noise are the main sources of imprecision in light conversion.

All of these noise sources impose a lower bound on the dynamic range, as they limit the minimal photocurrent that the device is able to report. Dark current is an artifact due only to the photodetector. Dark current limits the response to a minimum light intensity. Thermal and shot noise is generated by fundamental limits of the matter and devices by which the circuit is constructed. These noise sources are generated by the pixel follower and the column and global amplifiers. Reset noise is generated by the sampling of the reset photodiode voltage. Sampling noise is a reflection of the thermal noise on sampling capacitances in the signal path of the APS array.

In addition to intrapixel sources of noises, the differences between the output of multiple pixels generate image artifacts. Fixed pattern noise (FPN) is the most typical example and is the result of mismatch between individual photodetectors and photosensitive circuits. An array of pixels, all stimulated with the same light intensity, will produce different outputs (FPN in voltage or current) and thus generate noise in an image meant to be uniform.

We describe here the general theory behind photodetector noise and how to compute each noise contribution.

8.3.1 Temporal Noise

Temporal noise is one of the limitation of dynamic range and performance of a phototransduction circuit, particularly at lower lighting or high frame rates. Assessing temporal noise translates into the determination of integration and reset noise. For a typical APS image sensor, integration and reset noise are, respectively, expressed in Eq. (8.2) (Tian et al., 2001; Faramarzpour et al., 2006) for long integration times compared to the reset time.

$$\overline{V_n^2} = \overline{V_n^2(t_{reset})} + \overline{V_n^2(t_{int})} = \frac{1}{2}\frac{kT}{C_{int}} + q\frac{I_{ph} + I_{dc}}{C_{int}^2}t_{int} \qquad (8.2)$$

In Eq. (8.2), I_{ph} is the input photocurrent, I_{dc} is the dark current, t_{int} is the light integration time, and C_{int} is the integrating capacitance of node A in Fig. 8.2.

The first term corresponds to the reset noise power due to the reset phase of the pixel due to thermal noise. This term is significantly less than the value of kT/C usually reported in the literature (Tian et al., 2001), due to nonadiabatic charge transfer during the reset phase. The second term refers to integration noise due to dark current I_{dc} and photocurrent I_{ph} with the reset transistor turned off.

Dark currents are process, layout, and device dependent and cannot be easily quantified. Usually, dark current are measured and heuristics can be given for each specific device as a function of size and operating temperature.

Equation (8.2) gives an estimate of the total noise at the output of a pixel as presented in Fig. 8.2. The column circuits and global amplifiers also contribute to noise, but because they are peripheral circuits, their size can be increased and they can be optimized for low-noise operation. In general, Eq. (8.2) gives a good estimate of the total noise in an APS array.

8.3.2 Pixel and Photodiode Design and Noise Optimization

The signal-to-noise ratio (SNR), or output voltage dynamic range of an APS circuit, is defined as the ratio in decibels of the minimum to the maximum light intensity observable at required design precisions. The minimal detectable light is just above the dark current of the photodiode. The maximum light intensity is the intensity that brings the APS to saturation just at the end on the integration time t_{int}.

Once the expected photocurrent per unit area is calculated, a photo-diode can be sized for a particular application. In an integrating pho-todetector, the important quantities are the integrating capacitance size and the photodiode size. These two values determine the integration slope.

$$C_{int} = \frac{t_{int} I_{ph}}{V_{reset}} \tag{8.3}$$

$$v_n^2 = \frac{1}{2}\frac{kT}{C_{int}} + q\,I_{ph}\frac{t_{int}}{C_{int}^2} \tag{8.4}$$

$$\text{SNR} = \frac{1}{v_n^2}\frac{I_{ph}^2}{C_{int}^2}t_{int}^2 \tag{8.5}$$

Equation (8.3) calculates the maximum integration slope according to the integration time t_{int}, the photocurrent I_{ph}, and the reset voltage V_{reset} (maximum when $V_{int} = V_{reset}$). Equation (8.4) calculates the volt-age noise power at the integration node (see Sec. 8.3). Equation (8.5) computes the SNR. Note that dark current has not been considered in this calculation because its value is much smaller that the photocur-rent, and shot noise is dominated by the photocurrent.

Given a certain SNR and integration time, the values of C_{int} and I_{ph} can be calculated. Start from the SNR figure. For example, 8bits precision means that the noise has to be smaller than $1/\sqrt{12}$ of an LSB, thus in this case SNR = $1/\sqrt{12}\,2^{-8}$. Substituting Eqs. (8.3) and (8.4), into Eq. (8.5), we can compute the photocurrent I_{ph} as a function of the integration time. This pixel design methodology is optimal for noise performance of the sensor.

8.4 Design of an SOS APS

In this section, we present the design of a back-illuminated 32 × 32 image sensor array implemented in the 0.5-μm SOS (Peregrine, 2008a) process capable of ultraviolet imaging. The image sensor array was designed to perform at high frame rates and with low light intensities. The sensor we present here can performs "snapshot" image acquisition and analog readout at a continuous rate of 1000 frames/s and consumes less than 1 mW. Each pixel consists of a photodiode and a memory capacitor in a 40 μm × 40 μm area with a fill factor of 43%. This sensor takes advantage of the transparent substrate in SOS in order to operate with back illumination. Because sapphire is transparent to UV light (wavelengths of less than 400 nm), whereas silicon and silicon dioxide are not, this design is suited for hyperspectral imaging at high speeds. The sensor can image light from both the front and back sides of the die, making it one of the first of its kind.

The image sensor presented here is also the first fully functional, high-SNR array designed in the SOS process. In recent years, SOI circuits have been used for a variety of applications. Several groups have shown working photodetectors (Zheng et al., 2000; Uryu and Asano, 2002). However, there is no working sensor array in existence— only characterizations and simulations of full image arrays are available (Kucewicz et al., 2004; Brouk et al., 2007; Choa et al., 2007).

The image sensor presented in this chapter features the following innovations: (1) it is the first fully functional array implemented in SOI/SOS technology, (2) it can be operated with back illumination, (3) the pixels feature a reflector that allows light to pass twice through the photodetector, (4) it was designed to be able to operate at thousands of frames per second, and (5) it is intrinsically radiation tolerant.

The sapphire substrate in SOS is optically transmissive from the UV (200 nm) to the infrared (5500 nm) spectrum (Andreou et al., 2001), and makes this device favorable for hyperspectral imaging in multiple bands with the use of back illumination. The back illumination also allows the pixels to have higher fill factors than traditional CMOS photodiodes because the metalization layer will not block the light. The tolerance to ionizing radiation makes this SOS image sensor suitable for outer-space instrumentation.

8.4.1 SOS APS System Overview

The block diagram of the SOS image sensor is shown in Fig. 8.4. The array is composed of 32 × 32 pixels with SOS PN photodiodes as detectors. Row-and-column scan shift registers are used to address individual pixels and output their integration voltage. A global operational amplifier is used as a buffer to convey the pixel integration voltages to

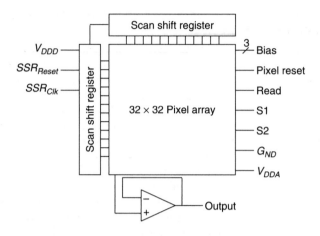

FIGURE 8.4 Block diagram of the image sensor. The system is composed of a 32 × 32 photodiode array. The two scan shift registers are used to access the individual pixels. A global operational amplifier is used as a buffer to the output.

the output. The image sensor array features 12 output pins, presented in Fig. 8.4. Power supply of analog and digital components (V_{ddd}, V_{dda}) were separated to avoid coupling of digital switching noise to the sensitive analog components. SSR_{Reset} and SSR_{Clk} are, respectively, the shift register reset signal and clocking signal. $S1$, $S2$, $PixelReset$, and $Read$ are pixel signals. Three bias voltages are used for pixels and the global amplifier.

The pixel is shown in Fig. 8.5. The size of each pixel is 40 μm × 40 μm with a fill factor of 43%. The capacitor is used as a memory element to store the integrated voltage. The $Read$ signal connects/disconnects the photodiode from the rest of the pixel circuitry. $S1$ and $S2$ signals connect the storage capacitor to the photodiode. The $Reset$ signal shorts the capacitor, resetting the capacitor voltage to a predefined value (voltage at the $Bias$ pin). $ColSel$ is a signal from the scan shift register to connect the storage capacitor to the Out pin. The V_{dda} pin is set to 3.3 V and Out is connected to the input pin of the global operational amplifier.

A normal pixel operation consists of three phases, shown in Fig. 8.6. First, the storage capacitor is reset to an initial value (voltage at the $Bias$ pin) during the reset phase by setting the $Reset$ line, which shorts the storage capacitor. Then, during the integration phase, $Read$ is set, connecting the photodiode to the rest of the pixel circuitry. $S1$ is set to V_{dda}, $S2$ is set to G_{ND}, and $Reset$ is set to G_{ND}, which connects the photodiode to the capacitor. The current from the photodiode discharges the capacitor as a function of the light intensity. Finally, during the

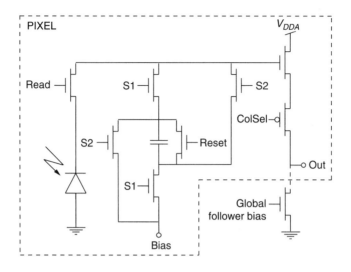

FIGURE 8.5   Pixel schematic. The *Read* signal connects/disconnects the photodiode from the pixel. The *Reset* switch shorts the storage capacitor to *Bias* voltage. *S1* and *S2* signals connect the capacitor to the photodiode and the output. *ColSel* is controlled by the scan shift registers to select the pixel.

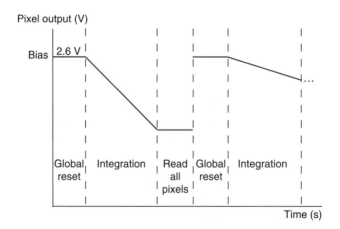

FIGURE 8.6   The three phases during a normal operation of the pixel. (1) During *Reset*, the storage capacitor is set to the voltage at the *Bias* pin. (2) During Integration, S1 is set to V_{dda} and S2 is set to G_{ND}, which discharges the storage capacitor relative to the light intensity at the photodiode. (3) During the *Read* interval (very short, not drawn to scale), the diode is disconnected by setting the *Read* signal to G_{ND}, and the capacitor stops discharging. This process is repeated for video streams.

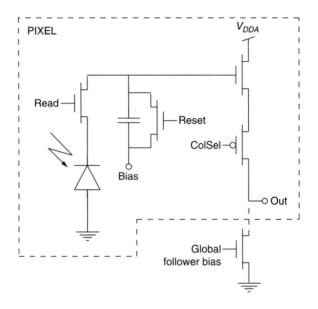

FIGURE 8.7   Simplified pixel schematic with capacitor switching circuit removed. The *Read* signal connects/disconnects the photodiode from the pixel. The *Reset* switch shorts the storage capacitor to *Bias* voltage. *ColSel* is controlled by the scan shift registers to select the pixel.

readout phase, *Read* is set to G_{ND} and the scan shift register sets *ColSel* to G_{ND}. This connects the storage capacitor to *Out*, which is connected to the global operational amplifier.

The *S2* switch was designed to allow frame differencing on a pixel. The first frame would be stored in the capacitor, whereas *S1* is set V_{dda} and *S2* to G_{ND}. Then, the second frame would be stored by setting *S1* to G_{ND} and *S2* to V_{dda}. However, due to capacitive coupling between the capacitance of the photodetector and the storage capacitor during switching, there is a voltage offset between the two frames related to light intensity, and the frame subtraction feature cannot be used. Notice, however, that the pixel performs as a normal APS circuit when S1 is always on and S2 always off or vice versa. In this case, the pixel schematic simplifies to the one presented in Fig. 8.7.

The image sensor performs a snapshot image acquisition (Kleinfelder et al., 2001) because all the pixels are reset at the same time and integrate over the same interval before readout. Unlike image sensors with rolling shutters (Krymski et al., 1999), motion artifacts are not a problem for the image sensor.

The global operational amplifier is connected in a feedback loop and has a gain of one. The primary purpose of the operational amplifier is to act as a voltage buffer between the output of the pixels and the circuitry external to the imager chip.

8.4.2 SOS APS Testing and Characterization

In this section, we present data collected and measured from a fabricated version of the photodiode-based SOS APS image sensor array. The photodiode used in this design was a PN type and has been characterized and designed as reported in sec. 7.4. The sensor was connected to a 12-bit commercial ADC with a maximum conversion rate of 3 Ms/s. The time to integrate an output swing of 1 V at different intensities were measured to determine the integration characteristics of the photodiode. A dispersing filter was placed between a white light source (Genesys LS-150) and the image sensor. The light intensities were measured using a commercial photographic digital lux meter.

The integration time (T_{int}) and the light intensity (L_I) have an inversely linear relationship for the photodiode, as shown in Fig. 8.8. Because the change in voltage (1 V) and the capacitance of the storage element (200 fF) is known, the current drawn by the photodiode relative to the light intensity can be calculated using the measured relationship between integration time and intensity [Eq. (8.12)]. This relationship between the current drawn by the photodiode (I_{in}) and the intensity

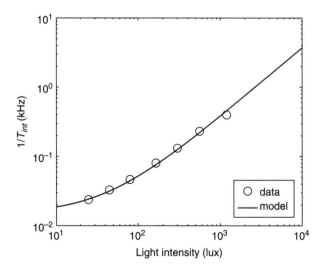

FIGURE 8.8 Linearity of the photodiode and model fit. The data was gathered by measuring the integration time with an output swing of 1 V at different light intensities. The equation of the linear model is $T_{int}^{-1} = 0.37 \cdot L_I + 15$.

is shown in Eq. (8.13), where the coefficient is $7.4 \cdot 10^{-14}$ A/lux and the dark current is $3 \cdot 10^{-12}$ A.

$$I_{in} = 200\,\text{fF} \cdot \frac{1\,\text{V}}{T_{int}} \tag{8.6}$$

$$T_{int}^{-1} = \frac{I_{in}}{(200\,\text{fF})(1\,\text{V})} \tag{8.7}$$

$$= 0.37 \cdot L_I + 15 \tag{8.8}$$

$$I_{in} = (7.4 \cdot 10^{-14}) \cdot L_I + (3 \cdot 10^{-12}) \tag{8.9}$$

The quantum efficiency of the photodiode was measured by using the model derived from the integration characteristics [Eq. (8.10)]. The quantum efficiency is reported in Sec. 7.4.1.1.

Various frames were captured at different light intensities and frame rates to measure the properties of the system. A SUNEX DSL107 lens with an f-stop of 2.4 was used to focus the images when applicable. In order to quantify the fixed-pattern noise in the system, 64 frames were captured at the different conditions, and the standard deviations of the pixel in each frame was computed. The standard deviation for images at different light intensities and frame rates are shown in Fig. 8.9. The average standard deviation is 3 mV at low light intensities. The deviation saturates at high illuminations because the pixel output voltage

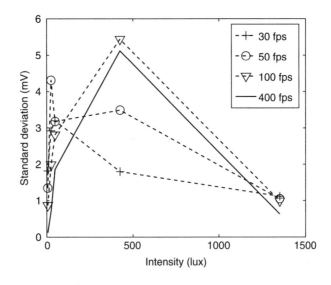

FIGURE 8.9   Measured average pixel standard deviation as a function of light intensity and frame rate. This is the fixed-pattern noise over 64 images.

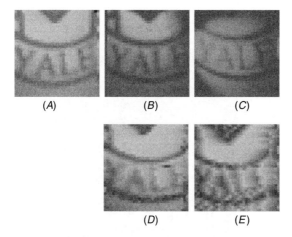

FIGURE 8.10   Images captured with lens at various frame rates: (*A*) 30 fps with 2.7 mW power consumption, (*B*) 500 fps, (*C*) 1000 fps, (*D*) 30 fps with 647 μW power consumption, and (*E*) 30 fps with 181-μW power consumption.

saturates to the minimum level, and thus compresses the standard deviation as well. At midscale illumination, the standard deviation increases with the frame rate, as expected, because the mismatch between pixels is compressed with longer integration times as more and more pixels saturate.

The output of the image sensor at different frames rates and power consumption are shown in Fig. 8.10. The photodiode operates correctly with an f 2.4 lens even at 1000 fps (200 μs integration time).

Another three sets of images (of 64 frames each) were captured to measure the fixed-pattern noise of the image sensor. First, the image sensor was placed in the dark and the photodetector signal was integrated for 100 μs to calculate the reset noise. Next, the image sensor was placed in the dark and the photodetector signal was integrated for 100 ms to measure the dark current. Finally, light was focused on the image sensor and adjusted to be near the saturation point for a given integration time of 30 fps to calculate the gain noise. For noise calculations, the standard deviations of each pixel over the 64 frames were calculated and the mean of the standard deviations of the 1024 pixels were calculated for the above three cases. The standard deviation was 0.8 mV for a short integration time (10 μs) in the dark with a swing of 3.8 mV, 2.2 mV for a long integration time (100 ms) in the dark with a 368 mV swing, and 3.8 mV when the image sensor was close to saturation while running at 30 fps with a 1.3-V swing. The SNR was calculated by taking the mean swing of each pixel at near

FIGURE 8.11   Picture of the fabricated SOS image sensor die.

saturation over 64 frames and dividing it by the mean of the standard deviation of each pixel over 64 frames.

The power consumption was measured by observing how much current the image sensor was drawing through the V_{ddd} and V_{dda} lines. The global operational amplifier and the pixels are powered by V_{dda} and the scan shift registers are powered by V_{ddd}. A Keithley 2400 General-Purpose SourceMeter was used to measure the current. The scan shift register consumes 45 μW, whereas the photodiode array and the global operational amplifier consume 200 μW. The power consumption remained constant for all frame rates tested.

Although the data in Fig. 8.10 shows the image sensor operating up to 1000 fps, the system is capable of much higher frame rates. The operational amplifier is able to operate up to 10.5 MHz, or 10,250 fps and is the limiting factor in the system. Also, the ADC used in the system is only capable of 3 Ms/s, thus no images above 3000 fps can be obtained from the current setup unless a faster ADC is used.

A picture of the SOS APS die is given in Fig. 8.11. The summary of the sensor performance is given in Table 8.1.

We engineered the fully functional camera pictured in Fig. 8.12 based on the SOS image sensor. The camera can be operated by any computer with an USB port.

Technology	0.5-μm SOS
Array size	32 (H) × 32 (V)
Total size	1.5 mm x 1.5 mm
Pixel size	40 μm x 40 μm
Fill factor	43%
Power consumption	650 μW (@ 3.3 V)
Fixed-pattern noise	0.8 mV (dark, 10-μs integration) 2.2 mV (dark, 100-μs integration) 3.8 mV (light, before saturation)
Output voltage swing	1.5 V
Dark current	3 pA
Conversion gain	0.8 μV/e-
SNR	48 dB
Max frame rate	1000 fps

TABLE 8.1   Photodiode SOS APS array properties

FIGURE 8.12   Camera designed for our SOS image sensor. The image sensor is mounted on a custom-built PCB with external components connected to the USB-based FPGA board.

8.4.3 SOS APS Array with Phototransistor Detector

We have also designed an SOS APS array with a phototransistor as detector. The phototransistor characteristics were reported in Sec. 7.4.2. We report in this section data collected and measured from the fabricated version of the phototransistor SOS APS image sensor array.

The nonlinearity of the phototransistor was measured by observing the photocurrent integration curve at different light intensities. The photocurrent generated by the phototransistor is not initially linear but settles to a linear curve after a certain amount of time. The initial nonlinearity is shown in Fig. 8.13. The nonlinearity reduces the total possible integration time. At 256 lux, the phototransistor saturates after 25 μs. Under indoor office illumination conditions, anything below 200 fps (greater than 4 ms integration time) results in a saturated image.

The time to integrate 1 V at different intensities was measured to determine the integration characteristics of the phototransistor. A dispersing filter was placed between a white light source (Genesys LS-150) and the image sensor. The light intensities were measured using a commercial photographic digital lux meter.

The integration time (T_{int}) and the light intensity (L_I) have an inversely linear relationship for the phototransistor at certain light

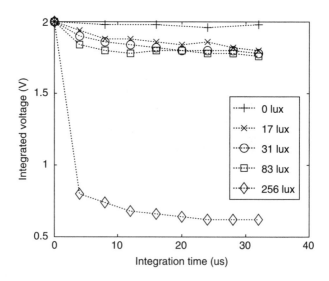

FIGURE 8.13 Initial nonlinearity of the phototransistor. The data was gathered by measuring the integration voltage at different light intensities. Due to the nonlinearity, the phototransistor saturates after 25 μs at 256 lux.

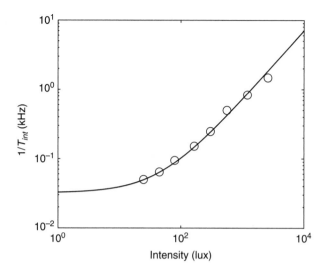

FIGURE 8.14 Linearity of the phototransistor and model fit. The data was gathered by measuring the integration time of 1 V at different light intensities. The equation of the linear model is $T_{int}^{-1} = 0.7 \cdot L_I + 32$.

intensities, as shown in Fig. 8.14. Because the change in voltage (1 V) and the capacitance of the storage element (200 fF) is known, the current drawn by the phototransistor relative to the light intensity can be calculated using the measured relationship between integration time and intensity [Eq. (8.12)]. This relationship between the current drawn by the phototransistor (I_{in}) and the intensity is shown in Eq. (8.13), where the coefficient is $1.4 \cdot 10^{-13}$ A/lux and the dark current is $6.4 \cdot 10^{-12}$ A.

$$I_{in} = 200\,\text{fF} \cdot \frac{1\,\text{V}}{T_{int}} \tag{8.10}$$

$$T_{int}^{-1} = \frac{I_{in}}{(200\,\text{fF})(1\,\text{V})} \tag{8.11}$$

$$= 0.7 \cdot L_I + 32 \tag{8.12}$$

$$I_{in} = (1.4 \cdot 10^{-13}) \cdot L_I + (6.4 \cdot 10^{-12}) \tag{8.13}$$

The quantum efficiency of the phototransistor was obtained using the model derived from the integration characteristics [Eq. (8.10)]. The quantum efficiency of the backside phototransistor is shown in Sec. 7.4.2. The SOS phototransistor has a higher quantum efficiency in near UV. However, at longer wavelengths, the quantum efficiency drops off because the thickness of silicon in the SOS process is only

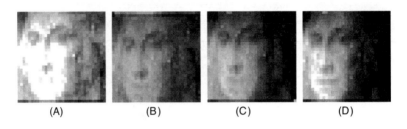

(A)   (B)   (C)   (D)

FIGURE 8.15   Image captured with lens at various frame rates: (*A*) 200 fps, (*B*) 500 fps, (*C*) 1000 fps, and (*D*) 1200 fps.

100 nm. This results in a depletion region thinner than traditional CMOS processes, and photons at longer wavelengths do not generate electron-hole pairs.

The output of the image sensor at different frames rates and power consumption are shown in Fig. 8.15. The phototransistor operates correctly with a fF 2.4 lens even at 1200 fps (200-μs integration time).

Another three sets of images (64 frames each) were captured to measure the fixed-pattern noise of the image sensor. First, the image sensor was placed in the dark and the photodetector signal was integrated for 10 μs to calculate the reset noise. Next, the image sensor was placed in the dark and the photodetector signal was integrated for 100 μs to measure the dark current. Finally, light was focused on the image sensor and adjusted to be near saturation point for a given integration time for 30 fps to calculate gain noise. The fixed-pattern noise was calculated using the standard deviations of each pixel over the 64 frames, and the mean of the standard deviations of the 1024 pixels were calculated for the above three cases. The standard deviation was 4.3 mV for a short integration time (10 μs) in the dark with a swing of 17.7 mV, 7 mV for a long integration time (100 μs) in the dark with a 51.6-mV swing, and 39.8 mV when the image sensor was close to saturation while running at 200 fps with a 1.4-V swing. The SNR was calculated by taking the mean swing of each pixel at near saturation over 64 frames and dividing it by the mean of the standard deviation of each pixel over 64 frames.

The power consumption was measured by observing how much current the image sensor was drawing through the V_{ddd} and V_{dda} lines. The global operational amplifier and the pixels are powered by V_{dda} and the scan shift registers are powered by V_{ddd}. A Keithley 2400 General-Purpose SourceMeter was used to measure the current. The scan shift register consumes 45 μW, and the phototransistor array and the global operational amplifier consume 200 μW. The power consumption remained constant for all frame rates tested.

Technology	0.5 μm SOS
Array size	32 (H) × 32 (V)
Total size	1.5 mm × 1.5 mm
Pixel size	40 μm × 40 μm
Fill factor	43%
Power consumption	250 μW (@ 3.3 V)
Fixed-pattern noise	4.3 mV (dark, 10-μs integration) 7.0 mV (dark, 100-μs integration) 39.8 mV (light, before saturation)
Output voltage swing	1.5 V
Dark current	6.2 pA
Conversion gain	0.8 μV/e-
SNR	31 dB
Max frame rate	1200 fps

TABLE 8.2   Phototransistor SOS APS array properties

Although the data in Fig. 8.15 show the image sensor operating up to 1200 fps, the system is capable of much higher frame rates. The operational amplifier is able to operate up to 10.5 MHz, or 10,250 fps, and is the limiting factor in the system. Also, the ADC used in the system is only capable of 3 Ms/s, thus no images above 3000 fps can be obtained from the current setup unless a faster ADC is used.

A summary of the performance of the phototranistor based SOS APS array is given in Table 8.2.

8.4.4 Ultraviolet Testing

As seen from Sec. 7.4.2, and in particular in Figs. 7.13 and 7.14, the back-illuminated photodiode has a high quantum efficiency in the near-UV spectrum. Also, the integration of UV light by the image sensor was tested using a 60-W Hamamatsu Xenon Flash Lamp (L6604). A UV bandpass filter (Newport FSR-UG11) was placed on top of the image sensor. The light source saturated the image even at a 500-μs integration time. When a visible bandpass filter (Newport FSR-KG3) was used in conjunction with the UV bandpass filter with the same integration time, the image was dark as expected. No meaningful images are available, as we did not have access to a quartz lens.

8.5 Design of an SOS Digital Image Sensor Array

In this chapter, we report on the design and characterization of a 16 × 16 pixels SOS digital photosensor array fabricated in the Peregrine SOS 0.5-μm process. This image sensor is based on a PIN photodiode detector and is one of the first standard CMOS photosensor array, capable of transducing light simultaneously from both sides of a die. The design features a time-domain digital output with per-pixel AD conversion.

Conventional imagers integrate the photocurrent for a fixed time, usually dictated by the desired frame rate. Therefore, the integrated voltage is output according to a raster scan. On the other hand, this digital sensor integrated the photocurrent to a fixed voltage threshold. When the threshold is crossed, a 1-bit pulse is generated by the pixel. The magnitude of the photocurrent is represented as the time interval between two successive pulses. This interpulse interval is inversely proportional to the intensity. Our system is also different from conventional methods because the read-out of each pulse is initiated by the pixel itself. That is, each pixel requests access to the output bus when the integration threshold has been crossed. Such digital representation of light intensity yields devices capable of large dynamic range imaging and low-power dissipation (Yang, 1994; McIlrath, 2001; Culurciello et al., 2003). A few frequency-modulated and/or address-event imaging systems have been previously reported (Yang, 1994; McIlrath, 2001), however, the one presented here in Fig. 8.16 is the first to combine a conventional APS with a fully arbitrated address event system (Culurciello et al., 2001; Culurciello et al., 2003; Culurciello and

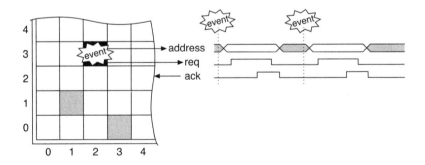

FIGURE 8.16 Address–event generation in an array. Events are 1-bit communication packets sent with an asynchronous protocol over a serial bus. Request (req) and acknowledge (ack) signals modulate and time the transfer from the transmitter to one or multiple receivers.

Andreou, 2004; Culurciello and Etienne-Cummings, 2004; Culurciello and Andreou, 2006). Pulse-modulated output has also been reported for biomimetic image gradient sensors from Barbaro et al., 2002.

The imager presented here mimics the octopus' retina by converting the light intensity directly into a pulse train; most other biological retinas represent light intensity as an analog signal. This biologically inspired readout method simultaneously favors brighter pixels, minimizes power consumption by remaining dormant until data are available and offers a pixel-parallel readout. In contrast, a serially scanned array allocates equal portion of the bandwidth to all pixels independent of activity and continuously dissipates power because the scanner is always active. Here, brighter pixels are favored because their integration threshold is reached faster than darker pixels (i.e., the request-acknowledge-reset-integrate cycle operates at a higher frequency). Consequently, brighter pixels request the output bus more often than darker ones. Also, virtually no power is used by the pixel until an event is generated; therefore, low-intensity pixels consume little power. Encoding the data as a stream of digital pulses provides noise immunity by quantization and redundancy. Furthermore, representing intensity in the temporal domain allows each pixel to represent a large dynamic range of outputs. The integration time is, in fact, not dictated by a regular scanning clock, and therefore a pixel can use the whole bus bandwidth by itself or can remain silent. This provides a simple and efficient way of obtaining dynamic range control without the use of additional circuitry that varies the integration time of each pixel based on the light intensity. Pixel-parallel automatic gain control is an inherent property of this time-domain imaging and address–event representation (AER) readout scheme.

Events from individual pixels are multiplexed on an asynchronous digital serial bus that transmits their addresses to one or multiple external receivers, shown in Fig. 8.16. The bus is arbitrated with digital circuits that ensure that only one pixel at a time is using the single-output bus. The digital arbitration circuits also prevent bus collisions; a pipeline controls the sequence of output events. A digital asynchronous interface is employed to minimize the design complexity and the power consumption. The analog pixel value is encoded as a pulse density stream of address events. The frequency-modulated signal can be reconstructed by integration or simply by counting the number received events over a predetermined window of time. The multiple threshold MOS transistors available in the SOS technology enable circuit optimizations for low-power and low-voltage operation.

Section 8.5.2 characterizes the SOS photodiode used in the image sensor array. Section 8.5.1 reports on the operation of the pixel. Section 8.5.3 is an overview of the address–event architecture of the imaging system. Section 8.5.4 summarizes the results obtained after testing the SOS digital image sensor array.

8.5.1 Digital Pixel Design

The key components in an address–event digital image sensor is the digital–output pixel. The pixel is responsible for requesting access to the output bus when a pixel has reached the integration threshold. Generally, a prototypical CMOS imager employs a photodiode as a photosensitive element. The relatively small photocurrent is integrated on a capacitor and subsequently read out, as seen in Sec. 8.4.1. An Address–event imager will convert light into events by integrating photocurrent up to a fixed threshold. The integrated voltage changes very slowly if the light intensity is low. The event generator inside the pixel must convert this slow-changing voltage into a fast-changing signal in order to minimize the delay between when the threshold is passed and when output bus access is requested. Furthermore, the fast transition also limits an individual pixel's power consumption. Because large arrays of pixels are generally employed, it is necessary to reduce the power of each pixel as much as possible.

The pixel used in the imager solves both the transition speed and power consumption problem with an elegant current positive feedback circuit (Culurciello et al., 2003; Culurciello and Etienne-Cummings, 2004). Power consumption and transition speed are closely related because CMOS digital circuits only consume power during switching. Hence, reducing the transition time will also reduce the power consumption. Our event generator has simultaneously a large gain, large bandwidth, and minute power consumption. This circuit can be used for various other applications where high-speed and low-power consumption are required. Fig. 8.17 is a schematic of the pixel.

Event generation occurs as follows. Initially, the inverter input voltage, V_{in}, is high (after the reset pulse). Transistor Q2 is off and so is the feedback switch Q6. In addition, the inverter output voltage, V_{out}, is low. As the capacitor C is discharged by the photocurrent, V_{in} decreases and transistor Q2 begins conducting. Slightly before V_{in} reaches the threshold of Q2, a subthreshold current flows through the inverter and is fed back to the input through transistors Q4 and Q6. Notice that V_{out} starts to rise before the feedback circuit is activated, which subsequently switches Q6 on and starts the current feedback. The mirror pair Q4, Q5 is sized for current gain. The feedback current mirror operates in subthreshold initially but increases exponentially as V_{in} decreases further. We approximate the start of the switching process as the value of V_{in} where the fed-back current equals and surpasses the photocurrent. At this point, V_{in} accelerates toward the

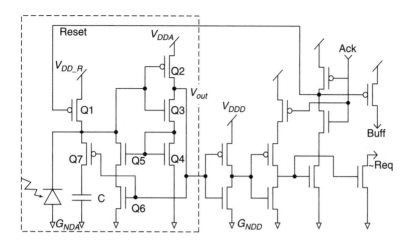

FIGURE 8.17 Pixel schematic of the digital image sensor. The circuit in the dashed area is the analog circuit portion of the pixel, responsible for generating events. The right side of the circuit is the digital interface to asynchronously communicate the event and prevent race conditions.

ground, V_{out} accelerates towards V_{dd}, and the switch transistor Q7 turns off, which disconnects the integration capacitor from V_{in} and causes V_{in} to accelerate further. Furthermore, as V_{in} plunges below the threshold voltage of Q3, it shuts off the feedback mirror, which cuts off the current in the Q2, Q4 branch and causes V_{out} to accelerate further toward V_{dd}. As can be seen, the transition takes place just before the threshold voltage of Q2 is reached, the capacitance at the V_{in} node is suddenly decreased, and Q3 and Q4 cut off for a low-current yet high-speed circuit. This circuit is unique in this respect (Culurciello et al., 2003; Culurciello and Etienne-Cummings, 2004).

Because the proposed imager measures the time to integrate photon-generated charges to a threshold voltage, the consistency of this threshold voltage, which is set by the event generator in each pixel, plays an important role in the image quality. From an analysis of the circuit in Fig. 8.17, we can say that the switching transition begins when the feedback current becomes comparable to the photocurrent. This definition is justified by the fact that at the switching point the input slew rate doubles because of feedback. As this happens, the positive feedback quickly switches the output. The input voltage at the start of switching, $V_{in,switch}$, is given by Eq. (8.14).

$$V_{in,switch} = \frac{nKT}{q} \ln \frac{i_{ph}}{\frac{W}{L}_{Q4} \frac{L}{W}_{Q5} \frac{W}{L}_{Q2} I_{Q2}} \qquad (8.14)$$

In Eq. (8.14), I_{in} is the input photocurrent, I_{Q2} is the weak inversion transistor Q2 current for zero bias. The subthreshold current through transistor Q4 causes the current feedback to start operating, and the inverter's output voltage also starts increasing. At the same time, transistor Q7 disconnects the integrating capacitors from the input of the inverter, thus reducing its load. The fast-increasing positive current feedback can then quickly drain the inverter's input capacitance. The magnitude of this positive feedback is at all times directly related to the current generated by Q2 and the gain of the feedback current mirror. Once the input of the inverter reaches the ground, the inverter current goes to zero and so does the feedback, because the NMOS transistor, Q3, turns the diode-connected transistor Q4 off. Thus, at the initial and final states there is no power supply current.

8.5.2 Photodiode Design and Characterization

The fabricated SOS digital imager employs a native PIN photodiode as photosensitive element. Spectral and temporal characterization of such structures have been reported previously (Abshire, 2001; Apsel et al., 2003) and more recently (Uehara et al., 2003). Refer to Sec. 7.4.3 for more information on the SOS PIN photodetector. Using ultrathin silicon photodiodes has advantages and disadvantages. Photon absorbtion in the ultrathin (100 nm) silicon layer is small, thus severely degrading the quantum efficiency to wavelengths other than blue and ultraviolet. This is in contrast to bulk CMOS photodetectors that are sensitive to red and infrared and have weak response to blue. Blue and ultraviolet sensitivity requires ultrashallow junctions that are hard to achieve in standard bulk CMOS processes. Using a PIN photodiode decreases junction capacitance and thus yields devices with bandwidths in excess of 5 GHz (Apsel et al., 2003).

The photodiode used in our pixel has a horizontal structure 100 nm thick (the thin-film Si layer in the SOS process) and 16.4 μm long, with a 1.2 μm intrinsic silicon layer between the anode and the cathode. The photocurrent is integrated on a 250-fF capacitor. In the proposed application of adaptive wavefront correction, the poor absorbtion in the PIN photodiodes is an advantage, as we are only employing the photosensor array in the feedback loop and the PIN photodiode is present just to sample the light from the laser beam.

When the PIN photodiode is reverse biased with the nominal supply voltage of 3.3 V, we measured high levels of dark current. The dark current was 17 pA in the diode mentioned above or approximately 1 pA/μm of diode length (refer to Chap. 7). This imposes a lower bound on the lower limit on the sensor's dynamic range. Assuming a conservative threshold of 0.5 V for triggering an event, in the dark a pixel takes 250 fF · 0.5 V / 17 pA = 7 ms to integrate enough light to produce an event. This is clearly a small time constant compared to other image sensors based on the same architecture as (Culurciello et al., 2003), and ultimately limits the quality of the images.

Integrating for longer period of time reduces noise in integrating pixels. A high level of dark current results in decreased image contrast. Events from the sensor arise with time constant of 27 μs (7 ms/256, where 256 is the number of pixels). This is the minimum event rate at the output of the image sensor. Ambient light generates a photocurrent of 30 to 40 pA; therefore the image sensor's SNR is only 2–3. A low-power (5 mW) laser pointer light generates 100 pA and an integration time of 1.25 ms. A high-intensity illumination lamp generates a maximum of 600 pA with an integration time of 0.2 ms. The problem with excess leakage currents in the the PIN photodiodes can be solved by redesigning the photosensitive structure, or using a lower-leakage PN junction as the ones reported in Sec. 7.4.1.

Figure 8.18 shows a plot of the photocurrent per unit length of the SOS PIN photodiode when illuminated from both front and back sides of the die. Light intensity was measured with a calibrated photometer and a variable high-intensity source. Note that light integrated from

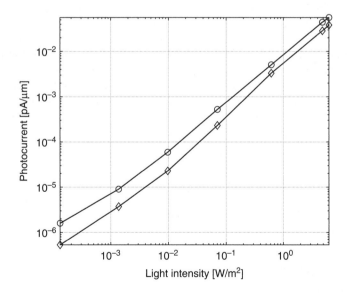

FIGURE 8.18 Difference between the front-illumination (diamond) and back-illumination (circle) photocurrent of the SOS PIN photodiode used in the digital image sensor. The input light wavelength was a broadband white halogen lamp with a median wavelength of 555 nm.

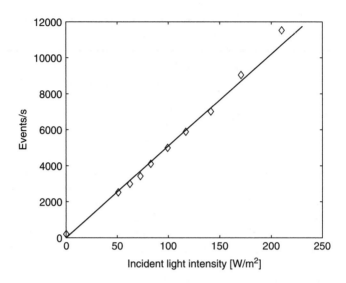

FIGURE 8.19 Event rate versus light intensity.

the back side generates higher photocurrents than the front side. This occurs because the front side of the die is covered by metal and SiO_2 layers that filter some of the incident light.

Figure 8.19 shows the event frequency versus incident light intensity at 555 nm for a single pixel in the array. For the data in this figure, the light was focused on a single pixel on the array using a lens, and the light intensity from the laser was varied using neutral density filters. The event frequency f_{ev} is linear with respect to the incident light intensity I_{in}. A relationship for f_{ev} is given in Eq. (8.15) where the empirical parameter L_s is equal to $51[Hz \cdot m^2/W]$.

$$f_{ev}[Hz] = L_s \cdot I_{in}[W/m^2] \tag{8.15}$$

The output event frequency spans approximately 2 orders of magnitude (200 to 11,500 Hz), and thus the array is capable of encoding data with 4 to 6 bits of precision, which is sufficient for the intended application.

8.5.3 System Architecture

The digital image sensor uses address-event (AE) representation output format. An AE communication channel is a model of the transmission of neural information in biological systems (Mead and Mahowald, 1988; Mead, 1989). Information is presented at the output in the form of a sequence of pulses, where the interevent interval or the event frequency encodes the analog value of the data being communicated. The AER model trades the complexity in wiring of the biological systems for the processing speed of integrated circuits. Neurons in the human brain make up to 10^5 connections with their neighbors, a prohibitive number for integrated circuits. Nevertheless, the latter are capable of handling communication cycles that are 6 orders of magnitude smaller than the interevent interval for a single neuron. Thus it is possible to share this speed advantage among many cells, and create a single communication channel to convey all the information between two neural populations. AER uses an asynchronous protocol for communication between different processing units (Boahen, 2000; Boahen, 2005).

As exemplified by Fig. 8.16, the information, divided into 'events', is sent from a unique sender to a unique element in a receiving population. Events are generally in the form of a pulse, therefore only their address is the important data to reconstruction and the time of occurrence. The information packet is therefore the address of the transmitter cell. In the case of our imager, events are individual pixels reaching a threshold voltage and requesting the bus for communication with a receiver. As a result, the system represents light intensity on a pixel as a frequency-modulated sequence of addresses where the time interval between identical addresses (pixels) is inversely proportional to the intensity. An AE system is generally composed of a multitude of basic cells or elements either transmitting, receiving, or transceiving data. Reconstruction of data necessitates storage, as events must be counted or accumulated to reassume the form of intensity signals.

Individual pixels integrate light on a local capacitor and when a threshold is reached, they request access to the output bus. After an event has been generated, an additional AER infrastructure in the pixel is required to communicate the event to the output bus by means of the boundary arbitration circuitry. A detail analysis and comparison of synchronous and asynchronous readout schemes can be found in (Culurciello and Andreou, 2003). The pixel circuit implementing the pulse density modulation is shown in Fig. 8.17. The pixel address (X, Y locations) appears at the output after arbitration (refer to the arbiter tree in Fig. 8.20) in the form of an event. Figure 8.17 shows a schematic caption of the pixel, where the right portion is the digital circuitry responsible for communicating the event to the outer array circuitry. This digital portion of the pixel generates a row request, Req. To provide robust noise immunity between the analog and digital portions of the pixel, the output of the event generator is buffered before passing it to a row-wise wired OR. The wired OR indicates that a pixel in that row has requested access to the output bus. The second inverter in the pixel digital buffer (see the digital portion of Fig. 8.17) has an additional PMOS transistor controlled by the returning acknowledge

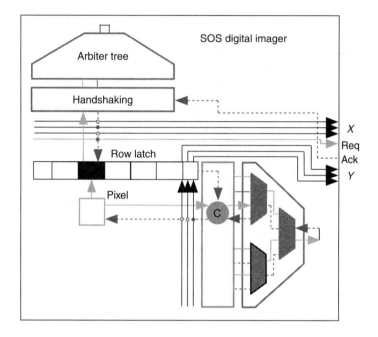

FIGURE 8.20 SOS digital image sensor output communication circuits. Individual pixel communicate events to row and column arbiter trees. These digital circuits resolve contention while multiple pixels access the bus and also pipeline multiple requests. The output of the sensor is the address of each transmitting pixel, with an asynchronous interface as reported in Fig. 8.16.

Ack signal. The additional transistor blocks any other request that might arise if the Ack signal has not been previously reset (i.e., a communication cycle has been completed). Analogously, an additional NMOS in the Ack signal path prevents racing conditions by only acknowledging a pixel whose request has been allowed to reach the boundary circuits. Hence, a handshaking protocol is initiated by the pixel that requests the output bus provided it has previously been acknowledged; also, the pixel acknowledges provided it has previously issued a request and gained access to the bus. This forms a four-phase handshaking sequence, which is also repeated at the row and column level, and as exemplified in Fig. 8.15.

The boundary circuits are used to arbitrate between active pixels (i.e., pixels that have generated events). This arbitration is executed in two steps (Culurciello et al., 2003). First a row arbitration tree selects one row from which at least one request has been generated. Next, the column arbitration tree selects and outputs the individual pixels within the row. When a row is selected, the entire row is copied into a buffer located above the array (Row Latch). This buffering step provides a pixel access speedup and improved parallelism by realizing a pipelined readout scheme. Simultaneously, the address of the row is also decoded and placed on the output bus Y. When a row request, (i.e., the wired OR signal) is asserted, many active pixels may exist within the row. The Buff signal indicates which pixel in the row has issued a request. Once copied, the entire row is acknowledged/reset (signal Ack), and photon integration starts anew. Column arbitration is performed on the buffered row. The arbitration tree selects the active elements in the buffer, computes and outputs their X addresses before clearing the buffer. A new active row is obtained when the buffer is clear. Performing column arbitration on the buffered row also improves readout speed by eliminating the large capacitance associated with the column lines. This capacitance is encountered when arbitration is performed within the whole array.

The imager power consumption can be reduced even further by using more elaborated circuits that eliminate the wired OR Req and Buff lines. The AE architecture employs pseudo-CMOS logic, which can be substituted with fully static or dynamic logic for larger power savings. On the other hand, the use of pseudo-CMOS logic greatly simplified the design of the large number of inputs OR gate required per each row and column.

Reconstruction of the images can be performed by counting the number of events in a given window of time (*histogram imaging*), or by computing the interevent time between successive events from the same pixel (*interevent imaging*) (Culurciello et al., 2003). To obtain a pixel intensity image, the interevent interval must be converted into light intensity. The photocurrent is inversely proportional to the interevent interval or directly proportional to the event frequency. To perform these transformations, each event is time indexed relative to a global clock and the time between successive events computed (instantaneous interevent interval or interevent imaging), or the number of events over a fixed interval is to be counted (average interevent interval or histogram imaging). In either case, the AER data must be stored or accumulated in a memory array. This can be in the form of an analog storage (capacitive storage, for example) or in the digital domain. A workstation computer was used to accumulate events and generate the images presented here. An interface program was responsible for collecting up to 1-M events, and then reconstructing an image histogram in memory. Real-time medium-quality images can be displayed every 1 to 5 K events. We also associated each event with a time index to analyze the temporal characteristics of the imager. The timing circuitry was a programmable Altera FPGA acting as a 28-bit counter.

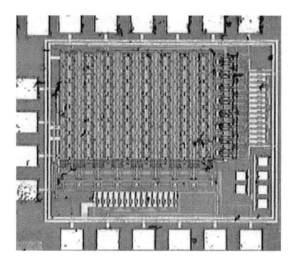

FIGURE 8.21 Die micrograph for the 16 × 16 sensor array. The array of identical pixel cells is clearly visible inside the die. The circuits at the bottom, and right side of the die are respectively, the column and row digital arbiter circuits. The output of this image sensor is a digital word of 5 + 5 bits indicating the address of the transmitting pixel.

The main drawback of this approach is the complexity of the digital frame grabber required to count all the events produced by the array. A high-resolution timer (up to 24 bits for hundreds of picosecond resolution) and a large frame buffer are required [up to 15 MB for a full VGA array (640 × 480 pixels)] would be required to obtain an instantaneous image for every event. The timer indexes each event and compares it with the last time an event at that pixel was recorded. The difference is inversely proportional to the light intensity. The buffer must hold the latest pixel time index and the intensity value. An analysis of optimal receiver design for similar image sensors is discussed in the paper by (Apsel and Andreou, 2000).

8.5.4 SOS digital image sensor testing and characterization

The die area for the sensor array is $1.2 \times 1.2 \text{ mm}^2$ with pads. The pixel size is $29.6 \times 42 \text{ } \mu\text{m}$. A micrograph of the die is shown in Fig. 8.21. The SOS wafer was polished only on the front side after fabrication. To obtain a clear die on both sides, we have polished the back side of the die using a mechanical lapping machine. Lapping was performed up to a surface roughness of $1 \text{ } \mu\text{m}$. The mechanical polishing resulted in

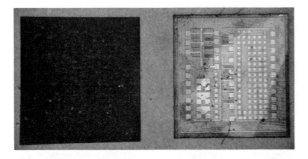

FIGURE 8.22   An SOS digital image sensor test die before (left die) and after (right die) mechanical polishing. Notice as the backside was originally opaque, and it becomes transparent after polishing.

an optically clear die on both sides. Alternatively, index matching fluid can be employed to fill in the asperities of the backside as in (Andreou et al., 2001). The effectiveness of the mechanical polishing is evident in Fig. 8.22.

Sample image sequences given from the sensor are given in Figs. 8.23 and 8.24. Figure 8.23 is obtained by illuminating on the frontside, Fig. 8.24 on the backside of the die (BSI). In both figures, a moving spot is sampled four times and the output event rate was 0.48 Mevents/s. These images were reconstructed by creating a normalized histogram after collecting 10,000 events. The noise present in

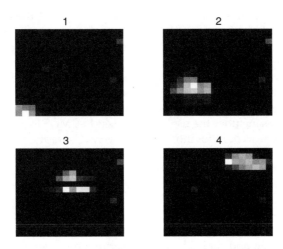

FIGURE 8.23   Output from a light spot moving up-right impinging on the front side of the sensor array. The poor quality of these images is due to the large dark current of the PIN photodiode employed in this sensor array.

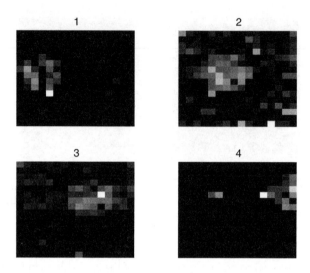

FIGURE 8.24 Output from a light spot moving upright impinging on the backside of the sensor array. The poor quality of these images is due to the large dark current of the PIN photodiode employed in this sensor array.

these images is due to errors in the collections of addresses and to the limited number of pixels. The output address bus was not properly buffered in this first version of the image sensor and thus was difficult to read reliably.

Power consumption of the analog array is 0.60 μW. Because there are 256 pixels in the array, this amounts to a consumption of about 2.32 nW per pixel. Low-power operation is a consequence of using the pulse frequency modulation pixel (Culurciello et al., 2003) as well as optimized circuit design employing MOS transistors with three different thresholds. The energy consumption for each event is on the order of 4.83 fJ. Digital power consumption was measured to be 1.1 mW. This includes the use of three pseudo-CMOS logic signals in the array that account for most of the power consumption. In the next version of the chip, the wired OR will be replaced with a tree-based fully CMOS design, thus dramatically reducing the power dissipation. The circuit design both for the pixel and periphery uses the available zero threshold transistors (Andreou et al., 2001) to reduce the complexity, minimizing the number of transistors and wires necessary for bias circuits.

Both analog and digital power consumption measurements were conducted using a supply voltage of 3.3 V and with an output event rate of 0.48 Mevents/s. The average amount of energy spent by the array to communicate an event to an external receiver is 2.3 μJ. This

Technology	0.5-μm SOS
Array size	16 (H) × 16 (V)
Total size	1.2 mm × 1.2 mm
Pixel size	29 μm × 42 μm
Fill factor	17%
Power consumption	1.1 mW (@ 3.3 V)
Pixel energy consumption	4.8 fJ/event
SNR at ambient light	2–3 limited by dark current
Dark current	17 pA
Max data rate	0.48 Mevents/s

TABLE 8.3   SOS digital pixel sensor array properties

figure takes into account the switching of nine pads (4 + 4 addresses and a request signal) and the power consumption of the pad drivers, that in this technology is a 40-pJ/event (from simulation data). The summary of the sensor performance is given in Table 8.3.

8.6 Summary

SOS technology has some advantages over a standard CMOS technology: reduced short-channel effects, reduced parasitic capacitances, reduced body effects, reduced latchup, reduced leakage currents due to lower area parasitic junctions, and ultra-low-noise figure. These advantages are very useful for the design of image sensors in SOS. In particular, there are a few features that have been mentioned in the chapter worth repeating: (1) the possibility of backside illumination due to the transparent substrate property of the SOS technology, (2) the implementation of PMOS transistors in SOS does not require an N-well, allowing their use as reset transistors in APS pixel, (3) the floating body of SOS NMOS transistor allows implementation of APS using a high-gain lateral bipolar transistor (LBT) as a sensing element, and (4) six types of MOSFET transistors, having multiple threshold voltages are available in SOS. This allows a more efficient APS design, widening the dynamic range of the sensor. (5) SOS image sensors are also radiation tolerant and can be used in extreme space environments.

SOS has the disadvantage of reporting lower quantum efficiency in the visible spectrum because of the thin silicon film thickness, making the implementation of a sensing element more difficult than in standard CMOS technology.

We have presented the first fully-working back-illuminated image sensor in an SOS 0.5-μm process that is also capable of detecting UV light. The image sensor has a resolution of 32 × 32 pixels. Each 40 μm × 40 μm pixel contains a photodiode or phototransitor and memory capacitor. Pixel reset, integration, and output occur in full-frame "snapshot" fashion. The image sensor is able to achieve continuous 1000 frames/s operation and can consume as low as 650 μW of power. A better operational amplifier and ADC will greatly improve both the analog readout and digital conversion speeds.

We have also presented a 16 × 16 pixel SOS-CMOS APS array with a digital interface. We reported on the design, fabrication, and testing results of the individual components of the array and demonstrate functionality for both front- and backside imaging. This sensor can be used as a secondary light sensor to image light in the same focal plane of the main sensor. Using a single focal plane saves additional beam splitter and optical elements that are generally used to monitor the main focal plane.

CHAPTER 9

SOS Biosensor Interfaces

9.1 Introduction

The advancements in medicine of the last century, and the resulting benefits in disease diagnosis and patient recovery, has been made possible by the availability of electrical biosensor interfaces. Biosensors have been used to detect a variety of endogenous and exogenous signals, as chemical concentrations, ionic currents, and field potentials. In electrical biosensor interfaces, biological signals are converted into electrical signals in the form of either currents or voltages. We therefore refer here to current-mode and voltage-mode biosensor interfaces, referring to the input electrical signal measured by the biosensor. Electrical input noise is an important parameter in the design of biosensors because it influences the signal-to-noise ratio (SNR) achievable with the device. During the design of a biosensor interface it is important to compute the input-referred electrical noise in order to quantify the minimum detectable signal and also compute the SNR. The SNR of electrical biosensors is affected by the source noise, fundamental electrical noise sources, and by the desired input bandwidth. Biosignals generally have a bandwidth of a few kilo-Hertz. As an example, neural potentials are generally monitored at a bandwidth of 10 to 20 kHz (Wise et al., 1970; Wise 1984; Harrison and Charles, 2003; Wise et al., 2004; Harrison et al., 2007; Weerakoon et al., 2008b; Weerakoon et al., 2008c) and patch-clamp ionic currents are monitored at 1 to 10 kHz (Sigworth and Klemic, 2005; Laiwalla et al., 2006a; Weerakoon and Culurciello, 2007a; Weerakoon et al., 2008a).

In addition, when venturing in the scale of nano and submicron devices and materials, technological progress is strictly related to the availability and sensitivity of the instrumentation used to characterize the performance of the fabricated structures. Typically, the instrumentation required measures small currents and voltages, generally on

265

the order of femto- to nanoamperes, and microvolts. Noise and performance of the measurement systems are affected by the relative size of the measurand and the measuring system. In particular, nano- or microscale devices operation should not be affected by the undesired loading of the measurement system. Large measurement systems provide versatile performance with high sensitivities at the expense of bandwidth (Wise et al., 2004; Weerakoon et al., 2008b; Weerakoon et al., 2008c). In order to obtain measurements with both high sensitivity and large bandwidth, measurement system dimensions should be comparable to the sample. Integrated circuits become attractive as instrumentation for biomedical and physical small-scale devices, both for performance and reduced size, but also for the low added electronic noise and undesired capacitive loading, a common problem deriving from the cables of large discrete components systems.

This chapter presents an overview of the current state-of-the-art SOS biosensor interfaces. We present both current-mode biosensor interfaces (generally referred as potentiostats) and voltage-mode biosensor interfaces implemented in the SOS process. We present numerical techniques to compute theoretical noise performance and also present measurement systems and measurements on actual noise data for both voltage-mode and current-mode biosensors implemented in SOS technology.

9.2 Noise and Sensing Limits

Low-amplitude voltage and current measurements are often complicated by the presence of electronic noise. Electronic noise can be *fundamental* or *man made*. Fundamental noise sources derive from the physical laws and the property of condensed matter. Man-made noise is induced by circuit design, layout, topological placement, and system architecture.

There are also fundamental limits on the SNR that can be attained with biosensor and electrical interfaces measuring small currents and voltages. In this section, we review these limitations so that designers can quickly assess the feasibility of measuring systems. The SNR physical limitations also allow us to compare state-of-the-art biosensors. Man-made noise source can be avoided with experienced and robust engineering and can ultimately approach the physical limits.

We begin by examining the fundamental limits of noise due to source noise and electrode or biosensor physical interfaces. It is good practice to first calculate the total contribution of the input noise sources and then design electronic interfaces that contribute by adding only a fraction of those input noise. This design procedure reduces the requirements of electronics so that noise performance can be traded for reduced power consumption, reduce circuit size, and larger recording bandwidths.

9.2.1 Low-Voltage Measurements Limits

Voltage measurements are limited by the source impedance of the biosignal. The source impedance is the sum of the electrode impedance and the tissue impedance, generally on the order of tens of $k\Omega$ to a few $M\Omega$ (Wise et al., 2004). The electrode and tissue resistance generates thermal noise. Its noise power $V_{n,in}^2$ is given by Eq. (9.1), where k is the Boltzmann constant, T the temperature, and R_{el} the electrode resistance.

$$\bar{V}^2{}_{n,in} = 4kT R_{el}df \qquad (9.1)$$

Equation (9.2) computes the SNR, where V_{in}^2 is the input signal power and df is the input signal bandwidth.

$$\text{SNR}_v = \frac{V_{in}^2}{4kT R_{el}df} \qquad (9.2)$$

By plotting the voltage SNR function as a function of the input voltage amplitude and the electrode resistance, we obtain the data reported in Fig. 9.1.

Figure 9.1 is indispensable to designers, as it shows the maximum attainable SNR as a function of signal intensity and also electrode resistance.

As an example, in typical neural recording biosensor interfaces the signal bandwidth is 10 kHz (Wise et al., 2004) and signal amplitudes are on the order of 100 μV RMS. The figure shows that a typical neural signal of 100 μV RMS can be recorded with 6 bits of precision when the electrode impedance is 1 $M\Omega$. This electrode reports a noise of 12.8 μV RMS, and sets the limits for any recording amplifier. As a result of the above calculations and plots, we can see that the electronic interface noise performance only need to slightly better than the electrode-source limitations, so its added noise is negligible. A neural recording amplifier using this electrode only needs to be designed with an input-referred voltage noise of approximately 1 to 10 μV RMS. The state of the art is currently a few μV RMS (Harrison and Charles, 2003). If better interface electrodes to the nervous system are available, the SNR will increase as the electrode impedance decreases. For, example a 10-KΩ has a voltage noise of only 1.28 μV RMS.

9.2.2 Low-Current Measurements Limits

The sensitivity of silicon-based electronic current measuring systems is limited by the discrete nature of the charge and the shot noise. Shot

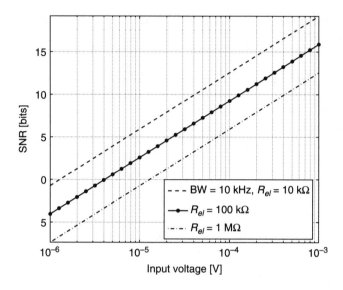

FIGURE 9.1 SNR for low-voltage measurements as a function of source resistance and input voltage amplitude. The measurement bandwidth is considered to be 10 kHz. This is the typical SNR obtainable in neural recording systems for measuring cellular field potentials. In these systems, the electrode impedance is on the order of 100 kΩ to 1 MΩ, and the input voltage amplitudes are on the order of 100 μV RMS.

noise can be calculated with Eq. (9.3), where q is the elementary charge, I_{in} is the input biocurrent, and df is the measurement bandwidth.

$$\bar{I}^2_{n,in} = 2q\, I_{in} df \tag{9.3}$$

Current noise is influenced by the value of the input current and the desired bandwidth. As an example, the current noise component of a 1-pA signal measured at 10 kHz is 56.6 fA RMS. The noise component becomes 176 fA RMS at 100 kHz. A 1 fA current signal measured at 10 kHz results in a noise of 1.79 fA RMS. Notice that this limitation is only present in specific materials such as silicon. In special kinds of materials that exhibit liner macroscopic resistance, noise limits are lower than the value predicted by shot noise limitations (Gomila et al., 2004).

The trade-off between the amplitude of the input current and the bandwidth impose a trade-off on the attainable SNR. The value of the SNR translates into an equivalent and effective number of bits that

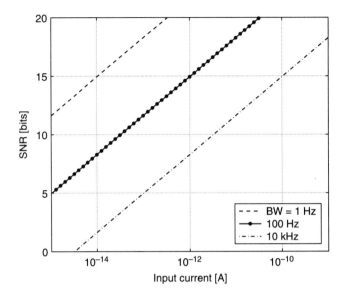

FIGURE 9.2   SNR for low-current measurements as a function of the input current amplitude and the recording bandwidth.

each recording sample can provide. The SNR can be computed with Eq. (9.4).

$$\text{SNR}_i = \frac{I_{in}}{2q\,d\,f} \tag{9.4}$$

The SNR of Eq. (9.4) is plotted in Fig. 9.2 as a function of bandwidth and input current amplitude. This SNR is the upper limit on low-current measurements that can be obtained by an electrical system. From the figure we can see that the fundamental limits of current measurements allow us to obtain 8 bits of resolution at 10-kHz bandwidth for an input current of 1 pA.

Low-current measurements can be performed with passive components in shunt or feedback configuration with respect to an active amplifier. For practical reasons, low-current measurement are conducted with capacitors or resistors as passive sensing elements.

Resistive-feedback circuits are more noisy than capacitive feedback ones. A large feedback resistor of several tens of megaohms is needed to turn a nanoampere of currents into a few tens of millivolts outputs. Feedback resistors also limit the sensor's bandwidth. Resistive-feedback is adequate for biosensor measuring nano-amperes of current at 10-kHz bandwidth, and are used in patch–clamp biosensors for

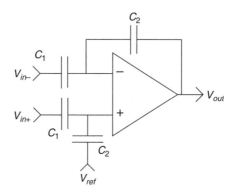

FIGURE 9.3 Schematic of an AC-coupled voltage amplifier implemented in SOS. This amplifier configuration is typical in continuous-time voltage recording for field potential biosignals. The amplifier gain is given by $A_v = C_1/C_2$.

whole-cell recording (Weerakoon and Culurciello, 2007a; Weerakoon et al., 2008b; Weerakoon et al., 2008c).

9.2.3 Voltage-Mode Biosensor Interfaces and Noise Performance

In this section, we compute the electrical noise contribution of a SOS biosensor interface for voltage measurements. The most typical voltage-mode interface is an AC-coupled operational amplifier head-stage as seen in Fig. 9.3. The operational amplifier in Fig. 9.3 is the one seen in Sec. 4.1.

Voltage-mode biosensors detect small input field potentials. Neural field potentials have signals with a peak amplitude of 100 μV up to 1 mV. These signal are generally amplified approximately 100 to 1000 times before being digitized and further processed. This corresponds to a choice of $C_1 = 100 - 1000 \cdot C_2$ in Fig. 9.3. These biosensors require input-referred noise levels of 1 to 10-μV RMS (Harrison and Charles, 2003; Wise et al., 2004).

The circuit model used to compute the input-referred voltage noise of the biosensor interface of Fig. 9.3 has also been reported in Secs. 3.6 and 4.1. The model for noise calculation in Fig. 9.3 is identical to the noise of the voltage follower seen in Sec. 4.1. In both configurations, in fact, the input differential pair of the operational amplifier is the dominant additive electronic noise source. The flicker noise due to the trapping of charges below the gate of the transistor is a significant problem in making low-noise SOS voltage-measuring systems using

MOSFETs. In bulk CMOS processes, this problem is alleviated using P-channel transistors which have been shown to have one or two orders lower K_f than N-channel devices (Harrison and Charles, 2003) in bulk CMOS processes. On the other hand, in SOS process, there is no significant difference between NMOS and PMOS noise performance, as reported in Secs. 3.6 and 4.1.

A typical voltage noise power spectral density plot for the amplifier described in Fig. 9.3 is shown in Sec. 4.1. At low frequencies the contribution of flicker noise is visible in a declining linear slope. At higher frequencies, the thermal noise of the circuit prevails over the flicker noise and the power spectral density flattens (see Fig. 4.8).

9.2.4 Current-Mode Biosensor Interfaces and Noise Performance

In this section we compute the electrical noise contribution of an SOS biosensor interface for current measurements. Low-current measurements can be performed with passive components in a shunt or feedback configuration with respect to an active amplifier. For practical reasons, low-current measurement can be conducted with capacitors or resistors as passive sensing elements. Inductors can also be used for large AC currents but are not generally employed in biosensor interfaces for their size and the small currents involved. Table 9.1 summarizes the advantages and disadvantages of each measuring technique. For nonintegrated high-precision measurement resistive feedback circuits are preferred. This is because of the high-offset voltage errors of shunt systems and the high noise introduced by the shunt resistor. The use of resistors is generally not viable in CMOS integrated circuits due to the high resistance value necessary for low-current measurements, generally on the order of 1 to 10 GΩ (in the picoampere range). The layout of a multi-gigohm resistor requires a very large silicon area, due to the relatively low resistance layers of common CMOS processes. Capacitive shunt and feedback measurement systems are very compact and appropriate for low-current measurements, but the capacitor

Type	Shunt	Feedback
Resistive	Low sensitivity and high offset voltage error— small resistors required	High sensitivity and low offset voltage error— large resistors required
Capacitive	High sensitivity—easy integration	High sensitivity—easy integration

TABLE 9.1 Shunt and feedback CMOS circuits

	Current Conveyor	Feedback Integrator	Follower Integrator
Type	Feedback	Feedback	Shunt
Bidirectional currents	Yes	Yes	No
Current amplification	Yes	No	No
Voltage clamp	Yes	Yes	No

TABLE 9.2 CMOS circuits for low-current measurements

size is proportional to the input current range and will be large if the current is above milliampere levels.

An ideal low-current measurement system provides the following features:

- measure low current with the maximum bandwidth and sensitivity available
- allow measurement of bidirectional currents (sinking/sourcing)
- allow control of the voltage at the input node (voltage clamp)

Measurement circuitry will always introduce additive noise in the system, and not all the circuit topologies allow bidirectional current flow. In the following sections we address three different kinds of integrated low-current measurement systems: current conveyor, capacitive feedback integrator, and follower integrator. Table 9.2 summarizes the properties of these three circuit topologies.

9.2.4.1 Resistive Feedback Current-Measuring Headstage

The most typical continuous-time current-mode interface is a resistive feedback transimpedance amplifier based on a operational amplifier headstage. The transimpedance amplifier is shown in Fig. 9.4 together with its noise sources, and it is based on the operational amplifier design reported in Sec. 4.1.

The operational amplifier input configuration allows us to concomitantly record input currents and also clamp the input voltage. Input is provided through the negative operational amplifier terminal, and the voltage clam is provided by the virtual short between the two input terminals of the operational amplifier. The resistive feedback transimpedance output can measure the input current by means of the relation in Eq. (9.5).

$$V_{out} = I_{in} \cdot R_f \qquad (9.5)$$

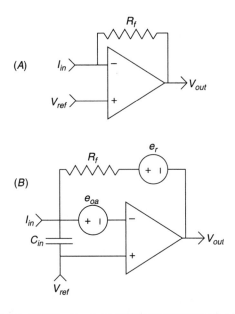

FIGURE 9.4 Schematic of a current measuring trans impedance amplifier with explicit noise sources. (*A*) Typical transimpedance stage with output $V_{out} = I_{in} \cdot R_f$. (*B*) The noise source e_r is the voltage noise from the large feedback resistor, and e_{oa} is the input-referred voltage noise of the operational amplifier. C_{in} is the total input capacitance. This schematic is a model for the calculation of the input-referred current noise in current-mode biosensors based on operational amplifiers.

The input-referred current noise of the the current measuring transimpedance amplifier is given by Eq. (9.6). Notice the the input shot noise contribution (first term) is generally much smaller than the added noise, so it is usually not included in the equations. The circuit model used to compute the input-referred current noise of the biosensor is reported in Fig. 9.4. The noise source e_r is the voltage noise from the large feedback resistor, and e_{oa} is the input-referred voltage noise of the operational amplifier. C_{in} is the total input capacitance.

$$S_I = \frac{S_r}{R_f^2} = 2q\,I_{in} + \frac{e_r^2}{R_f^2} + \frac{e_{oa}^2}{R_f^2} + \frac{e_{oa}^2}{X_C^2} \left[\frac{A^2}{Hz} \right] \tag{9.6}$$

The first term is the shot noise of the input current, and the second term is the current noise of the feedback resistor R_f. The third term is the operational amplifier voltage noise reflecting on the feedback resistor R_f. Notice that the operational amplifier voltage noise is a simplification of the equation given in Sec. 4.1. The fourth term is the operational amplifier voltage noise reflected on the input capacitance. Equation (9.6) can be turned into Eq. (9.7) by expressing each of the terms as a function of the design parameters. In Eq. (9.7), K is the Boltzmann constant, T is the absolute temperature, g_m is the transconductance of the operational amplifier's input transistors, K_f is the process-dependent flicker noise parameter of these transistors, C_{ox} is the oxide capacitance at the gate of the input transistors.

$$S_I = 2q\,I_{in} + \frac{4KT}{R_f} + \left[\frac{8KT}{3g_m} + \frac{K_f g_m^2}{C_{OX}L^2 f}\right]\frac{1}{R_f^2}$$

$$+ \left[\frac{8KT}{3g_m} + \frac{K_f g_m^2}{C_{OX}L^2 f}\right]4\pi^2 C_{in}^2 f^2 \qquad (9.7)$$

The critical parameters for low-noise current amplifiers design are input capacitance, voltage noise, and input leakage current. Input transistors are vital in establishing these characteristics. The gate capacitance C_g of the input MOSFET is proportional to the area of the transistor, whereas the thermal noise e_n decreases as the square root of the area (assuming constant gate length). The noise of the recording system is proportional to $C_{in}e_n$ where the total input capacitance is $C_{in} = C_g + C_{el}$ with C_{el} being the capacitance of the electrode and any other capacitance on the input node. By differentiating Eq. (9.6) with respect to C_g, the optimal gate can be shown to be the gate width corresponding to $C_g = C_{el}$. The transimpedance amplifier multiplies the input current by R_f to produce a voltage proportional to the input. The current noise at the input can be shown to be inversely proportional to R_f. However, the parasitic capacitance and the physical size of R_f in layout limits its maximum usable value. It can be demonstrated that a small capacitance of just a few hundred femtofarads can create undesirable oscillations in the headstage (Weerakoon et al., 2008c). A typical plot of the input-referred current noise from an headstage as the one shown in Fig. 9.4 is shown in Fig. 9.37.

9.2.4.2 Linear Integrator Current-Measuring Headstage

A current integrator circuit can be used in low-current integrated measurement systems with large bandwidth. The integrator is a high-gain amplifier with a shunt-integrating capacitor C_f between its input and output terminals. The capacitor voltage is proportional to the integration time and the input current. The integrator is usually followed by a comparator that stops the integration once the voltage across C_f reaches a fixed value. This circuit can perform a sigma–delta conversion of the input current with very little additional circuitry, as we report in Sec. 9.3.1.

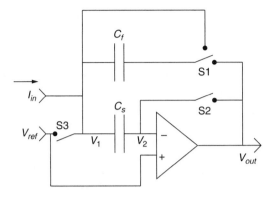

FIGURE 9.5 Linear integrator current-measuring circuit. An operational amplifier with capacitive feedback C_f integrates the input current i_{in}. Offsets and flicker noise of the operational amplifier can be reduced by using a switched-capacitor technique and sampling the input offset and flicker noise and records it on a capacitor C_s.

Capacitive low-current measurement circuits operate by integrating the input current on a small capacitor. A typical small-size integrated capacitor of 100 fF and an input current of 1 pA result in a maximum readout rate of 10 Hz, as a result of Eq. (9.8).

$$f_{int} = \frac{1}{T_{int}} = \frac{I_{in}}{C_f \cdot V_{int}} \tag{9.8}$$

Figure 9.5 shows an implementation of the integrator. Capacitor C_f is the integration capacitor of the input current i_{in}. Offsets and flicker noise of the operational amplifier can be reduced by using a switched-capacitor technique historically called *autozero*. This technique samples the input offset and flicker noise and records it on a capacitor C_s. The capacitor is then used in the signal path to subtract the unwanted quantities. To reset the stage, in step 1, switch S2 and S3 are closed and S1 shorts the capacitor C_f. Both terminals of capacitor C_s are set at voltage V_{ref}. During current integration, in step 2, S1 connects to the output of the operational amplifier; S2 and S3 are open.

Details on an implementation with cascoded inverters instead of an operational amplifier will be provided in Sec. 9.3.1. This switched capacitor implementation of the integrator circuit is very useful for sigma-delta conversion of the current. Conversions of 12 bits can be obtained at sampling frequencies of 10 KHz or more on picoamperes of current (Murari et al., 2004; Laiwalla et al., 2006a; Stanacevic et al.,

2007). Notice also that this circuit exhibits the best noise performance for integrating circuits because the flicker noise is reduced by the autozero technique.

9.2.4.3 Follower Integrator Current-Measuring Headstage

Another circuit that can be used to monitor small currents is the follower integrator current-to-voltage converter. This converter, typically used in active pixel sensor (APS) CMOS cameras (Culurciello et al., 2003), is able to measure picoamperes of current when the source is connected to ground and when there is no necessity of controlling the input voltage. Figure 9.6 is a circuit schematic of the follower integrator current measuring headstage. Node A is initially reset to the the supply voltage, and its parasitic (or explicit) capacitance is discharged by the input current source. The voltage at node A will decrease linearly in time and is buffered by a voltage follower circuit. The output voltage can then be read in analog mode or can be converted to digital using pulsewidth modulation (PWD) converter or a conventional ADC.

The disadvantage of using the current integrating sensor of Fig. 9.6 is that one node of the nanodevice must be connected to ground.

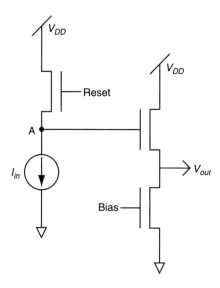

FIGURE 9.6 Follower integrator current measuring headstage. Node A is initially reset to the the supply voltage, and its capacitance is discharged by the input current source. The voltage at node A will decrease linearly in time and it is buffered by a voltage follower circuit to provide an output voltage.

On the other hand the noise level of this circuits can be calculated with Eq. (9.9).

$$\overline{V_n^2} = \overline{V_n^2(t_{reset})} + \overline{V_n^2(t_{int})} = \frac{1}{2}\frac{kT}{C_{int}} + q\frac{I_{in} + I_{dc}}{C_{int}^2}t_{int} \qquad (9.9)$$

In Eq. (9.9), I_{in} is the input current to be measured, I_{dc} is the DC current component (or leakage), t_{int} is the current integration time, and C_int is the integrating capacitance of node A in Fig. 9.6.

The first term corresponds to reset noise power due to the reset phase due to thermal noise. This term is significantly less than the value of the kT/C usually reported in the literature, due to a nonadiabatic charge transfer during the reset phase. The second term refers to integration noise due to undesired current I_{dc} and input current I_{in} with the reset transistor turned off. Note that for this circuit, flicker noise can be reduced by using correlated double sampling. This technique is ubiquitous in APS-based image sensors.

9.2.4.4 Current Conveyor Current-Measuring Headstage

A current conveyor circuit is a suitable circuit element for amplifying low currents and apply voltage biases for biosensor interfaces. The current conveyor circuit performs decoupling and linear operations in current mode in the same way that the operational amplifier performs in voltage mode. The conveyor can decouple the input current and give current amplification, allowing to relax both noise and performance specification for the measuring stages. The current conveyor can be used as a preamplifier coupled with the integrative or continuous headstages presented in the previous sections.

A schematic of the current conveyor is given in Fig. 9.7. For simplicity all devices have the same size. If a potential is applied to terminal X, the same potential appears on terminal Y. The current flowing into terminal X will be conveyed into terminal Z with high output impedance. The potential at X is independent of the current flowing and any current flowing into X will flow into Y as well. By inspecting Fig. 9.7, one can easily determine that all the above condition are satisfied by nature of the input operational amplifier and the current mirrors. The circuit will be operated with X as the input port and Z as the output port. Z will act as a current source mirroring the value of the current into X. The output current from Z can be then measured by means of the integrative or continuous headstages presented in the previous sections. Notice that the input current can be amplifier by this stage by making the output transistors (leftmost transistors in Fig. 9.7) M times the size of the other transistors. Multiple current mirror stages can also be cascaded to obtain large amplifications of the input current.

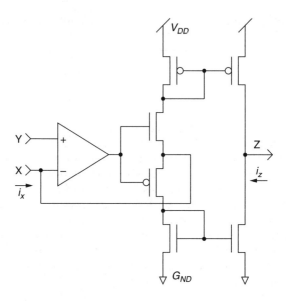

FIGURE 9.7 Current conveyor current-measuring headstage. This circuit can provide amplification of an input current, while at the same time providing input voltage bias. X and Y are the inputs, and Z is the output. The input current can be provided at the X terminal, while Y can be used as a bias voltage for X.

The noise performance of the current conveyor circuit will determine a lower bound on the measurable current levels. The total input referred noise S_I at the input node X is given by the sum of the operational amplifier input-referred noise and the thermal noise of the four input transistor of the conveyor. This assumes that the first stage amplifier the current, so that subsequent stages do not need to be accounted for in the noise calculations. The input-referred current noise of the the current conveyor is given by Eq. (9.10). The noise source e_{cc} is the voltage noise from the current conveyor four transistors input branch, and e_{oa} is the input-referred voltage noise of the operational amplifier, as seen in Secs. 3.6 and 4.1. X_C is the total input shunt impedance.

$$S_I = 2q\,I_x + \frac{e_{cc}^2 + e_{oa}^2}{X_C^2}\left[\frac{A^2}{Hz}\right] \qquad (9.10)$$

The first term in Eq. (9.10) is the input current shot noise contribution. The other term of the equation is due to the current conveyor input branch and the operational amplifier reflecting on the input impedance (capacitance) at node X.

$$S_I = 2q\,I_x + \left[4 \cdot \frac{8KT}{3g_{m,cc}} + \frac{8KT}{3g_{m,oa}} + \frac{K_f g_{m,oa}^2}{C_{OX} L^2 f} \right] 4\pi^2 C_{in}^2 f^2 \quad (9.11)$$

Equation (9.10) can be turned into Eq. (9.11) by expressing each of the terms as a function of the design parameters. In Eq. (9.11), K is the Boltzmann constant, T is the absolute temperature, $g_{m,cc}$, and $g_{m,oa}$ are, respectively, the transconductance of the current conveyor branch and the input transistors of the operational amplifier. K_f is the process-dependent flicker noise parameter of these transistors, C_{ox} is the oxide capacitance at the gate of the operational amplifier input transistors, I_X is the input current to be measured, K_f is the flicker noise coefficient and, and finally C_{in} is the total input capacitance at node X. In Eq. (9.11), the first term is the shot noise of the input, generally the smallest term. The second term is the operational amplifier noise and also the conveyor first branch of four transistor contributing to thermal noise because of their channel conductance $g_{m,cc}$.

Figure 9.8 shows a current conveyor implementation that does not require the use of an input operational amplifier while providing the same benefits and operation of the circuit in Fig. 9.7. X and Y are the inputs, and Z is the output. The input current can be provided at the X terminal, while Y can be used as a bias voltage for X. The output current can be input to any of the current-measurement circuit we showed in the previous sections. In this case, because of the lack of the operational amplifier, the total input referred current noise is given by Eq. (9.12).

$$S_I = 2q\,I_x + \left[4 \cdot \frac{8KT}{3g_{m,cc}} \right] 4\pi^2 C_{in}^2 f^2 \quad (9.12)$$

Figure 9.9 shows the test results from a fabricated SOS CMOS current conveyor like the one in Fig. 9.8.

9.2.5 Measurement Setup

We designed and implemented a low-noise amplifier to be able to record the electrical noise in our fabricated biosensor interfaces. The noise amplification circuit is shown in Fig. 9.10.

Table 9.3 reports the type and values of each component used in the design of the noise measurement circuit in Fig. 9.10.

The first stage of amplification is performed by means of a ultralow noise operational amplifier Linear Technologies model LT1128. This amplifier has a gain–bandwidth product of 13 MHz. In this first stage, the input signal was amplified 300 times, allowing us to visualize

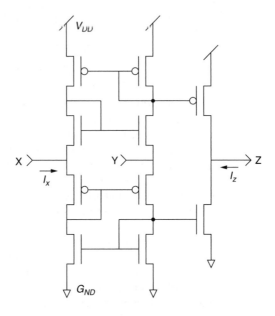

FIGURE 9.8 SOS CMOS current conveyor circuit without operational amplifier input. This circuit configuration provides the same features and operation as the circuit in Fig. 9.7, but without the use of an operational amplifier. X and Y are the inputs, and Z is the output. The input current can be provided at the X terminal, while Y can be used as a bias voltage for X.

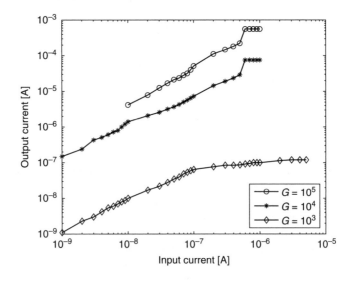

FIGURE 9.9 Current conveyor circuit without operational amplifier input test.

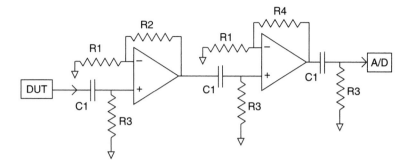

FIGURE 9.10 Noise amplification circuit used to measure and record electrical noise parameters from an operational amplifier used in the design of a biosensor.

and record the input noise signal with a bandwidth of up to 40 kHz. The second stage amplifies the signal a further five times. The second stage uses a National Semiconductors operational amplifier model LF411CN. This amplifier has a gain–bandwidth product of 1 MHz. The amplifier circuits are AC coupled to the inputs to avoid amplifying any DC bias or offset. AC coupling was performed by means of a RC high-pass filter. The cutoff frequency of this filter was set to 15 Hz.

The performance of our noise amplifier is given in Fig. 9.11. This figure shows the amplifier input-referred noise power spectral density. Integrated over a 15-Hz to 10-kHz bandwidth the total noise is 400 nV RMS. Notice that this instrument has the equivalent noise power spectral density of a 1-K Ω resistor. This is equivalent to a 0.4-μV RMS noise integrated over a 10-kHz bandwidth.

Device Type	Value
R1	1 KΩ
R2	300 KΩ
R3	100 KΩ
R4	5 KΩ
C1	0.1 μF

TABLE 9.3 Circuit components type and values for the noise measurement circuit in Fig. 9.10

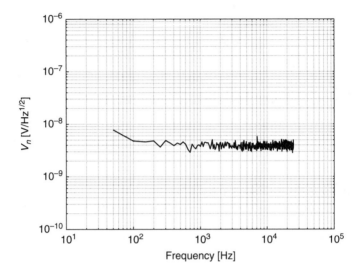

FIGURE 9.11   Noise performance of the noise amplification circuit used to measure and record electrical noise parameters from operational amplifier used in the design of a biosensor.

9.3 SOS Current-Mode Biosensor Interfaces

In this section we present and discuss the design of CMOS low-current measurement systems for sensors and biosensor interfaces. We will discuss both a integrating current sensor and a continuous-time current sensor implemented in the SOS technology. Low current measuring systems are an important biosensor interface due to the low levels of current of biological compounds and their small size. One of the most accurate and performing current measuring biosensor interface is the patch-clamp amplifier (Hamill et al., 1981; Sakmann and Neher, 1995; Sigworth, 2003), which requires the measurement of bidirectional currents of 1 to 50 nA with a RMS noise of less than 5 pA and bandwidth of 10 kHz for whole-cell measurements. We will use patch-clamp recording as an example application of the current measuring biosensor interface. We will use this example to discuss noise performance and electrode interface issues that are directly related to the ultimate performance that can be obtained from a SOS CMOS current measuring system.

9.3.1 Integrating Current-Measuring Interfaces

In this section, we present an integrating SOS current measurement system. The prototype was fabricated on a 0.5-μm SOS process. The device employs an integrating headstage with a pulse frequency modulated output, ranging from 3 Hz to 10 MHz. A digital interface produces a 16-bit output conversion of the input currents. This device was designed as an integrated patch-clamp current amplifier capable of recording from pico- to tens of microamperes of current. The high-dynamic range of seven decades and the picoampere sensitivity of the instrument was targeted to whole-cell patch-clamp biosensor recordings. We report on the electronic characterization of the fabricated device, dynamic performance, and examples of measurements on biological cells for patch-clamp applications. The device can be used in an advanced planar high-throughput patch-clamp screening system for testing medicines.

We designed the current measuring system based on asynchronous sigma-delta ADCs (McIlrath, 2001; Laiwalla et al., 2006a). The sensor is based on a pulsed-output current integrator circuit with reset frequency proportional to the input current (Murari et al., 2004; Genov et al., 2006; Stanacevic et al., 2007), as described in Sec. 8.5. This architecture permits high oversampling ratios at the bandwidths of interest. A block diagram of our current-measuring system is shown in Fig. 9.12, and is similar to the one presented in Sec. 8.5.

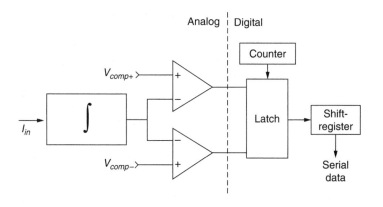

FIGURE 9.12 Integrating current measurement amplifier overview. The analog portion of the circuit integrates the input current up to a positive or negative threshold V_{comp+} and V_{comp-}. After integration, a digital pulse latches an oversampled 16-bit counter to produce a digital output.

The input current is integrated over a capacitor until the integrator output reaches either of the compare voltages V_{comp+} or V_{comp-}. At the end of integration, the change in the comparator's state generates a pulse. The digital components of the system comprise a 16-bit counter, latch, and shift register. The counter is free running, and its value is latched when a pulse from the analog circuitry is detected. This value is then transferred to the shift registers, and serially communicated to a computer-based data-acquisition system. Since the output data is oversampled and synchronized by means of an external clock, the time between two integration pulses can be measured accurately. The difference between two latched values thus yields the integrator's reset frequency. This frequency can be used along with the system transfer function in Eq. (9.13) to calculate the input current.

$$I_{in} = C_f \Delta V f_{pulse} \tag{9.13}$$

I_{in} is the input current, C_{int} the integration capacitance used in the integrator, and ΔV is the voltage swing of the integrator.

The switched-capacitor headstage integrator was used in place of a conventional operational amplifier to reduce the noise reflecting on the headstage input capacitance (Laiwalla et al., 2006a). If an operational amplifier is used in the design, the input flicker noise level require the input transistor to be large. This in turn forms a large input capacitance where other noise sources would reflect and reduce the overall performance of the amplifier.

9.3.1.1 System Components

Low current measurements are very sensitive to noise (Bandyopadhyay et al., 2002). To minimize the impact of the noise sources in our integrated system, we employed a switched-capacitor implementation (Murari et al., 2004). This realization also reduces power consumption. There are two main components in the analog circuitry of our chip: the *integrator* and the *comparator*. All amplifiers are implemented using single-stage cascoded inverters which offer high gain and low noise when operated in the subthreshold region (Vittoz, 1994). The cascoded inverter schematic is shown as an insert in Figs. 9.13 and 9.14. The active transistor loads on the input transistor increase the gain of the inverter. The gain of the amplifier can be controlled by tuning V_{bias} and operating the circuit in the high-gain subthreshold region. From simulation, the cascoded inverter has a gain of 2000. The gain varies only 2% with a change of 10 K in temperature. This variation is not an issue in cellular experiments, where the temperature is kept constant by the measurement setup.

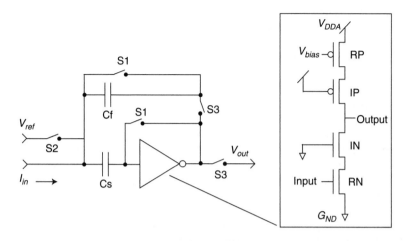

FIGURE 9.13 The headstage of the amplifier is a current integrator. During the reset phase, switches S1 and S2 are closed. During the integration phase, switch S3 is closed. The cascoded inverter is portrayed in the insert.

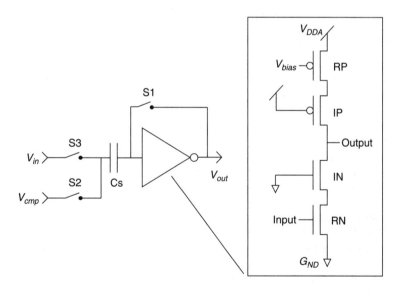

FIGURE 9.14 Comparators used in the current measurement amplifier to detect the end of integration cycles. During the reset phase switches S1 and S2 are closed, while S3 is closed during the comparison phase. The cascoded inverter is portrayed in the insert.

Headstage Integrator   A schematic of the headstage integrator is shown in Fig. 9.13 (Laiwalla et al., 2006a). The integrator uses a user-selectable 60- or 600-fF feedback capacitor C_f to sense the input current. The integrator is initially reset to a reference voltage V_{ref}, chosen to be approximately at the middle of the integration voltage swing. Using a 3.3-V supply, V_{ref} is set to 1.6 V to maximize the integration range. Because the system uses correlated double sampling (CDS), two samples are taken for each measurement. The first sample is taken during the reset phase of the system, when V_{ref} is connected to the input of the amplifier and switches S1 and S2 are closed. This voltage is stored on capacitor C_s of 1 pF, together with any correlated input noise. The input and output of the inverter are shorted, forcing both nodes to the inverter's logic threshold and highest gain. The second sample is collected during the operating phase of the circuit (switch S3 is closed) when the device is integrating the input current. Because the voltage noise is stored on C_s, the current seen by the integrator is the difference between the samples. Time-correlated noise such as flicker noise is thus partially subtracted from the integration voltage. The integrator's three-phase reset is designed to minimize the charge injection due to simultaneous switching. The switching sequence is S1, S2, S3 in Fig. 9.13. We used compensated switches with dummy half-size transistors. We used transistors of the intrinsic kind for all the current sources and to eliminate the need of biases in the cascode. The input transistor is of the RN kind to decrease the amplifier current and keep it in the subthreshold region. The transistor length and width was 2 μm.

Voltage Comparator   The comparator design uses the same principles as the integrator headstage (Laiwalla et al., 2006a). The schematic for the comparator is shown in Fig. 9.14. The input node is switched between the integrator output V_{in} and the compare voltage V_{cmp}, and the amplifier is operating in open-loop configuration. The operation is divided in two phases. During the reset phase (switches S1 and S2 are closed), the compare voltage V_{cmp} is stored on capacitor C_s, together with any correlated noise. The inverter is also initialized to its logic threshold. During the comparison phase, switch S3 is closed. The input V_{in} is connected and the comparator changes state when this voltage exceeded V_{cmp}. We used two separate comparators in our circuit, one with a positive compare voltage and the other with a negative one with respect to V_{ref}, as portrayed in Fig. 9.12. The value of $|V_{ref} - V_{cmp}|$ was set to 0.5 V to obtain a pulse frequency of approximately 10 kHz with an input current of 3 nA. The device logic circuitry was designed so that each comparator generated a positive output when it reached the comparison point. This pulse latched the counter and served as a reset signal for the entire system.

9.3.1.2 Input-Referred Noise Analysis

Noise in the current measurement amplifier can be categorized with respect to its place of origin into two main sources: the input transistor and the reset switches in the headstage integrator (Laiwalla et al., 2006a). The contributions of these sources are:

- *Input current shot noise:* the current to be measured has an intrinsic shot noise associated with it due to granularity of carriers and, in case of biosensor interfaces, ionic noise. Figure 9.2 shows the value of this limit. Input current shot noise ($I_{in,shot}$) was one of the lowest sources of noise in this circuit.

- *Input resistance current noise:* any input impedance between the current source and the amplifier input will contribute to the total current noise ($I_{R,in}$). This noise depends on the value of the input resistance, and at gigohm values, the noise is much lower than other sources of noise.

- *The input transistor:* The input transistor of the headstage integrator is the major noise source in the current measurement amplifier. The input transistor is the RN transistor in Fig. 9.13. The channel of the transistor gives rise to thermal noise. There are four transistors in the integrating headstage (including the input, as can be seen in Fig. 9.13), and each of these contributes to the total thermal noise of the device (all lumped into variable V_{themal}). Moreover, the entrapment of electrons in the transistor gate oxide leads to flicker noise ($V_{flicker}$), which varies as the inverse of the operating frequency. This noise source, unless canceled or reduced with CDS, is the dominant noise source in the circuit at the low frequencies of operation of the current measurement amplifier (less than 10 kHz).

- *Reset noise:* The switched capacitor architecture of the current measurement amplifier introduces reset noise (kT/C noise) into the circuit as the capacitors are discharged through the finite resistance of the MOSFET switches (V_{reset}). This is one of the major sources of noise in this design.

- *Quantization noise:* This noise (I^2_{quant}) is due to the conversion from current to frequency of the converter. The operating frequency of the counter in Fig. 9.12 determines the quantization resolution of the converter. This, depending on the pulse frequency and counter clock, can be one of the major sources of noise in the circuit.

The total noise in the current measurement amplifier is the sum of the above-mentioned sources, and is given by Eq. (9.15).

$$I_{noise}^2 = I_{in,shot}^2 + I_{R,in}^2 + \frac{V_{reset}^2 + V_{themal}^2 + V_{flicker}^2}{Z_{in}^2} + I_{quant}^2 [A^2] \quad (9.14)$$

Z_{in} is the input impedance of the headstage. The total system noise is dominated by flicker noise at lower frequencies and by reset noise at high frequencies. Because the current measurement amplifier works in the kilohertz range, flicker noise would be the largest constituent of noise observed in these amplifiers. If the headstage was implemented as a regular operational amplifier, the only way to reduce flicker noise is to increase the size of the input transistor, at the expense of a significantly higher input capacitance. In order to provide sensitivity in the picoampere range, we have employed an established technique to reduce flicker noise in our design. A noise correlated in time, flicker noise levels can be lowered using CDS (Enz and Temes, 1996). This yields a lower noise system with the total input noise described in Eq. (9.15) (Laiwalla et al., 2006a).

$$I_{noise}^2 = I_{in,shot}^2 + I_{R,in}^2 + \frac{V_{reset}^2 + V_{themal}^2}{Z_{in}^2} + I_{quant}^2 [A^2] \quad (9.15)$$

Substituting the definitions of the noise sources into Eq. (9.15) yields Eq. (9.16).

$$I_{noise}^2 = 2I_{in}q + \frac{4kT}{R_{in}} + \left(\frac{kT}{C_f} + \frac{8}{3}kTg_m \right)(2\pi f_{pulse}C_f)^2 + C_f \Delta V \frac{f_{pulse}^2}{f_{clk}}$$

$$(9.16)$$

The parameters of Eq. (9.16) are R_{in} is the input resistance used to provide the current (1 GΩ), I_{in} is the input current, $C_s = 1$ pF is the input sampling capacitor, $g_m = 4.7 \times 10^{-7}$ and $L = 2$ μm are the transconductance and length, respectively, of the RN input transistor, and $C_f = 600$ fF is the integration capacitance. See Fig. 9.13 for details on these components and parameters. The process parameters used for evaluation are given in Table 9.4.

The noise performance and modeling of Eq. (9.16) are given in Sec. 9.3.1.3. When these parameters are taken into account, the only important noise sources are the reset noise and the quantization noise. In this case, Eq. (9.16) turns into Eq. (9.17). The parameters of Eq. (9.17) are $\Delta V = \| V_{ref} - V_{minus} \| = 0.5$ V, $C_f = 600$ fF is the integration capacitance, f_{clk} is the counter clock used to measure the pulse frequency,

g_m	$4.51 \cdot 10^{-7}$ S
C_{OX}	$2.47 \cdot 10^{-3}$ F/m^2
A	$19.4 \cdot 10^{-12}$ m^2
T	298 K
I_d	$300 \cdot 10^{-12}$ A
k	$1.28 \cdot 10^{-22}$
C_f	$60/600 \cdot 10^{-15}$ F
C_{in}	$1 \cdot 10^{-12}$ F
K_f	$1.28 \cdot 10^{-22}$
A_f	0.889
R_{in}	1/10 GΩ

TABLE 9.4 Process parameters for
noise calculations

k is the Boltzmann constant, and T is the circuit temperature (room
temperature 25°C).

$$I_{noise}^2 = \frac{kT}{C_f}(2\pi f_{pulse}C_f)^2 + C_f \Delta V \frac{f_{pulse}^2}{f_{clk}} \tag{9.17}$$

9.3.1.3 Results

We tested an integrated current measurement amplifier by sourcing
a range of input currents from a few microamperes to a picoampere
while recording the frequency of the output pulse (Laiwalla et al.,
2006a). The amplifier was powered at 3.3 V with an Agilent 3631A DC
power supply. The voltage noise on the power supply and the bias
voltages was measured as less than 2 mV RMS. The input currents
were sourced by applying a voltage across a mega-ohm resistor. A
Keithley 2400 Source Meter was used to source and measure the input
current. The frequency of the output pulse was measured using a
Tektronic TDS2014 Four Channel Digital Storage Oscilloscope. We
obtained a linear transfer function across the entire range of tested
currents in a range 3 pA to 10 μA, as shown in Fig. 9.15. The output
pulse frequency was in the range 3 Hz to 10 MHz, and was observed
to increase in discrete steps when very high currents were sourced
as quantization noise due to low oversampling became dominant.
We operated the device with clock frequencies of up to 50 MHz, and
were thus able to extend the upper limit of the current measurements.
The use of fast clocks and the SOS process make this device one of the

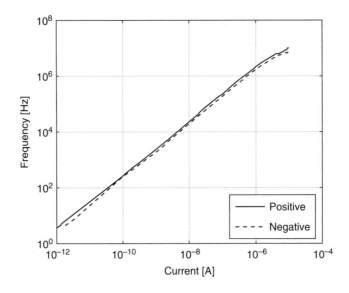

FIGURE 9.15 Measured output pulse frequency of the headstage integrator as a function of positive and negative input currents, as integrated on a 600-fF capacitor. The dynamic range of the current measurement amplifier was measured to be 7 decades.

largest dynamic-range current measuring system reported (Murari et al., 2004).

The power consumption of the current measurement amplifier was estimated for both positive and negative currents and for both analog and digital supplies. The results for the nanoscale currents range of interest are plotted in Fig. 9.16. We observed that the power consumption increases at very low currents and low output pulse frequencies. The slow integration of low currents causes the digital interface to spend a long time near the logic threshold. This causes short circuit currents in the digital supply. Also notice the dependency of the analog power consumption on the direction of the current, due to charge injection. For a higher current (not in Fig. 9.16), both analog and digital power consumption increases with current as the reset frequency rises. The analog power consumption is mostly due to the short circuit current during the resent phase.

The dynamic performance of the current measurement amplifier was measured by digitizing a 400-Hz input sine current with an amplitude of 50 nA. The sampling rate was 2 kHz. Figure 9.17 shows a plot of the recorded sine current and demonstrates the ability of our current measurement amplifier to record data at high-speeds. Notice that the glitches in Fig. 9.17 are due to readout errors of the current

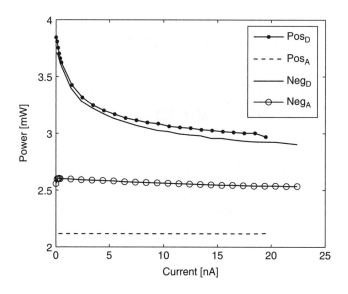

FIGURE 9.16 Measured power consumption of analog and digital circuitry of the SOS integrated current measurement amplifier. The power consumption was estimated for both positive and negative currents and for both analog and digital supplies.

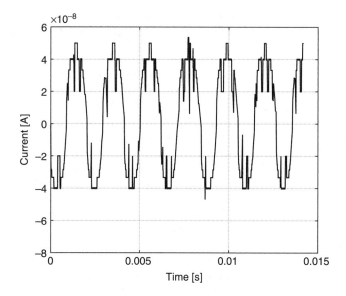

FIGURE 9.17 Measured dynamic performance of the SOS integrated current measurement amplifier. The input is a 400-Hz sine current of 50 nA amplitude sampled at 2 kHz.

Digital interface   Comparators   Headstage

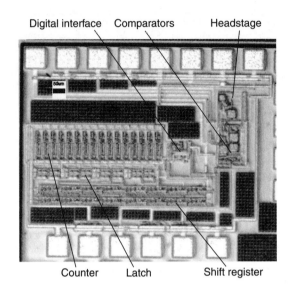

Counter   Latch   Shift register

FIGURE 9.18 Die micrograph of the SOS integrated current measurement amplifier. The die size is 1260 μm × 1040 μm.

measurement amplifier data. We are working to eliminate this problem in the next version of the custom data acquisition software we are developing to operate the device.

The die size of the integrated current measurement amplifier measures 1260 μm × 1040 μm with pads and 1140 μm × 560 μm without pads. The headstage integrator measures 150 μm × 225 μm. A micrograph of the fabricated die is reported in Fig. 9.18.

To test the device with picoampere input currents, we used a very large resistance (1 GΩ and 10 GΩ) to convert a large (volt-level) input voltage into a very low current. The external resistors were soldered to the device input pin, which is set at the V_{ref} voltage by the circuit (see Fig. 9.13). The device was placed in an aluminum box with BNC connectors for external instruments. V_{ref} was set to 2 V and the input voltage was varied from 3 to 1 V to provide a ±1 nA input current swing. This technique is common practice when measuring ultralow currents.

We used an Opal Kelly XEM3001 field-programmable gate away (FPGA) board to collect the digital pulsed data from the current integrator, and we implemented a graphic user interface to display and save the data. The current integrator settings used in the experiment are: $V_{bias} = 1.876$ V, $V_{ref} = 2$ V, $V_{minus} = 1.5$ V, $V_{plus} = 2.5$ V. All testing were conducted with $C_f = 600$ fF, and the FPGA timing clock running at 0.2 MHz (at least 200 times the maximum integrator bandwidth).

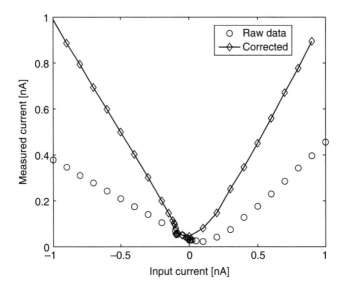

FIGURE 9.19 Transfer function of the current measurement system with an input current of ±1 nA. The raw data can be compensated for offsets and gain mismatch to report the corrected transfer function (solid line). The input current was injected using a 1-GΩ resistor. The measured current is an absolute value, as the device also provides a digital sign bit.

Figure 9.19 shows the transfer function of the current measurement system with an input current of ±1 nA. Each point in the transfer function has been measured 100 times. Figure 9.19 shows the mean value. These data were collected using a 1-GΩ input resistance. For ease of view, the absolute value of the currents is given in Fig. 9.19, as the device also provides a digital current sign bit. In order to compensate for the device's imprecise components, we have corrected the transfer function curve to show the linearity of the device. The raw data collected from the device show an offset of 0.1 nA toward positive currents. This offset translates into a 0.1-V offset on the comparators threshold voltage. This offset is expected, as the comparators we used were not designed for low offset voltage. In addition, there is a gain mismatch between the theoretical values [Eq. (9.13)] and the final results. The positive current raw data was 0.51 times smaller than the measured value, and the negative 0.35 times. This gain mismatch is due to a size discrepancy in the integration capacitance C_f. We have compensated the offset and gain errors, and the resulting transfer function is the "corrected" data set in Fig. 9.19.

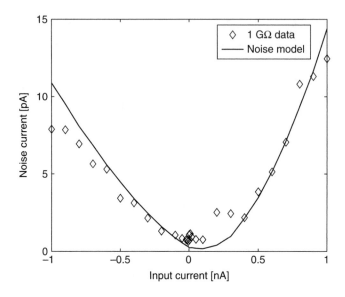

FIGURE 9.20 Measured RMS noise of the integrator circuit. The input current was injected using a 1-GΩ resistor. The measured current is an absolute value, as the device also provides a digital sign bit.

Figure 9.20 reports the measured total rms noise from the device. The noise is measured by computing the standard deviation of the 100 samples collected for Fig. 9.19. We have computed a mathematical model of the RMS noise of the current integrator. The RMS current noise value $I_{n,RMS}$ can be computed with Eq. (9.17).

Quantization from the counter clock (f_{clk}) is the major source of noise for large currents (>100 pA), together with the kT/C_f component. Low current (<100 pA) noise is dominated by the integrator thermal noise: $I_{n,thermal}^2 = e_{n,OA}^2(2\pi f_{OA})^2(C_f + C_{in})^2$. $e_{n,OA}^2$ is the integrator thermal noise, $f_{OA} = 10$ kHz is the integrator bandwidth, $C_{in} = 50$ pF is the input capacitance. The flicker component is reduced by the correlated-double sampling process.

Figure 9.21 shows a close-up of the noise data in Fig. 9.20 (open diamonds) but also reports the data recorded with a 10-GΩ resistor (open circles). The noise model does not change when a large resistor is used because the resistor noise is negligible. The model closely matches the data with the larger resistor (see circles on the left side of Fig. 9.21), which helps to reduce external noise and lower the thermal noise component. The matching is very adequate for the negative currents, whereas the noisier data collected for positive currents does not provide a good match with the model. This is also visible in the lower

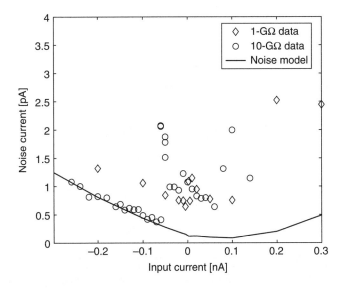

FIGURE 9.21 Measured RMS noise of the integrator circuit. Current were injected with a 1-GΩ resistor (diamond) and a 10-GΩ resistor (circles). The measured current is an absolute value, as the device also provides a digital sign bit.

right portion of Fig. 9.20. The model, however, matches the data well in the whole range of Fig. 9.20, confirming its validity. Finally, note that the results measured in this section should be compared with the interpretation given in reference (Stanacevic et al., 2007), which states that the major cause of noise is shot noise from a reference current source. The SOS system does not use a current source, so this offers even better noise performance. As can be seen in Fig. 9.21, at least for negative currents, the RMS noise is less than 350 fA with an input current of 50 pA and a bandwidth of 110 Hz. This is one of the highest sensitivity values obtained with an integrated current integrator circuit and the first one designed in the SOS process.

9.3.1.4 Experiments

The patch clamp is the gold standard in electrophysiology and is a fundamental technique for screening drugs and medical compounds, and ultimately ensuring consumer safety (Hamill et al., 1981; Roden, 2004). Electrophysiologists use the patch-clamp technique to measure the currents flowing through the membranes of living cells. The currents measured by the patch-clamp is used to study the effect

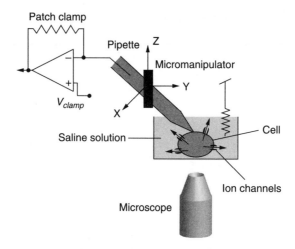

Patch clamp

Pipette Z

Micromanipulator

Y

V_{clamp}

X

Saline solution

Cell

Ion channels

Microscope

FIGURE 9.22 A conventional whole-cell patch-clamp recordings, a pipette filled with electrolyte solution is used as the electrode. It takes considerable skill to attach the pipette to the cell using micromanipulators and a microscope.

of medical compounds and the behavior of ion channels, the structures responsible for cell membrane conductivity (Sigworth, 2003). The patch-clamp technique reports the highest SNR ratios available in a biosensor but is labor intensive when performed manually. In a whole-cell patch-clamp recording, the ion-channel current across the whole cell membrane is measured while clamping the membrane at a known potential. Two whole-cell recording setups are shown in Fig. 9.22. In conventional patch recordings, a glass pipette is used as the electrode. The pipette is filled with an ionic solution and is attached to the cell using micromanipulators under a microscope. Suction is first applied so that a tight seal ($\sim 10^{10}$ Ω) is formed between the pipette tip and the cell membrane. Further suction is applied to break the cell membrane and gain access to the interior of the cell.

Ion channel currents range from a few picoamperes for single-channel recordings to tens of nanoamperes for whole-cell measurements. Voltage steps between 10 and 100 mV are applied to the membrane during an experiment to activate ion channel proteins and permit ionic currents to flow across the membrane. Currents are bidirectional, depending on channel type and membrane potential, and the bandwidth of interest is between a few hertz to 10 kHz in bench-top systems. Higher sampling rates and bandwidth are desired for more precise characterization of ion channels. The patch-clamp amplifier

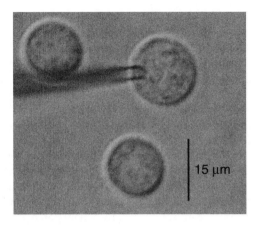

FIGURE 9.23 The patch-clamp protocol on a RBL cell. A glass pipette is attached by gigaseal to one RBL cell approximately 15 μm diameter.

must have a large dynamic range in order to record the large transient currents after the stimuli and must be highly sensitive to currents in the pico- to nanoampere range.

To verify the usefulness of the SOS current measuring amplifier in lifescience applications, we used the patch-clamp amplifier to measure the ionic conductances of rat basophilic leukemia cells (RBL) (Laiwalla et al., 2006a). Figure 9.23 shows the gigaseal of a glass pipette to a RBL cell used in our measurements. The data collected from the RBL cells is due to inward rectifier potassium channels, which conduct only in the inward direction. The pipette solution used in the experiment was 130 mM KCl, 10 mM NaCl, 1 MgCl2, 2 CaCl2, 10 HEPES, 1 EGTA, pH 7.4 with NaOH; the bath solution was 130 mM KCl, 4.4 mM NaCl, 2 CaCl2, 2 MgCl2, 10 HEPES, 5 dextrose, pH 7.4 with NaOH. Fig. 9.24 show the time response as step voltages are applied to RBL cells. The

FIGURE 9.24 Measured patch-clamp protocol on the RBL cells. The input current is converted to a voltage using the integrating headstage. The bath and pipette solutions both contained 130 mM of K+.

y-axis shows the measured voltage (at the output of the analog filter), which relates linearly to the measured current.

A conventional patch-clamp experiment involves using a glass micropipette (with an aperture of 1 μm), and maneuvering it to make contact and seal it to the membrane of a cell residing in a culture solution. The pipette contains the recording electrode (also known as the patch electrode), and a command electrode is placed in the culture solution (known as the bath electrode). Suction is applied through the pipette to cause the cell membrane to rupture (whole-cell configuration), thus shorting the electrode to the intracellular potential. Step voltages in the range of 10 to 120 mV are then applied between the patch and the bath electrode, and the corresponding ionic flow is measured. These voltage steps are applied to the V_{ref} terminal of Fig. 9.13. We have connected our device in series to a commercial patch-clamp amplifier, so that we could use its data acquisition software and real-time visualization. DACs driven by the commercial HEKA amplifier software were used to provide voltage steps to an external resistor connected to the amplifier's input node. We decoded the amplifier response by taking the digital reset pulse train at the output of the amplifier and RC filtering it with a time constant of 100 μs. The filtered positive and negative reset spikes were then passed through a difference amplifier and low-pass filtered using an analog eight-pole Bessel filter. This was equivalent to converting the variable-frequency reset spikes to scaled analog voltages and provided an efficient way for quantifying circuit current. The filter output was digitized at 10 kHz by standard patch-clamp software Digidata and visualized using Clampfit version 8.2. Our measurements were made using a standard patch clamp setup, demonstrated in Fig. 9.25.

The inward rectifier measurements with the pipette clamped to the RBL cell are shown in Fig. 9.24. To obtain this figure, we applied step voltages from 200 mV down to −200 mV in 10-mV steps. With the pipette in the bath solution there is the equivalent of a 100-MΩ resistor between the electrodes. The analog filter 3-dB bandwidth was set to 500 Hz. Cell measurements were obtained by means of a gigaseal clamp implying that the seal resistance between the glass electrode and the cell membrane is larger than a gigaohm, and that current leaks are minimized. Notice that the ion channel is an inward rectifier potassium channel in the RBL cells, therefore it does not pass current in the positive direction. For this reason the I-V characteristic of the RBL cell measurement is one sided, as can be seen in Fig. 9.26.

The results of Fig. 9.24 obtained with our patch-clamp system are comparable in terms of noise performance to commercial automated systems (Axon Instruments OpusXpress, 2008) (Nanion Technologies, 2008). These are also designed for whole-cell recordings and have noise levels of 1 to 10 pA RMS. Furthermore, the results of Fig. 9.24 are

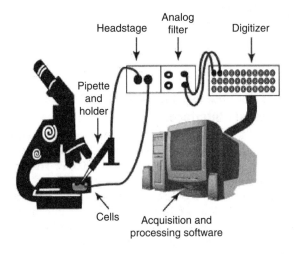

FIGURE 9.25 The patch-clamp experimental setup.

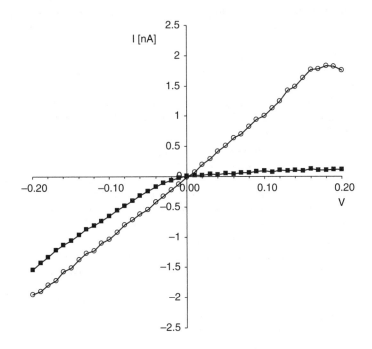

FIGURE 9.26 Measured I-V characteristics on the RBL cells. This measurement is the mean current during the voltage steps. Notice the rectifying behavior of the RBL cells. The filled squares represent RBL cell current, and the open circles represent current through a 100-MΩ resistor.

virtually identical to the recording with a commercial high-resolution patch-clamp amplifier as the Molecular Devices Axopatch 200B (Laiwalla et al., 2006a).

9.3.2 SOS Continuous-Time Current-Mode Biosensor Interfaces

In this section, we present a continuous-time SOS current measurement system implemented in the 0.5-μm SOS CMOS process (Weerakoon et al., 2008b; Weerakoon et al., 2008c). The system is capable of recording currents up to ±20 nA, with a RMS noise of 5 pA at 10-kHz bandwidth using a transimpedance amplifier with resistive feedback. In our design, we use a large resistor in conjunction with an operational amplifier to obtain linear amplification of current. Using SOS technology, we minimized the significant parasitic capacitance of resistors made in conventional bulk CMOS processes (Peregrine, 2008a). This leads to wider bandwidth and increased stability in the current-measuring amplifier. The system can compensate for the capacitance and the resistance of the electrode, up to 20 pF and up to 70% of the series access resistance respectively. This is the first fully integrated implementation of a patch-clamp measurement system with series access resistance and parasitic capacitive compensation capability. The integrated patch-clamp system will be used to fabricate high-throughput planar patch-clamp systems. Our design can be used as a fully integrated patch-clamp amplifier and is 10 million times smaller in volume than commercial bench-top systems. Other than minimizing space requirements, an integrated patch-clamp amplifier reduces noise and offers better electrical performance by decreasing cabling and other parasitic capacitances that lower the measurement bandwidth (Weerakoon et al., 2008b; Weerakoon et al., 2008c). Although several integrated circuits able to measure low current levels have been previously realized, they lack the essential ability to compensate for the resistance and the parasitic capacitance of the electrode (Pandey and White, 2001; Genov et al., 2006; Laiwalla et al., 2006b; Ayers et al., 2007; Stanacevic et al., 2007; Weerakoon and Culurciello, 2007a).

In the following section, we introduce the patch-clamp system, present the theory of series resistance compensation, and analyze the design of circuitry to remove the parasitic capacitive transients. We analyze the noise of the patch-clamp system, present the hardware test bed developed to record current measurement data, and conclude by providing experimental results on the use of the system for patch-clamp measurements.

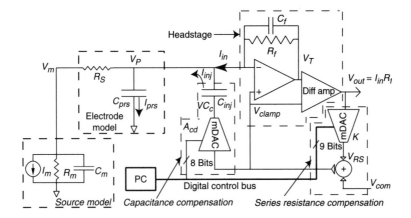

FIGURE 9.27 Block diagram of an integrated current measuring system with electrode compensation. A transimpedance amplifier followed by a difference amplifier produces an output proportional to the input current. Two compensation circuits provide electrode capacitance and resistance compensation. The amount of compensation is controlled by two multiplying DACs. The device was designed to measure currents of ±20 nA. The current source model and electrode model are marked. All operational amplifiers are identical to the ones described in Sec. 4.1.

9.3.2.1 Continuous-Time Current-Measuring System Overview

Figure 9.27 is a block diagram of our current-measuring system headstage with electrode compensation circuitry. The currents are typically a few nanoamperes in amplitude with a bandwidth of 5 to 10 kHz. Our current-measuring amplifier system consists of three main components: the *transimpedance headstage* to monitor the input current, a *resistive compensation circuit* to compensate for the series resistance of the electrode, and a *parasitic capacitive compensation circuit* to compensate for the current drawn by the electrode parasitic capacitance (Weerakoon et al., 2008b; Weerakoon et al., 2008c).

The headstage monitors the input current while clamping the electrode at V_{com} with an operational amplifier. This voltage clamp allows us to finely control the voltage at the current source terminal and is fundamental in current-measuring systems. The input current I_{in} is recorded by an input current-to-voltage transimpedance amplifier that uses resistive feedback (R_f). The nonconductive substrate of SOS technology makes it possible to fabricate large resistors with minimal parasitic capacitance, which increases speed, sensitivity and stability (Cristoloveanu and Li, 1995; Kuo and Su, 1998; Andreou et al., 2001). C_f is the shunt capacitance across the resistor. This shunt

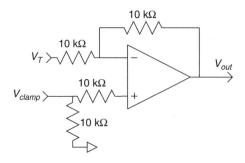

FIGURE 9.28 Schematic diagram of the integrated current-measuring system's difference amplifier.

capacitance is necessary to make the transimpedance amplifier stable and to increase the bandwidth of the voltage clamp. A difference amplifier subtracts the command voltage from the tranimpedance amplifier's output. The difference amplifier was implemented as shown in Fig. 9.28. The resulting output voltage is proportional to the input current and is low-pass filtered by the transconductor time constant $\tau_z = R_f C_f$, as shown in Eq. (9.18). All operational amplifiers used in the design of this continuous-time current-measuring system are identical to the ones described in Sec. 4.1.

$$V_{out} = \frac{I_{in} R_f}{1 + \tau_z s} \tag{9.18}$$

The voltage clamp transfer function relating the electrode potential V_P to V_{clamp} is derived in Eq. (9.19). The operational amplifier is assumed to have a single pole located at τ_{Amp}. C_t is the total capacitance at the input of the current-measuring amplifier.

$$V_p = \frac{V_{clamp}}{1 + \tau_{clamp} s} = \frac{V_{clamp}}{1 + \frac{C_t}{C_f} \tau_{Amp} s} \tag{9.19}$$

Using the parameters in Table 9.5 we calculate τ_z and τ_{clamp} to be 7.5 and 0.8 µs, respectively, resulting in a transconductor bandwidth of 20 kHz and a voltage clamp bandwidth of 180 kHz, respectively. Command voltage steps (V_{com}) between 0 and 100 mV are applied to the cell membrane during experiments while recording the input currents. The device was designed to measure cell membrane currents up to ±20 nA.

Series Resistance Compensation The access resistance R_S to the cell is typically 5 to 20 MΩ due the series resistance of the electrode.

Feedback resistance, R_f	25 MΩ
Feedback shunt capacitance, C_f	0.3 pF
Total capacitance at the input, C_t (typical)	6 pF
Parasitic capacitance at electrode, C_{prs} (typical)	5 pF
Series resistance of the electrode, R_S (typical)	10 MΩ
Operational amplifier time constant, τ_{Amp}	0.05 μs
Transconductor time constant, τ_z	7.5 μs
Injection capacitance , C_{inj}	10 pF
Voltage clamp time constant, τ_{clamp}	0.8 μs
Source model resistance, R_m (typical)	10 GΩ
Source model capacitance, C_m (typical)	30 pF
Source model resistance	20 GΩ
Source access time constant, τ_a	300 μs
Source current amplitude, I_m	±1−20 nA

TABLE 9.5 Parameters of the current-measuring system, source, and electrode models

The capacitance of the source (in patch-clamp application, this is the cell membrane) is about 10 to 100 pF. The time constant associated with charging this capacitance when a potential step V_{com} is applied to the electrode is several hundred microseconds. There is also a voltage error of tens of millivolts between the electrode potential V_P and the source potential V_m due to the voltage drop across the series resistance, as the input currents I_m in the order of nanoamperes are typical. Series resistance compensation is used to minimize this error.

In the series resistance compensation circuit shown in Fig. 9.27, we estimate the voltage error caused by the electrode resistance and we make a correction to the command voltage, thereby effectively reducing the electrode resistance (Sakmann and Neher, 1995). This reduction in series resistance enables us to voltage clamp the source accurately. It also reduces the time needed to charge the membrane capacitance C_m and enables the circuit to monitor input current variations occurring immediately after the control voltage is applied. The series resistance compensation circuit takes the current monitoring signal and scales it by a variable factor and adds it to the command voltage (V_{com}) in positive feedback polarity. A 9-bit digital setting K of the compensated resistance is applied to an mDAC to control the amount of compensation. The series resistance compensation mDAC

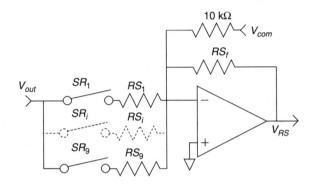

FIGURE 9.29 Schematic diagram of the integrated current-measuring system's series resistance mDAC. All operational amplifiers are identical to the ones described in Sec. 4.1.

was designed to scale the output voltage between 0.06 to 30 times with 9-bit resolution. A diagram of the 9-bit DAC is shown in Fig. 9.29. The total output at the summing amplifier when switch SR_i is closed is given by Eq. (9.20).

$$V_{RS} = -\left[\sum \frac{RS_f}{RS_i} \times V_{out} + V_{com}\right] \qquad (9.20)$$

In our design, RS_f was chosen as 10 kΩ and values of 612.5 Ω, 1.25 kΩ, 2.5 kΩ, 5 kΩ, 10 kΩ, 20 kΩ, 40 kΩ, 80 kΩ and 160 kΩ were chosen for resistors RC_1-RC_9 respectively.

In order to derive the behavior of V_m with V_{com} with series-resistive compensation, we make the following assumptions. First, because τ_{clamp} is much shorter than τ_a (see Table 9.5), we assume that V_{com} appears at V_P instantaneously. Second, we assume that the source model resistance R_m is very large compared to R_S and hence ignore it. The transfer function representation of this simplified system is shown in Fig. 9.31. The command voltage V_{com} is added to the compensation signal to yield the electrode potential V_p. A filtered version of this voltage with an access time constant $\tau_a = R_S C_m$ appears at the membrane. The electrode current is the product of C_m and the time derivative of V_m. This current is converted to a voltage according to the transfer function of the transimpedance amplifier. The resulting current monitoring voltage is then scaled by a factor K to yield the compensation voltage. The setting K can be written as in Eq. (9.21) where α is the fraction of R_S that is compensated. The entire

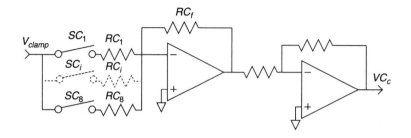

FIGURE 9.30 Schematic diagram of the integrated current-measuring system's capacitive compensation, mDAC. All operational amplifiers are identical to the ones described in Sec. 4.1.

closed-loop transfer function relating V_m to V_{cmd} can then be written as in Eq. (9.22).

$$K = \alpha R_S \tag{9.21}$$

$$\frac{V_m}{V_{com}} = \frac{\tau_Z s + 1}{\tau_Z \tau_a s^2 + (1 - \alpha)\tau_a s + 1} \tag{9.22}$$

$$\zeta = \frac{(1 - \alpha)\sqrt{\frac{\tau_z}{\tau_a}}}{2} \tag{9.23}$$

$$\alpha \leq 1 - 2\sqrt{\frac{\tau_z}{\tau_a}} \tag{9.24}$$

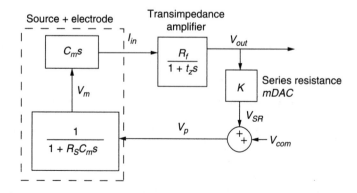

FIGURE 9.31 Block model of the current-measuring system with series resistance compensation. All operational amplifiers are identical to the ones described in Sec. 4.1.

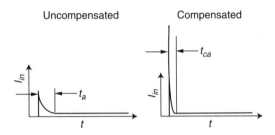

FIGURE 9.32   The series resistance compensation circuit increases the system's bandwidth.

The damping factor ζ of the second-order transfer function in Eq. (9.22) can be written as in Eq. (9.23). When ζ drops below unity, the output will overshoot. Therefore, α is restricted to the range given in Eq. (9.24). The maximum level of compensation is limited by the bandwidth of the transconductor and also by the parasitic capacitance of the electrode as we will show in the next section. By substituting the values in Table. 9.5 in Eq. (9.24) we would be able to compensate up to 70% of R_S ($\alpha_{max} = 70\%$) and thereby see a threefold increase in bandwidth. This increase in bandwidth is illustrated in Fig. 9.32. The time constant τ_{ca} associated with charging the membrane capacitance C_m through the compensated resistance ($R_S - \alpha R_S$) is given by Eq. (9.25).

$$\tau_{ca} = C_m(R_S - \alpha R_S) \qquad (9.25)$$

Therefore, when compensated, α approaches 1 and τ_{ca} decreases. When R_S is compensated, due to charge conservation, the overshoot will increase when τ_{ca} decreases.

Parasitic Capacitive Compensation   The capacitive compensation circuitry shown in Fig. 9.27 compensates for the current I_{prs} drawn by the electrode capacitance C_{prs}. The electrode capacitance is compensated by injecting a current I_{inj} through an integrated capacitor C_{inj} of 10 pF. An 8-bit digital setting of AC_d is applied to a mDAC so that a range of C_{prs} values can be compensated. The implementation of the parasitic compensation mDAC is shown in Fig. 9.30 and is identical to that of the series resistance compensation mDAC. The output of the inverting summing amplifier in the capacitive compensation mDAC is sent through an inverting amplifier to obtain VC_c. When switches

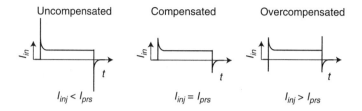

FIGURE 9.33 The capacitive compensation circuit eliminates parasitic overshoots.

SC_i are closed, the output of the capacitive compensation mDAC is given by Eq. (9.26).

$$VC_c = \sum \frac{RC_f}{RC_i} \times V_{clamp} = AC_d \times V_{clamp} \qquad (9.26)$$

In our design, RC_f was chosen as 10 kΩ and values of 75 Ω, 150 Ω, 306 Ω, 612.5 Ω, 1.25 kΩ, 2.5 kΩ, 5 kΩ, 10 kΩ were chosen for resistors RC_1 to RC_8, respectively, allowing parasitic capacitances up to 20 pF to be compensated.

When the value of AC_d is set as given in Eq. (9.27), no parasitic current is drawn from the headstage and the parasitic capacitance is fully compensated.

$$(AC_d - 1)C_{inj} = C_{prs} \qquad (9.27)$$

Figure 9.33 shows the predicted response of the parasitic capacitance compensation circuit. When uncompensated, the headstage provides the currents needed to charge the parasitic capacitance. These "fast" currents appear as narrow overshoots in the current-monitoring signal with the same polarity as the control step. The fast transients are superimposed on the slower transients associated with the time constant that charges the source voltage (i.e., τ_a). When properly compensated, $I_{prs} = I_{inj}$ and the overshoots do not appear. When overcompensated, $I_{inj} > I_{prs}$ and the overshoots appear negative.

Parasitic capacitance compensation prevents the electronics from saturating because it eliminates the need for the headstage to provide the large transient currents needed to charge the parasitic capacitances. An added benefit of capacitive compensation is that it reduces the effects of a noisy stimulus. In the same manner, that the parasitic currents caused by voltage steps are canceled, the compensation circuitry cancels out the currents caused by noise in the stimulus voltage (Sakmann and Neher, 1995). Furthermore, this technique increases the amount of applicable series resistance compensation. To understand why, let C'_{prs} be the residual parasitic capacitance. The sign

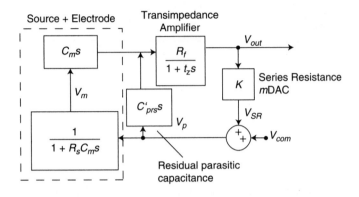

FIGURE 9.34 Block model showing the effect of residual parasitic capacitance on series resistive compensation.

of C'_{prs} could be either positive or negative, depending on under- or overcompensating, respectively. The transfer function representation of the patch-clamp system with this residual parasitic capacitance included is shown in Fig. 9.34. The transfer function derived in Eq. (9.22) can now be written as in Eq. (9.28). Positive C'_{prs} adds a phase lead to the R_S compensation loop, introducing a damping effect and thereby stabilizing the compensation. However, excessively large values of C'_{prs} increases the magnitude of the feedback and introduces oscillations. To prevent this, C'_{prs} is kept smaller than τ_Z / R_S (i.e., about 0.75 pF) (Sakmann and Neher, 1995). Therefore, it is imperative that one cancels out the parasitic capacitance before performing series resistance compensation.

$$\frac{V_m}{V_{com}} = \frac{1}{(\tau_a s + 1)(-K Z_A C'_{prs} s + 1) - \tau_a K Z_A s} \tag{9.28}$$

9.3.2.2 Noise in the Continuous-Time Current-Measuring System

Low-amplitude current measurements are often complicated by the presence of background noise. The background noise in these measurements arise mainly from the electronic circuitry and the source-electrode network (Weerakoon et al., 2008b; Weerakoon et al., 2008c).

Source-Electrode Noise in a Continuous-Time Current-Measuring System
In most low-current measurement (e.g. patch-clamp whole-cell recording), the dominant background noise contributor is the source-electrode network shown in Fig. 9.35 (Neher and Sakmann, 1995). The noise level of source plus electrode network, S_P can be calculated

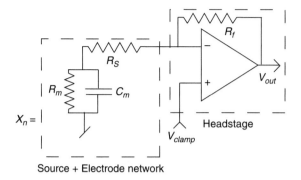

Source + Electrode network

FIGURE 9.35   Source-plus electrode network. The noise level of this network limits the performance of several low-current measurements.

using Eq. (9.29) (Johnson, 1928). Here, K is the Boltzmann constant, T is the absolute temperature, and X_N is the equivalent impedance of the source-plus electrode network.

$$S_P = \frac{4KT}{Re(X_N)} \qquad (9.29)$$

X_N can be calculated using Eq. (9.30).

$$X_N = \frac{R_m}{1 + R_m C_m s} + R_S \qquad (9.30)$$

The current noise of the series combination of R_S and the source capacitance dominates in Eq. (9.29). The current noise of the source resistance can be ignored (Neher and Sakmann, 1995). The current noise power spectrum of this simplified network is shown in Eq. (9.31).

$$S_P = \frac{4KT R_S (2\pi f C_m)^2}{1 + (2\pi f R_S C_m)^2} \qquad (9.31)$$

When designing a low-noise amplifier to measure low-amplitude currents, it is sufficient to ensure that the noise contribution of the amplifier is less or comparable to S_P in the recording bandwidth.

Electronic Noise in the Patch-Clamp System   The electronic noise in the current measurement system arises primarily from the feedback resistor and the operational amplifier used to implement the I–V converter. The operational amplifier noise performance has been already reported in Sec. 4.1. The circuit model used to compute the electronic noise of the patch-clamp system is shown in Fig. 9.36. The total output voltage noise spectral density S_V is given by Eq. (9.32). Y_c is the resulting impedance at the input due to X_n, C_{prs}, and C_{gs}. C_{gs} is the

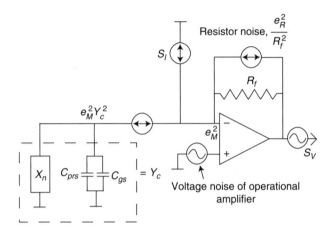

FIGURE 9.36 The model used to calculate the electronic noise of the current-measuring system.

gate capacitance of the input transistor. The input-referred current noise spectral density S_I of the current amplifier can be calculated by Eq. (9.33). e_M is the sum of the flicker noise and thermal noise components of the input transistor of the headstage and e_R is the thermal noise of the feedback resistor (Kansy, 1980; Sarpeshkar et al., 1993; Nemirowsky et al., 2001). e_M and e_R can be calculated using Eqs. 9.34 and 9.35, respectively.

$$S_V = e_M^2 + e_R^2 + e_M^2 Y_c^2 R_f{}^2 \tag{9.32}$$

$$S_I = \frac{S_V}{R_f{}^2} = \frac{e_M^2 + e_R^2}{R_f{}^2} + e_M^2 Y_c^2 \tag{9.33}$$

$$e_M^2 = C_T + \frac{C_F}{f} \tag{9.34}$$

$$e_R^2 = 4KTR_f \tag{9.35}$$

$$SI_{Tot} = S_I + S_P \tag{9.36}$$

Here, C_F is the process dependant flicker noise coefficient and C_T is the thermal noise coefficient of the input transistor (see Sec. 4.1 for details). The total input referred current spectral density, SI_{Tot} of the patch-clamp system is given by Eq. (9.36). Figure 9.37 shows S_I, the theoretical unloaded noise of the patch-clamp amplifier, plotted using Eq. (9.33). The flicker noise component of the noise dominates S_I at low frequencies. At higher frequencies, S_I is dominanted by

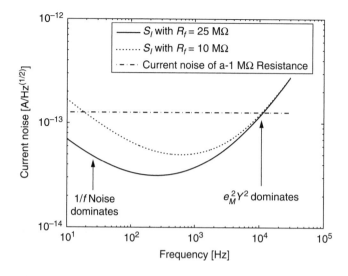

FIGURE 9.37 Electronic noise of the unloaded current measuring headstage. The current noise spectral density S_I is dominated by flicker noise at low frequencies and at higher frequencies by the voltage noise that is reflected on the input impedance. S_I has an inverse relationship with the feedback resistance R_f. When using a R_f of 25 MΩ, the value of S_I at 5 kHz is comparable to the current spectral density of a 1-MΩ resistance.

the voltage noise that is reflected, Y_c. The current noise at the input decreases with R_f. However, the physical size of of R_f in the layout as well the headroom limitations in the amplifier, limits its value. A R_f of 25 MΩ was chosen for our design. When using a R_f of 25 MΩ, the value of S_I at 5 kHz is comparable to the current spectral density of a 1-MΩ resistance. The current noise spectral density S_P of the cell–electrode network is compared with SI_{Tot} in Fig. 9.38. S_P is the dominant component of SI_{Tot} in the recording bandwith.

Noise in the Operational Amplifier The patch-clamp technique is extremely sensitive to noise due to the low amplitude of the membrane current and hence low-noise amplification is critical to our design. All amplifiers are implemented using the design described in Sec. 4.1. The input transistors are vital in establishing the noise characteristics of the operational amplifier. The gate capacitance C_{gs} of the input MOSFET is proportional to the area of the transistor, whereas the thermal noise e_n decreases as the square root of the area (assuming constant gate length). The noise of the recording system is proportional to $C_{in}e_n$ where the total input capacitance is $C_{in} = C_g + C_{prs}$. The characteristics of the operational amplifier are summerized in Sec. 4.1.

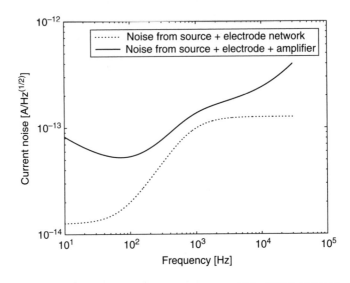

FIGURE 9.38 Current noise spectral density, S_P of the source plus electrode network compared with the total input referred current noise spectral density, SI_{Tot}. S_P is the dominant component of SI_{Tot} in the recording bandwith.

9.3.2.3 Hardware Test Bed

A diagram of the hardware test bed is shown in Fig. 9.39. The entire system is powered at 3.3 V using a USB bus, and the digital interface was provided using a field programmable gate array on an Opal Kelly 3001 board. The command voltage was provided by an Analog

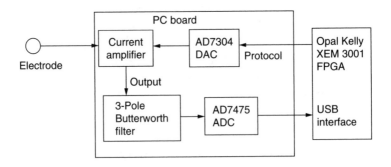

FIGURE 9.39 A block diagram of the harware test bed. The test bed consists of an ADC, a DAC, and an antialiasing filter. The current measurement system is controlled using a GUI-driven C++ program.

Instruments 8-bit AD7304 DAC. The output of the amplifier was digitized using a 12-bit Analog Instruments AD7475 ADC. The data was sampled at 62.5 kHz and low-pass filtered at 20 kHz using a three-pole Butterworth filter. A graphic user interface C++ program allows the compensation mDACs to be set, the stimulus pulses to be generated, and current signals to be plotted. The entire system consists of two stacked circuit boards and was packaged in a shielded metal box (Weerakoon et al., 2008b; Weerakoon et al., 2008c).

9.3.2.4 Experimental Results

Figure 9.40 shows the measured step responses of the current measurement system while using series resistance compensation. As predicted by Eq. (9.24) and Fig. 9.32 we were able to decrease the time constant needed to charge the source from 600 to 200 μs, obtaining a threefold increase in bandwidth. This corresponds to compensating 70% of a 4-MΩ series resistance.

Figure 9.41 shows the measured response of the current measuring system while using parasitic capacitance compensation. As predicted from Eq. (9.27) and Fig. 9.34, we were able to see positive, negative, and no overshoots corresponding to undercompensation, overcompensation, and proper compensation of the parasitic capacitance. The shown response is for approximately 10 pF of parasitic capacitance at the input. We were able to compensate up to 20 pF of parasitic capacitance.

Figure 9.42 shows the measured input–referred current noise spectrum compared with the theoretical noise as calculated in Eq. (9.33). Integrating the input-referred noise yields an RMS current noise of

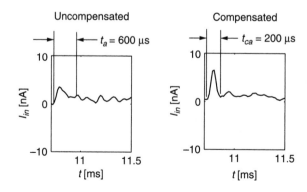

FIGURE 9.40 Measured response of the series resistance compensation circuit. The circuit is able to compensate for up to 70% of 4-MΩ series access resistance.

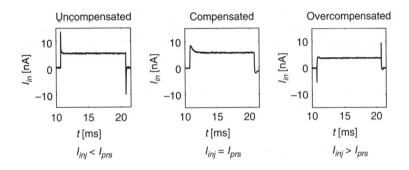

FIGURE 9.41 Measured response of the capacitive compensation circuit. This response was obtained while the circuit compensated a 10 pF parasitic capacitance.

5 pA at 10 kHz bandwidth. This corresponds to a SNR ratio of 250 or approximately 8 bits when the input is 1 nA. This result is comparable to state-of-the-art commercially available bench-top amplifiers made with discrete componenets. For example, the ionWorks Quattro amplifier from Molecular Devices (now MDS Analytic Technologies)

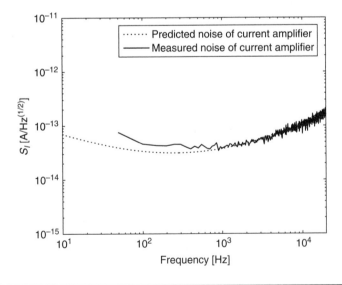

FIGURE 9.42 Measured amplifier input-referred current noise spectral density compared with the theoretical value calculated in Eq. (9.33). Integrating the input noise curve yields a RMS noise current of 5 pA.

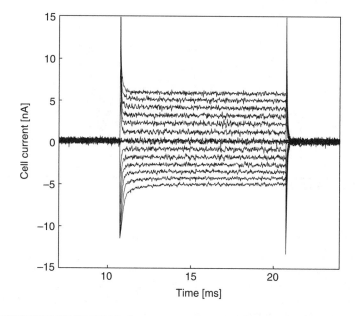

FIGURE 9.43 Time course of whole-cell measurements taken on human embryonic kidney cells expressing a high density of Slack channels carrying K+ current.

has noise levels of 10 pA of rms current at 10 kHz bandwidth (Molecular Devices Electrophysiology Instruments, 2008). The Triton-1 device from Tecella technologies has 0.5 pA RMS at 3-kHz bandwidth while using a R_f value of 1 GΩ (Tecella Electrophysiology Instruments, 2008).

Figure 9.43 shows recordings made from our current measuring system when operated as a patch-clamp amplifier. The measurement was conducted on human embryonic kidney (HEK) cells expressing a high density of Slack channels carrying K+ current. The time response of the patch-clamp amplifier was recorded as control voltage steps (V_{com}) are applied to the cell membrane from −80 to 80 mV in steps of 10 mV. The measurements were made using a conventional patch-clamp setup.

Figure 9.44 shows the glass micropipette electrode (1-μm diameter tip) and the HEK cells used to make the recordings from our current-measuring system operated as a patch-clamp amplifier. The parasitic compensation was set at 15 pF and the resistive compensation was set at 70%. The noise level on the output was 0.1 nA RMS. This excess noise is due to the noise generated in the hardware test bed and the compensation circuitry.

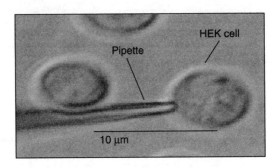

FIGURE 9.44   HEK cells and glass micropipette.

The power consumption of the patch-clamp system was measured as 300 µW with a 3.3-V power supply. Power consumption is dominated by the five operational amplifiers used in transimpedance and different amplifiers and the resistive and parasitic capacitive compensation mDACs. Each operational amplifier consumes 20 µA of bias current. The performance of the current-measuring system is summarized in Table. 9.6. A micrograph of the system die is shown in Fig. 9.45

9.4   SOS Voltage-Mode Biosensor Interfaces

In this section, we present the design of a voltage-mode biosensor interface designed in the SOS fabrication process. The circuit was designed to record biopotentials on the order of 100 µV to 1 mV RMS with

Process technology	SOS 0.5-µm CMOS
Input referred current noise at 10 kHz	5 pA$_{rms}$
SNR with 1-nA input	250
Resistive compensation capability	70% of 4 MΩ
Capacitive compensation capability	20 pF
Power consumption at 3.3 V	300 µW
Chip area (with pads)	1150 µm × 700 µm

TABLE 9.6   Performance of the SOS CMOS current-measuring system

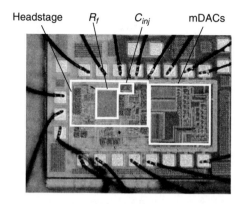

FIGURE 9.45 Die micrograph of the integrated patch-clamp amplifier.

a input noise of less than 10 μV RMS. This voltage amplifier biosensor interface was designed as a delta amplifier circuit. Figure 9.46 presents the operation of the delta amplifier. An input voltage biosignal V_{in} is amplified by a AC amplifier like the one presented in Fig. 9.3.

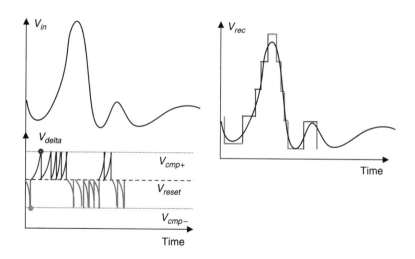

FIGURE 9.46 Operation of the SOS biosensor interface voltage amplifier. An input voltage biosignal V_{in} is amplified by a fixed gain amount. The amplifier output is initially reset to a V_{reset} voltage, and when it changes more than a threshold V_{cmp+} or less than V_{cmp-}, a pulse is generated and the output is once more reset to V_{reset}. A sign bit is generated with an output 1 if the output is above V_{cmp+} or 0 if it is below V_{cmp-}.

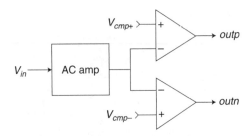

FIGURE 9.47 Block diagram of the SOS biosensor interface voltage amplifier. An input voltage V_{in} is amplified by an AC amplifier. Its output is compared against the two threshold voltage V_{cmp+} and V_{cmp-}. The input and output of the amplifier are reset to a voltage V_{reset} after passing one of the thresholds. Signals *outn* or *outp* generate a step response that is used to record the time of threshold crossing.

The amplifier output is initially reset to a V_{reset} voltage. When a change of the amplified input signal changes by more than V_{delta}, and is thus above a threshold V_{cmp+} or below V_{cmp-}, a pulse is generated and the output is once more reset to V_{reset}. As a pulse is generated, its time of appearance and its sign bit (1 if above V_{cmp+} or 0 if below V_{cmp-}) are recorded. This corresponds to a signed delta analog-to-digital conversion. The pulses and their signs are enough to reconstruct the original signal V_{rec}, as can be seen in the right of Fig. 9.46.

Figure 9.47 shows a block circuit diagram of an implementation of the delta amplifier. An input voltage V_{in} is amplified by an AC amplifier and then its output is provided as input to two comparators against the two threshold voltage V_{cmp+} and V_{cmp-}. The input and output of the amplifier are reset to a voltage V_{reset} after passing one of the thresholds. Signals *outn* or *outp* generate a step response that is used to record the time of threshold crossing.

Figure 9.48 is a circuit schematic of the entire delta amplifier implemented in the SOS process. The input is an AC amplifier with fixed gain also reported in Fig. 9.49. Pulse signals *outn* or *outp* are generated after comparisons against V_{cmp+} and V_{cmp-}. Signal pulse is the OR-ed version of *outn* and *outp*, and it is used to convey the time of the pulse, whereas *outp* can be used as a sign signal for the pulse polarity. An inverter chain delay provides feedback to reset the amplifier after a pulse has been generated. Notice that the amplifier input and output are reset to a voltage V_{reset} by means of transistors NMOS with sizes $[W,L]$ of [5 μm, 0.5 μm].

Figure 9.49 reports the schematic of the circuit used to implement the voltage-mode AC amplifier. Table 9.7 shows the size of the devices

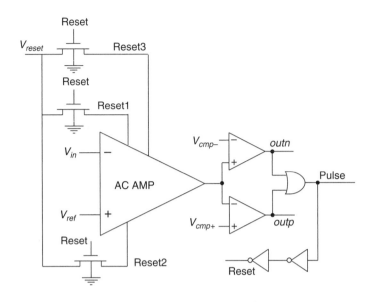

FIGURE 9.48 Circuit schematic of the SOS biosensor interface voltage amplifier. *outn* or *outp* are the pulse signal generated after the comparisons against V_{cmp+} and V_{cmp-}. Signal *pulse* is the OR-ed version of *outn* and *outp*, and it is used to convey the time of the pulse, whereas *outp* can be used as a sign signal for the pulse polarity. An inverter chain delay provides feedback to reset the amplifier after a pulse has been generated.

used in the design. This stage had a fixed AC voltage gain of 700 (35 pF/50 fF). Transistors M1 to M4 are used to bias the input of the operational amplifier to V_{reset} by means of reverse-biased diodes with a large impedance. The amplifier is also periodically reset to voltage V_{reset} after each pulse. Both the reset and M1–M4 transistors keep the floating nodes input of the operational amplifier around V_{reset} and make sure that they do not drift during operation, causing malfunction of the amplifier. This technique is typical in AC-coupled CMOS amplifiers.

Figure 9.50 presents a comparison of the AC amplifier operational amplifier connected as a voltage follower and its noise performance. The comparison is for two identical amplifiers implemented in both SOS and 0.5-µm CMOS technology from American Microsystems Inc. (AMI). The operational amplifier used in this design is seen in Fig. 9.3 and reported in Sec. 4.1. The AMI bulk CMOS amplifier had identical sizes to the SOS amplifier. From Fig. 9.50, it is clear that the SOS amplifier provides a worse noise performance than a bulk

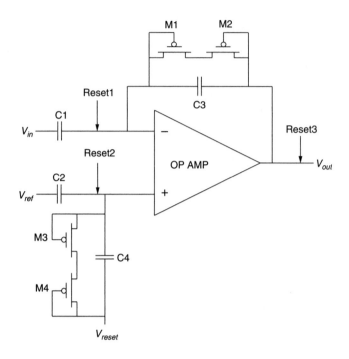

FIGURE 9.49 Circuit schematic of the AC amplifier used in the SOS biosensor interface voltage amplifier. Table 9.7 shows the size of the devices used in the design. This stage had a fixed AC voltage gain of 700.

CMOS implementation. The flicker coefficient is approximately 10 times worse in SOS ($5 \cdot 10^{-24}$ versus the value of bulk CMOS of $5 \cdot 10^{-25}$), and the thermal voltage is also 5 times higher than the bulk CMOS process. This difference is mostly due to the lower channel doping

Device Type	Size
C1, C2	35 pF
C3, C4	50 fF
M1–4	[5, 0.5] μm

TABLE 9.7 Parameters of AC amplifier used in the SOS biosensor interface voltage amplifier

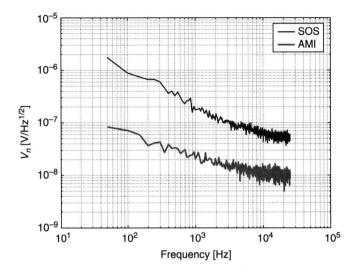

FIGURE 9.50 Comparison of the noise performance of the AC amplifier in Fig. 9.49 implemented in both the SOS and a 0.5 μm bulk CMOS technology from AMI.

of the SOS process, resulting in larger channel resistance. The higher flicker noise is due to the SOS interface and its traps.

Integrated over a 10-Hz to 20-kHz bandwidth, total RMS noise is 2 μV for the AMI process design, and 20 μV RMS for the SOS process design.

CHAPTER 10

SOS Design of Isolation and Three-Dimensional Circuits

10.1 Introduction

This chapter presents CMOS isolation circuits and three-dimentional circuits that can be obtained using the SOS fabrication process.

Single-chip isolation circuits can be obtained in SOS by taking advantage of the insulation of the sapphire substrate (Culurciello et al., 2005b; Culurciello et al., 2005c; Marcus et al., 2006; Culurciello et al., 2007). We present monolithic isolation techniques using SOS process and examine design and test results of digital isolation circuits in SOS. These devices can be used in a wide variety of applications that require passing signals across an isolation barrier: power supplies, remote sensing, and medical and industrial applications. These devices can also be used in every application when a common ground cannot be guaranteed to prevent ground loops from causing circuit damage and to offset errors at sensitive nodes.

Three-dimensional assemblies of multiples dies and circuits can be also obtained using capacitive coupling and alignment simplified by the transparency of the SOS dies (Culurciello et al., 2005a; Culurciello and Andreou, 2005). We explore the use of techniques for isolation to communicate power and data between two or more SOS dies. The circuits described here are one of the first examples of three-dimensional (3D) circuits obtained in the SOS process. Capacitive coupling can be used to transfer data and power in a 3D assembly of two SOS dies with no galvanic connections. This technique can be used to package multichip ensembles of sensors and processor without using galvanic connections. In addition, it can be used to advance the 3D fabrication technology of multichip modules. We also show that power exchange and communication can be achieved between sensors located in two different SOS dies. This technique is also extensible to bulk CMOS.

10.2 Isolation Circuits in SOS

An isolation buffer is an electrical circuit that communicates an input digital voltage from one region to an output digital voltage in a second region where the ground node is electrically isolated from the first one. A digital isolation amplifier takes a digital signal as input and transmits it to an output circuitry that will reproduce a delayed copy of the input. Input and output circuits are electrically isolated from their respective ground terminals. The isolation is desirable in harsh environments, for remote sensing in location with ground loops or where it is not possible to ensure a common ground signal between output and input nodes. Applications include biomedical equipment and instrumentation, control of high-voltage converters and circuits, and high-voltage or high-current factory environments.

An integrated version of an isolation circuit is conventionally an assembly of two separate dies packaged together (Waaben, 1975; Harper, 2000). The cost of the isolator can be high because of the expenses and the difficulties in packaging two dies with the desired isolation properties. In addition, bulk processes cannot provide isolation of two portions of a die because of the presence of a common substrate. The substrate is generally tied to a ground node to avoid CMOS latchup, and the ground resistance between two parts of the silicon die is generally a few ohms. Differences in grounds between two parts of a bulk CMOS circuit would generate large ground currents, circuit latchup, and subsequent damage.

The isolation properties of the SOS substrate make it a perfect candidate for high-performance monolithic isolation circuits (Culurciello et al., 2005b; Culurciello et al., 2005c; Marcus et al., 2006; Culurciello et al., 2007; Peregrine, 2008a). In fact, SOS circuits are isolated from each other by both the sapphire substrate below them and the field oxide around and on top of them. It is, therefore, possible to have two different circuits present on the same die and operating with different ground potentials.

Figure 10.1 illustrates the isolation property of two SOS metal layers (of a 2 nH transformer described in Sec. 10.2.4). Specifications for the isolation device described in this chapter were a ground-to-ground isolation of up to 100 V. We measured the isolation of the amplifier up to 110 V with a Keithley unit 236 and measured no significant current (Fig. 10.1), as evidence that the isolation in the SOS die is holding at least to the specifications. The actual *measured* breakdown of the device occurred in the proximity of 820 V between the grounds of input

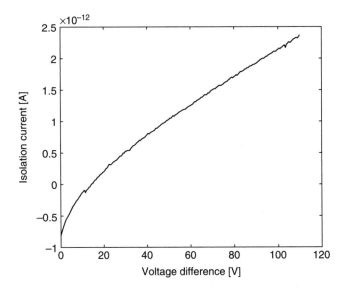

FIGURE 10.1 Isolation performance between coils in metal-1 and metal-3 in the SOS process. The two large coils are separated only by 3.6 μm, yet can provide much more than 100 V of isolation. Breakdown was measured at 820 V.

and output circuits. This isolation is guaranteed by the 3.6 μm separation from the metal-1 and metal-3 coils composing the interface between input and output. The breakdown measurements were conducted using FisherBiotech FB400 electrophoresis equipment. These isolation maximum ratings are not as high as the ones that can be obtained using a multidie isolator (more than 1000 V DC and 2000 V peak), but they are sufficient for many applications.

10.2.1 Monolithic Isolation Techniques

Several techniques are viable to electrically isolate two portions of a circuit or regions. The physical wired connection of the conduction electrical signal has to be converted to an electromagnetic field that can be transferred between two circuits without a physical connection. Alternatively, the input signal can be converted into other kinds of force fields using microelectromechanical systems (MEMS) or fluidic transducers, just to mention a few.

Conventional isolation circuits can be performed with optical couplers (optocouplers) (Waaben, 1975). The combination of a light source and a light sensor can transmit a signal by using amplitude modulation (turning the light on and off). To obtain optical couplers, the fabrication process needs to be of a special kind to be able to generate light. Light sources are not generally available in a bulk CMOS circuit because they require special doping levels and particular semiconductors combination (refer to Sze, 1981). The specialized nature of the process raises fabrication costs and still requires two dies in one package (Harper, 2000). To our knowledge, no light source was ever reported to date using the SOS process, therefore, this technique cannot be used in an SOS implementation.

Isolation can be performed using capacitive coupling between two parts of a circuit (Kuhn et al., 1995; Mick et al., 2002b). This technique requires that the two circuits be placed sufficiently close together to obtain a significant coupling. Capacitive coupling has been employed in bulk CMOS multichip modules to transfer data signals between multiple dies (Salzman and Knight, 1994; Gabara and Fischer, 1997; Mick et al., 2002a; Xu et al., 2004). On-chip isolation using a SOI substrate has been demonstrated for modem lines (Kanekawa et al., 2000) and for data only. We discuss one design of capacitively isolated circuits as an example and examine the tradeoffs in Sec. 10.2.2.

Another technique for isolation is using electromagnetic fields (refer to Kuhn et al., 2001). This is the ubiquitous case of a radio system where two transceivers can be located at very different ground potentials and still be able to communicate. It is possible to create a simple radio transceiver in a single CMOS die by using an oscillator and a transformer. We discuss one design of electromagnetically isolated circuits as an example and examine the trade-offs in Sec. 10.2.4.

10.2.2 Capacitively Isolated Circuits in SOS

Capacitive isolation circuits employ a capacitor C to transmit information from transmitter (circuit A) to receiver (circuit B) without a physical connection (Fig. 10.2). The communication is performed by

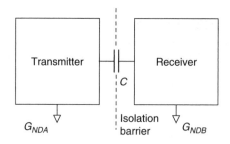

FIGURE 10.2 Principles of operation of a capacitively coupled isolation circuit. Two circuits with different grounds communicate though a capacitor. The transmitter circuit is A, and the receiver is B.

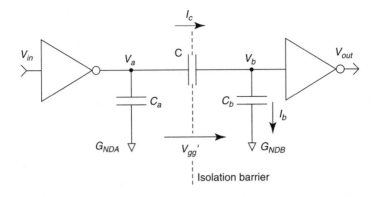

FIGURE 10.3 Model of operation of a capacitively coupled isolation circuit. Capacitor C is the coupling interface in Fig. 10.2.

using the electric field at the two capacitor plates. When the electric field changes, the receiving plate of the capacitor senses a current proportional to the charge removed or added to the capacitor. This charge is proportional to the capacitance and the voltage swing at the input terminal (Culurciello et al., 2005b; Culurciello et al., 2005c; Marcus et al., 2006; Culurciello et al., 2007).

To ensure proper operation, the isolation capacitance must be much higher than the receiver input capacitance. This is to avoid capacitive division of the transmitted voltage and for an additional reason that will be explained momentarily. Referring to Fig. 10.3, C is the isolation capacitor, and C_a and C_b are the parasitic capacitances at the two terminals of the isolation capacitor.

Consider now an AC model of the capacitive coupling. V_{in} is a digital signal whose value is always between V_{DD} and 0 V. Equivalently, V_a is the inverted voltage of V_{in}, and it is always between 0 and V_{DD} V. The voltage V_b can be calculated using Eq. (10.1).

$$V_b = V_a \frac{C}{C + C_b} \tag{10.1}$$

As can be seen from the capacitive divider Eq. (10.1), for proper operation the capacitance C must be much larger than the parasitic C_b. In the contrary case, the transmitted signal amplitude is going to be smaller than V_{DD} and errors will occur.

Let us consider the charge at nodes V_b and across the isolation capacitance C: $Q_b = C_b V_b$ and $Q_c = C(V_a - V_b)$. The maximum charge at V_b is $C_b V_{DD}$ and the minimum is 0 V. Assuming that the node V_b is protected toward the power supplies by two diodes of threshold V_{th}, then the maximum charge across C is $C(V_{DD} + 2V_{th})$ and the minimum is 0 V.

To assess whether the capacitive link is functional, we need to see if a swing of V_a can change the state of V_b (charge its terminal by a voltage swing of $\pm V_{DD}$). For example, consider the case when $V_b = -V_{th}$ and V_a is switching from 0 V to V_{DD}. We obtain that the initial charge $Q_{bi} = -C_b V_{th}$ and the initial charge across C is $Q_{ci} = CV_{th}$. The final value of the charge across the isolation capacitance is, therefore, $Q_{cf} = C(V_{dd} + V_{th})$. Therefore, the final charge at node V_b is $Q_{bf} = -C_b V_{th} + C(V_{DD} + V_{th})$. If we chose $C \gg C_b$ then the voltage V_b will be higher than V_{DD}, limited by the protection diodes. Thus operation requires that $C \gg C_b$.

A problem of capacitive isolation is that input voltage swings have to be detected and ground bounce swings have to be rejected. In fact, only the former is the desired signal, whereas the latter is a noise signal. If the ground bounces are much bigger and faster than the communication signal across the coupling capacitor, then communication might not be possible. But, in general ground, bounces are slow phenomena due to the large capacitance of ground nodes, so there is rarely a problem with this device. Let us consider the currents across the isolation capacitor C. The current i_c is expressed by Eq. (10.2).

$$i_c = C\frac{dV_a}{dt} - C\frac{dV_b}{dt} \tag{10.2}$$

By examining the second term on the right-hand side of Eq. (10.2), we can express it as in Eq. (10.3).

$$\frac{dV_b}{dt} = f\left(\frac{dV_{gg'}}{dt}, \frac{dV_a}{dt}\right) \tag{10.3}$$

The second variable on the right-hand side of Eq. (10.2), is governed by the input voltage V_{in}. The first variable is governed by the ground bounce (voltage $V_{gg'}$) and it is to be rejected. V_b can be influenced both by a change in V_a and $V_{gg'}$. Using superimposition of effects, we can write Eq. (10.4).

$$i_b = C_b\frac{dV_{gb}}{dt} - C_b\frac{dV_{gg'}}{dt} \tag{10.4}$$

With V_{gb} being the voltage between node b and the input ground (G_{NDA}). The first term on the right hand side of Eq. (10.4) is due to V_a or the input signal. The second term is the interference or noise due to ground bounce. We conclude that, for correct device operation, Eq. 10.5 has to be satisfied.

$$\frac{dV_a}{dt} \gg \frac{dV_{gg'}}{dt} \tag{10.5}$$

This imposes a constraint on the minimum signal slew rate. If the signal slew rate is much higher than the ground slew rate [Eq. (10.5)] then the ground bounces can be attenuated using a high-pass filter at the receiving node. This can be done by adding a leak current to the receiver-node, typically using a reverse-biased diode properly sized.

As an example, consider the case of $(dV_a/dt)_{min}$. This value must be bigger than $(dV_{gg'}/dt)_{max}$. With a power supply of 3 V and 10 Mbps communication, the value of $(dV_a/dt)_{min}$ is approximately 1.5×10^7, whereas the maximum grounds slew rate $(dV_{gg'}/dt)_{max}$ is about 1×10^6. These values satisfy Eq. (10.5).

In the following sections, we report on the design of two types of SOS digital isolation buffers. The two types of SOS capacitive isolation devices are differential transmission, and digitally modulated. Both implementations feature a monolithic single-chip isolation device with the following specifications (Culurciello et al., 2005b; Culurciello et al., 2005c):

- Data rate: higher than 10 Mbps.
- Temperature range:−55°C to +125°C operation.
- Isolation: minimum 100 V continuous ground isolation.
- Supply: can be powered only by the output/receiver side.
- Ground bouncing rejection: minimum 1 V/μs.
- Input rise/fall time: 10 ns max, 1.5 ns min.

We describe both types of implementation in the following sections.

10.2.2.1 SOS Capacitive Isolator
with Differential Transmission

In this section we report the design of a SOS digital isolation buffer with differential transmission at the baseband. The device is composed of a transmitter or input circuit A and a receiving or output circuit B (Fig. 10.2). Isolation is provided by capacitive coupling at baseband, avoiding modulation of the input signal.

A schematic of one isolation channel *isoCap2sc* is given in Fig. 10.4. The device is designed to withstand ground bouncing of more than 1 V/μs by using asynchronous circuitry to reject spurious transitions. This is done by employing a differential scheme at the input, before the capacitive isolation interface of Fig. 10.2.

A schematic of the asynchronous C-element circuit *celiso* is given in Fig. 10.5. The asynchronous C-element circuit *celiso* is a static logic cell that switches its output only when it detects a valid differential transition. A valid transition is a transition where one output has a logic level of 1 and the other a level of 0. If only one of its inputs

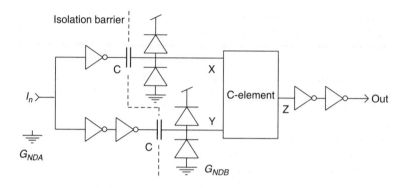

FIGURE 10.4   Capacitive isolator circuit: the isolation cell *isoCap2sc*. Differential transmission of the input signal through the capacitors allows to reject spurious transitions (transmitter circuit A, left of isolation barrier). An asynchronous C-element *celiso* demodulates the received signal (receiver circuit B, right of isolation barrier).

switches, because of a ground bounce or supply spike, the inputs will be both 0's or 1's, and the output will not commute. This allows us to reject ground bounces that cannot be differentiated from legitimate input transitions in a single-ended circuit scheme.

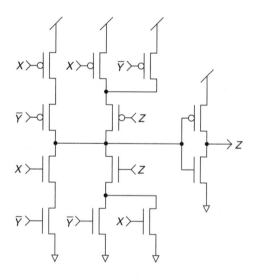

FIGURE 10.5   Asynchronous C-element cell for the capacitive isolator: *celiso*. The output changes state only if X and $\overline{Y}$ are asserted. Otherwise, the last output is maintained. The C-element implements the function in Eq. (10.6).

The *celiso* function can be expressed in concurrent hardware process (CHP) production rules by Eq. (10.6).

$$[\neg X \wedge Y] \rightarrow Z \uparrow, [X \wedge \neg Y] \rightarrow Z \downarrow \qquad (10.6)$$

The input signal is buffered by digital inverters at the transmitter/ input side of *isoCap2sc* (see Fig. 10.4) and communicated differentially to the receiver/output circuit using capacitive coupling. The coupling capacitors C have a capacitance of 150 fF, a value at least 10 times the parasitic capacitance C_b at the receiver floating node. The coupling capacitor have been designed using metal-1 and metal-3. The silicon area used by the capacitor is 175×60 μm^2. We mentioned in the previous section the reasons and calculations for the size of this capacitor.

The differential signal is buffered with inverters and recombined using the asynchronous circuit to reproduce a final digital output signal (*Out* in Fig. 10.4). At the receiver side, protection diodes (see Fig. 10.4 located at the receiver input node enforce that the voltage at the floating nodes always drifts to one of the supplies to prevent damage.

Both transmitter and receiver circuits operate at a nominal power supply of 3.3 V. The unit can be powered from input and output side or, alternatively, from the output side only. We employ a charge pump that operates on the output power supply and is capable of powering the input differential circuitry (Dickson, 1976; Culurciello et al., 2005a). A schematic of the charge pump circuit is given in Fig. 10.8. The pump generates the required 3.3-V input supply with enough current (1.5 mA) to drive the inputs at full speed. Protection diodes ensure that an unbound received voltage does not damage the input circuits. The charge pump has separate external supply connections so it can be disabled to save power when the input side is externally powered.

The final prototype of the capacitive isolation buffer named *isoCap2* is organized as an array of the four independent isolation channels *isoCap2sc* presented in Fig. 10.4. All channels are located in one single chip. All channels share the same input and output power supplies. The four-channel isolation buffer architecture is shown in Fig. 10.6.

The final prototype is organized as an array of four separate isolation channels (each made up of one *isoCap2sc* cell) and sharing the same input and output power supplies (Culurciello et al., 2005b; Culurciello et al., 2005c). A micrograph of the final device layout is shown in Fig. 10.7. The die has 16 bonding pads: the bottom 8 are (left to right) 2 input supplies (supply for the input isolation circuit and output of the charge pump), 4 data inputs, and 2 grounds (circuit and charge pump output ground), the top 8 are (left to right) the 2 output supplies (output circuit supply and charge pump input supply), and 4 data outputs, and 2 grounds (output circuit ground and charge pump input ground). The die size is 1×1.2 mm.

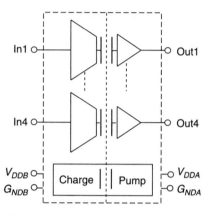

FIGURE 10.6 System architecture of a four-channel isolation buffer with differential transmission. Here only two channels are shown. The system is composed of four identical differential isolation channels *isoCap2sc* (presented in Fig. 10.4) and the charge-pump in Fig. 10.8.

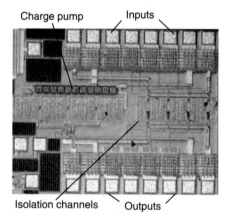

FIGURE 10.7 Micrograph of the fabricated SOS four-channel isolation amplifier. The die has 16 pins: four input (top), and four output channels (bottom), separate supply and ground for both input and output circuits and for the optional charge pump. The transmission scheme is internally differential. The die size is 1×1.2 mm.

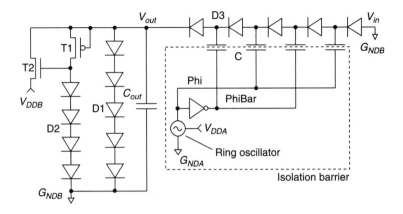

FIGURE 10.8 Charge pump circuit for the differential transmission SOS isolation buffer. The isolated circuit uses a ring oscillator to pump through four capacitive stages a voltage equal to the power supply. Diodes allow us to inject charge into the receiver reservoir capacitor. A series of transistors regulate the output voltage, whereas a series of diodes forces the maximum received voltage within the maximum supply range of the process (Culurciello et al., 2005a). The circuit inside the box is transmitter A and the circuit outside is receiver B.

10.2.2.2 SOS Isolation Charge Pump

The charge pump used in the isolator design is based on the Dickson charge pump design (Dickson, 1976; Culurciello et al., 2005a) and it is composed of four stages. A schematic is reported in Fig. 10.8. The circuit uses two square wave pumping clocks φ (Phi), and φ (PhiBar) to pump charge through coupling capacitors C. The output voltage V_{out} of the Dickson charge pump is given by Eq. (10.7):

$$V_{out} = V_{in} + N \left(\frac{C}{C + C_S} V_\phi - V_D - \frac{I_{out}}{(C + C_S) f_\phi} \right) - V_D \qquad (10.7)$$

C_S is the ground parasitic capacitance of each capacitor C (not shown in the figure). V_ϕ is the voltage amplitude of the clock φ, and f_ϕ is the clock frequency. With ideal diodes and capacitors, each stage increments the voltage V_{out} by a step equal to V_ϕ. The diode is employed in each stage to constrain the charge flow in one direction, allowing capacitors C to charge but not to discharge. The built-in diode voltage drop reduces the pumping voltage, subtracting V_D from the voltage pumped at each stage. When designing charge pump circuits using discrete components, Schottky or germanium diodes are often employed as they have lower built-in voltage than silicon PN junctions.

The open-circuit output voltage of the charge pump is reduced by the load current I_{out} [last term in parentheses of Eq. (10.7).]

In our isolation amplifier, we use the four-stage isolation charge pump shown in Fig. 10.8. An isolation charge pump architecture has the same functional blocks as a conventional charge pump but with the power supply rails of the oscillator galvanically isolated from the output supply rails. The charge pump was designed to provide an isolated power supply identical to the input supply to a galvanic isolation buffer. An 11-stage ring oscillator produces a 350-MHz square-wave clock signal to drive the pump. The output of the oscillator is buffered to drive the pumping capacitances C. Each capacitance has a value of 450 fF and forms an isolation barrier between the two regions of the circuit. These capacitors are designed using a parallel plate configuration of metal-1 and metal-3 (plate separation of 2 μm SiO_2 dielectric).

To optimize the design of the Dickson charge pump circuit, we employ MOS transistors with different threshold voltages that are available in the Peregrine SOS CMOS process (Andreou et al., 2001; Peregrine, 2008a). The following naming convention is employed to identify the device: regular Rx threshold ($V_{TH} \approx 0.7$ V), low xL threshold transistors ($V_{TH} \approx 0.3$ V, and intrinsic Ix, zero threshold transistors ($V_{TH} \approx 0$ V). The type of transistor is denoted by substituting x for either N or P. Diode D1 is the diode connected regular threshold transistors that is employed by the technology vendor as electrostatic discharge protection diodes.

Diode-connected NL MOS transistors are used as the rectifiers (D3) in each pump stage. The use of a low threshold MOS transistor in this part of the circuit minimizes the forward bias diode drop V_D to 0.3 V, and reduces the undesirable voltage drops on the rectifiers. A 9-pF capacitor at the output (C_{out}) together with a series of five diodes (D1) are used as the charge pump filter and constraint the maximum voltage within the maximum supply range of the process (3.3 V). Transistors T1, T2, and a diode chain of four diodes D2 form an active voltage regulator at the output of the charge pump to produce the regulated voltage V_{DDB} (Fig. 10.8). The series regulator transistor T2 is a zero-threshold IN device biased by a string of four regular threshold (RN) diode-connected MOS transistors (D2) and a current source. Transistor T1 is a self-biased current source implemented using an IP MOS transistor. Transistor T2 acts as an ideal voltage follower without a build-in voltage. The device type and sizes are given in Table 10.1.

The absence of substrate parasitics in the SOS CMOS process eliminates all the stray capacitance to ground C_S that further degrade the voltage gain of each stage [see Eq. (10.7)]. The stray capacitance in a 0.5-μm bulk-process can have a value of up to 10% of the nominal value (from extracted parameters MOSIS, 1999). In this case a capacitance of

Device	Type	Size [μm]
D1	RN	27.4/1.6
D2	RN	2/2
D3	NL	15/0.5
T1	IP	10/2
T2	IN	6/4

TABLE 10.1 Device type and size of MOS transistors used in isolation charge pump design

$C = 450$ fF can have a parasitic $C_S = 45$ fF. The capacitance C obtained with the SOS process has virtually zero parasitic capacitance, thus the ratio $C/(C + C_S)$ is 1 in SOS CMOS and 0.9 in a bulk CMOS. For a design with no parasitic capacitances, such as the one presented in this chapter, Eq. (10.7) simplifies to Eq. (10.8):

$$V_{out} = V_{in} + N \left(V_\phi - 0.3 - \frac{I_{out}}{C f_\phi} \right) - 0.3 \qquad (10.8)$$

Solving Eq. (10.8) results in a value of the unloaded V_{out} of 11.7 V, with a supply of 3.3 V, $V_{in} = 0$, $V_D = 0.3$ V, $N = 4$, $I_{out} = 0$. The equivalent result in a bulk process of Eq. (10.7) is V_{out} of 8.3 V with $V_D = 0.7$ V.

The measured unloaded charge pump voltage output V_{DDB} was approximately 2.5 V. The output voltage is limited to this value by the chain of protection diodes D1 in Fig. 10.8. Fig. 10.9 plots the measured charge pump output voltage (V_{DDB}) as a function of the input power supply (V_{DDA}) and the load. The charge pump was tested in three loading conditions: unloaded, loaded with 5-kΩ, and with 22-kΩ resistors. Using an input power supply V_{DDA} of 2 V the charge pump can supply 100 μA at 2 V for a 22-kΩ load or 0.6 V for a 5.3-kΩ load. The output of the charge pump falls rapidly when the current drawn is high or the input voltage is too low to counterbalance the losses in the charge pump diodes. The power consumption of the unloaded charge pump circuit is 2.5 mA at a 3.3-V supply. Current consumption is due to the oscillator circuit and the buffers to drive the pumping capacitors.

Figure 10.10 plots the scaling properties for the charge pump output current as a function of capacitors (pumping stages). The charge pump output is shown for an unloaded and unprotected pump. The charge pump output current can be increased to hundreds of milliamperes by sacrificing the isolation properties of metal-1 metal-3 capacitors and

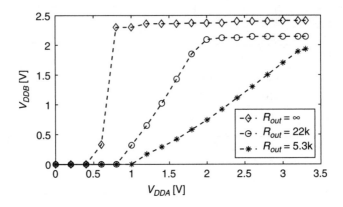

FIGURE 10.9 Received voltage output of the charge pump as a function of the load. Given the very modest requirements of the input circuit, the charge pump can generate both the voltage and currents required to operate the transmitter circuit A.

by using native MIM capacitors with 0.1-μm plate distances (SiO_2 dielectric).

The power consumption of the unloaded charge pump circuit is 2.5 mA at 3.3 V supply. The current consumption is due to the oscillator circuit and the buffers to drive the pumping capacitors. The charge pump provides continuous isolation between the power supplies in the excess of 800 V.

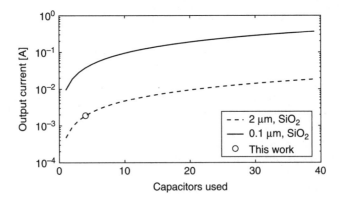

FIGURE 10.10 Scaling properties of the proposed charge pump (Culurciello et al., 2005a). The output current that the charge pump can source is a function of the number of coupling capacitors. The circle indicates the current design. A thinner insulation and a modest increase in coupling capacitors (10) can achieve 100 mA of current handling.

The unregulated power efficiency η of the charge pump can be calculated using Eq. 10.9.

$$\eta = \frac{I_{V_{DDA}} \cdot V_{DDA}}{I_{V_{DDB}} \cdot V_{DDB}} \tag{10.9}$$

Using a $I_{V_{DDB}}$ of 315 μA, a $I_{V_{DDA}}$ of 2.5 mA and a V_{DDA} of 3.3 V, the efficiency in the ideal case (no diode drop) is 32%, whereas it is 23% for the SOS CMOS implementation using multithreshold technology ($V_D = 0.3$ V), and it is 9% in case of a bulk CMOS implementation ($V_D = 0.7$ V). Note that this efficiencies are computed without output voltage regulator circuits. The measured power efficiency of the fabricated charge pump with voltage regulator is 4%. The current paths through the shunt-regulated diodes (D1 stack) and regulator bias (D2 stack) drains most of the charge pump charge to yield a low efficiency in the fabricated pump. The efficiency of the pump can be improved by sizing the current source transistor T1 to a lower branch voltage and by eliminating the shunt regulator (D1 stack), which is only included here for overvoltage protection.

The pump generates the required input supply voltage and currents in order to drive the inputs at 100 MHz. Outputs are buffered with digital inverters to be able to drive a 50 pF capacitive load. Protection diodes ensure that an unbounded received voltage does not damage the input circuits. The charge pump has separate external supply connections and can be disabled to save power when the input side is externally powered.

10.2.2.3 SOS Differential Capacitive Isolator Measurements and Results

We simulated the isolation buffer at the design corners for temperature, and transistors characteristics. We simulated at $-60°$C, $+130°$C for typical, fast, and slow transistors. We repeated the measurements at 10 Mbps and 100 Mbps. We tested the circuit in two configurations: powered from input and output and output only. The circuit performed correctly in all settings. We also tested the circuits with a power supply of $\pm 10\%$ of the nominal value at 10 Mbps. In both cases, the circuit was operational as in the nominal 3.3-V supply setting. We also measured device supply current (with and without the charge pump operating) versus supply voltage versus number and frequency of channels operating. These results are summarized in Table 10.2 (Marcus et al., 2006).

Static current consumption for both transmitter and receiver circuits at 3.3 V was about 34 μA, because the circuit is implemented with fully static digital logic. These data were collected with the charge pump circuit turned off. When the charge pump was turned on, the static

$V_{DD_{in}}$	$V_{DD_{out}}$	$I_{DD}(mA)$	Input Signal(s)
2.5	$V_{DD_{in}}$	0.1055	1 at 1 MHz
2.5	$V_{DD_{in}}$	0.2257	4 at 1 MHz
2.5	$V_{DD_{in}}$	4.789	1 at 80 MHz
2.5	$V_{DD_{in}}$	15.980	4 at 80 MHz
3.3	$V_{DD_{in}}$	0.034	1 at 0 Hz
3.3	$V_{DD_{in}}$	0.036	1 at 1 kHz
3.3	$V_{DD_{in}}$	0.041	1 at 10 kHz
3.3	$V_{DD_{in}}$	0.084	1 at 100 kHz
3.3	$V_{DD_{in}}$	0.2539	1 at 1 MHz
3.3	$V_{DD_{in}}$	0.6684	4 at 1 MHz
3.3	$V_{DD_{in}}$	5.956	1 at 80 MHz
3.3	$V_{DD_{in}}$	24.94	4 at 80 MHz
2.5	CP	1.620	1 at 1 MHz
2.5	CP	1.738	4 at 1 MHz
2.5	CP	6.302	1 at 80 MHz
2.5	CP	17.545	4 at 80 MHz
3.3	CP	2.5	1 at 0 Hz
3.3	CP	3.4978	1 at 1 MHz
3.3	CP	3.6751	4 at 1 MHz
3.3	CP	9.833	1 at 80 MHz
3.3	CP	25.280	4 at 80 MHz

TABLE 10.2 Measurements of SOS differential isolation circuit current for various power supplies and input signal configurations with a capacitive load of 25 pF

current consumption was 2.5 mA with an output supply of 3.3 V. The received power supply output of the charge pump is reported in Fig. 10.8. Setting the output power supply V_{DDB} to 1.5 V, the circuit worked perfectly with an input square wave of 30 MHz, and taking 3 mA from the power supply, with the charge pump circuit turned on. Isolation was verified experimentally, with the circuit operating with an input square wave of 30 MHz, $V_{DDB} = 3.3$ and $V_{GNDB-GNDA} = 25$ V. All these measurement were conducted with the isolation chip driving 2 ft of coaxial cable and a 25-pF load (oscilloscope load).

The simulated power consumption of the device with unloaded output pads is 1 mA when powered from both input/transmitter and output/receiver sides at 100 Mbps. The power rises to 2 mA when

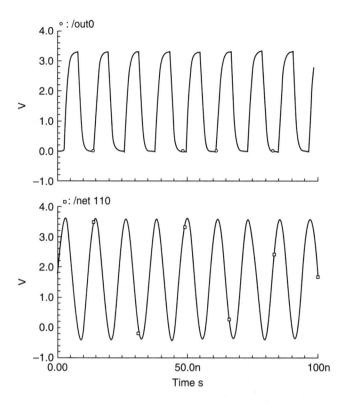

FIGURE 10.11   Simulation of the SOS differential isolation circuit with an 85-MHz input. The top trace is the output of one isolation channel, and the bottom trace is the input signal to the same channel.

powered only from the output/receiver side at 100 Mbps. These data are for typical devices at room temperature and with a 3.3-V supply.

$$\Delta t = \frac{C}{I} \Delta V \qquad (10.10)$$

Equation (10.10) calculates the time needed to transition across the isolation capacitor, given the isolation capacitance value C, the maximum drive current, and the ΔV voltage swing. If the voltage swing is the nominal digital supply voltage of 3.3 V and the maximum drive current is 100 μA, the transition time is 5 ns. The maximum current drive was hand calculated from a RN NMOS of size 4 × 0.5 μm in saturation (Peregrine, 2008a).

Figure 10.11 shows screen the simulation verifying functionality of the prototype with an input sine wave at 85 MHz. The input was

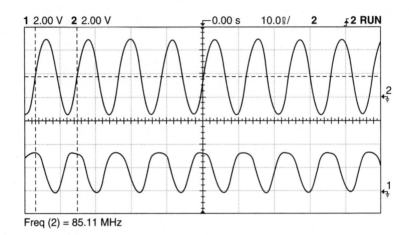

1 2.00 V **2** 2.00 V   ⌐−0.00 s   10.0⅋/   **2**   ⨍**2 RUN**

Freq (2) = 85.11 MHz

FIGURE 10.12 Oscilloscope traces verifying functionality of the SOS differential isolation circuit. The top trace is the 85-MHz input waveform. The bottom trace is the output of the isolation channel. Notice the similarity of the response of the circuit with the simulation in Fig. 10.11.

chosen to be a sine wave to be compatible with the measured data input in Fig. 10.12, which resembles a sine wave due to an impedance mismatch with the prototype's package. The bottom trace of Fig. 10.11 is the 85-MHz input square wave. The top trace is the output of the isolation channel. Fig. 10.12 shows screen shot from measured data captured with an oscilloscope. The top trace is the 85-MHz input square wave, the bottom trace is the output of the isolation channel. Notice the close match of simulations and measured data to stress the fidelity of the device models when operated above threshold (refer to Chap. 1 for details).

Next, we determined the minimum permissible input slew rate for error-free data transmission. We defined this as the lowest slew rate for which the output would still consistently toggle (Marcus et al., 2006). For fast edges, the isolation capacitor current i_C [Eq. (10.2)] is much larger than the leakage currents at the leakage node, which keeps V_b relatively unattenuated. But as the input edge rate slows, leakage currents become a larger fraction of i_C, and V_b begins to become attenuated. At some critical level of input edge rate, the received voltage will be too small to commute the output buffer, and a transmission error will occur.

To find this critical rate, we used the circuit of Fig. 10.13 and a 10-Hz 50% duty cycle square wave input to create an adjustable, slew-rate-limited input for the isolator. We then ramped the slew rate down until we found the minimum rate at which the device would still

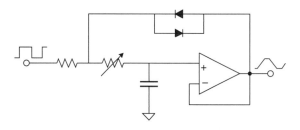

FIGURE 10.13   Adjustable slew rate limiter circuit used to determine the SOS differential isolation circuit minimum acceptable slew rate.

reliably operate. We found that this rate varied significantly among devices from a minimum of 0.01 V/μs to a maximum of 0.2 V/μs. The average measurement was 0.067 V/μs. These results are in good agreement with specifications of standard logic interfaces. For example, a typical interface device might be the HC14 Schmitt trigger inverter with minimum specified output slew rate of 18 V/μs. This is 90 times faster than the minimum rate recognized by the isolation amplifier and implies good margin in the interface.

We performed an end-to-end system test by calculating bit error rates through the device with operating temperature and ground bounce slew rate as parameters. The isolator was powered from 3.3-V input and output supplies with the charge pump disabled. We used a Fireberd 6000 bit error rate tester (BERT) to inject a pseudo-random data sequence into the device and record errors in the output sequence. The pseudo-random sequence was 2047 bits long and clocked at a frequency of 15 MHz. We connected a variable frequency sawtooth waveform between the isolator's input and output grounds to measure ground bounce rejection. The sawtooth had a 20 V_{pp} amplitude with 50% duty cycle and a variable frequency that we set to give ground bounce slew rates ($dV_{gg'}/dt$) from 1 V/μs to 12.5 V/μs. Because BERT has common input and output ground connections, we used pulse transformers to couple the clock and data signals between the tester and the isolator to avoid shunting the ground bounce signal. Average bit error rates were calculated after 1×10^9 transmitted bits. The complete setup is shown in Fig. 10.14.

The test was first performed at room temperature and then again at the military temperature extremes ($-55°C$ and $+125°C$). Fig. 10.15 shows the number of received bit errors as a function of ground bounce slew rate ($dV_{gg'}/dt$). It is readily apparent from these results that the device easily meets the 1-V/μs ground bounce rejection requirement over the full military temperature range.

A summary of the internally differential SOS isolation circuit performance characteristics is given in Table 10.3.

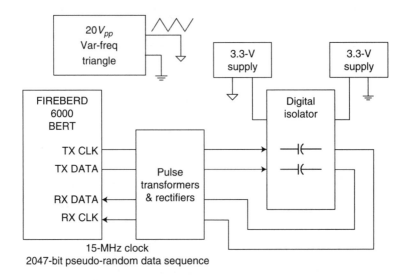

FIGURE 10.14 Test setup for measuring the bit error rate as a function of the ground bounce slew rate.

10.2.2.4 Application of the SOS Isolator: Isolated Power Supply Feedback

One of the most common applications requiring input/output galvanic isolation is power conversion. AC/DC and DC/DC converters accept a high-voltage (HV) bus input and convert it down to one or

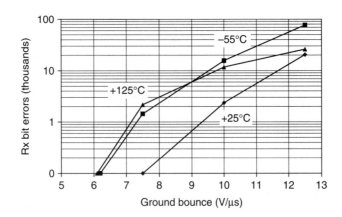

FIGURE 10.15 Number of received bit errors from the SOS differential isolator as a function of the ground bounce slew rate. The test was repeated at different temperature corners: 25°C (nominal) and military range +125°C and −55°C.

Isolation	100 V continuous, 800 V peak
Number of channels	four with charge pump
Signal bandwidth	85 MHz/channel
Supply voltage	2.5 V–3.3 V
Power consumption	Table 10.2
Silicon area (channel)	$230 \times 140 \ \mu m^2$
Silicon area (entire chip)	$1 \times 1.2 \ mm^2$
Charge pump	Receiver B
	can power transmitter A

TABLE 10.3 Summary of the SOS isolation circuit

more low-voltage (LV) secondary outputs, which are regulated against changes in line voltage and output loads (Marcus et al., 2006). Several converter architectures have been proposed to transmit control signals across the isolation barrier, either in the form of a DC error signal or an AC drive signal. An example of a typical power converter architecture is given in Fig. 10.16. In this scheme, a pulse-width modulator

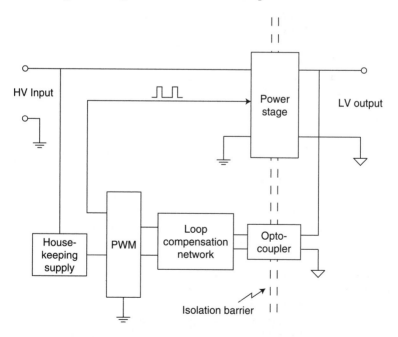

FIGURE 10.16 A typical isolated DC/DC converter architecture. The PWM senses the ground isolated output voltage through an optocoupler and drives the power stage as needed to keep the output voltage constant.

(PWM) controller located on the input side of the isolation barrier sets the power stage output voltage. A voltage feedback signal derived from the output side is transmitted across the isolation barrier to the controller. The PWM controller toggles the power to the power stage with a fixed-frequency train of variable width pulses to ensure that the power stage average output voltage equals the desired output voltage. As the output voltage changes as a result of input or load changes, the loop changes the pulse width to keep the average voltage constant and therefore regulate the output voltage. The traditional choice for feedback across the isolation barrier is optocoupling because of its simple implementation and linear transfer function. However, optocouplers suffer from variations in current-transfer-ration (CTR) and low bandwidth, which limits the performance of the resulting power converter. To improve performance and reliability, magnetic coupling can be used in place of optocouplers at a penalty of increased parts count and design complexity.

The SOS isolation amplifier provides a third alternative to these coupling schemes, offering simplicity, robustness, and excellent performance. To illustrate how this can be achieved, the traditional architecture of Fig. 10.16 has been modified as shown in Fig. 10.17 (Marcus et al., 2006). Here, the PWM is located on the output side where it directly–senses the output voltage and compares to a reference to generate a pulse width modulated drive signal. The isolation amplifier then transmits this signal back to the input side where it drives the main switch (through a high-power driver circuit), coupling energy through the main transformer to the output load(s). This topology optimizes implementation by placing the isolation amplifier (an inherently digital device) in a purely digital path and eliminates the need for DC–AC and AC–DC conversions of the analog error signal.

There are many advantages of this approach over traditional converter architectures. First, by eliminating the optocoupler (and its inherent CTR degradation issue), we have immediately improved the converter's reliability and long-term operation (Marcus et al., 2006). This is not a trivial result for space or military applications where 100% reliable operation in harsh environments is critical. Of special note is that this improvement in reliability does *not* come at the cost of additional components. On the contrary, eliminating the magnetics and chopper/rectifier circuits associated with the traditional high reliability magnetic coupling approach greatly reduces complexity and required board space of the final design. Another important advantage of the approach is that output voltage regulation tends to be superior relative to traditional feedback schemes. This is because the isolation barrier crossing has been relocated from the analog domain (error

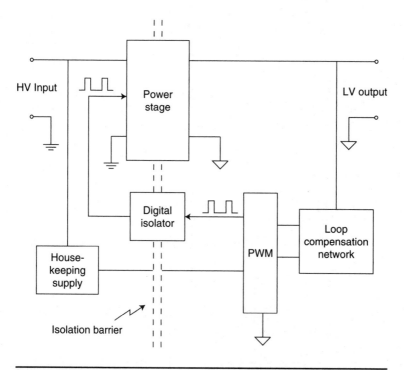

FIGURE 10.17 An alternative to the traditional DC/DC converter architecture. Here, the PWM is located on the output side of the isolation barrier for improved output voltage regulation. The isolation amplifier couples the PWM's output pulses back to the input side to drive the main power stage.

feedback path) to the digital domain (drive signal path). This reduces the number and type of components traversed by the error feedback signal, which correspondingly reduces the injected offset and bias errors at this critical node. As a result of these simplifications, output voltage regulation is significantly improved.

The scheme does suffer from one important disadvantage—that the housekeeping supply design becomes slightly more complicated because an isolated output is now required to supply the PWM bias. One simple solution is to use a chopper–transformer–rectifier combination to bring an unregulated supply rail to the secondary side. Local regulation (e.g., linear or switching regulators) are then used to provide a stable regulated supply for the PWM. Note that this scheme is appropriate only for applications where the dynamic loads are small

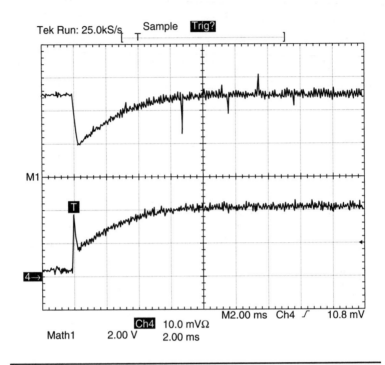

Tek Run: 25.0kS/s   Sample   Trig?

M1

4→

Ch4 10.0 mVΩ
Math1   2.00 V   2.00 ms

M2.00 ms   Ch4 ╱   10.8 mV

FIGURE 10.18   Output voltage transient response (top trace) and load step
stimulus (bottom trace).

and the input voltage range is narrow. Other applications may need
a more complicated design that could negate the benefits of using a
secondary side-referenced PWM in the first place.

To demonstrate the performance of the scheme shown in Fig. 10.17,
a simple prototype converter was built and tested. The circuit was a
basic flyback converter designed to supply 5-V output at several watts
from an isolated 28-V input bus. Two separate benchtop power supplies were used to power input and output circuits in this demonstration. The returns of each supply were tied together at a single point at
the input bench supply's chassis ground connection. Thus, input and
output circuits were galvanically isolated from each other with only
the main power transformer and the isolation amplifier crossing the
isolation barrier. The input voltage was allowed to vary from 22 to 36
V while the output was loaded from 100 mA to 1 A using variable
resistors.

The results were exceptional in terms of both line and load regulation: less than 0.1% and 0.45%, respectively over the full operating range. These results are two to three times better than comparable designs employing traditional optical or magnetic coupling methods (VPT Inc., 2005; LAMBDA Americas, 2005; Interpoint Corporation, 2005; Marcus et al., 2006). For completeness, we also performed a step-load test to demonstrate the converter's transient response with the isolation amplifier in the feedback path. We would expect no discernable effect on dynamic performance because the amplifier faithfully transmits the power pulses with negligible phase delay and, therefore, appears transparent to the control loop. Indeed, this appeared to be the case as the stable, overdamped response in Fig. 10.18 demonstrates. In this scope plot, the load-step current is the bottom trace at 500 mA/div and the output voltage transient response is the upper trace at 2 V/div. The load step was provided by momentarily switching a 5-Ω resistor in parallel with the nominal 50-Ω load. This resulted in a large current spike at the moment of switching as the output capacitance was quickly discharged. The input voltage in this test case was the minimum 22 V. We repeated the measurement at maximum input voltage of 36 V with similar results.

10.2.3 Digital Phase-Shift–Modulated Isolation Buffer in SOS

In this section, we present an alternative implementation of a digital isolation buffer. This design uses a digital phase-shift-keying modulation of the input signals and provides the advantage of a reduced silicon area and lower number of coupling capacitors per channel. We designed and fabricated the four-channel digital isolation amplifier depicted in Fig. 10.19, using the 0.5-μm SOS technology (Marcus et al., 2006; Culurciello et al., 2007). The isolation device was fabricated on a single die, taking advantage of the isolative properties of the sapphire substrate. In this section, we improve on previous designs by employing a digital phase-shift-keying modulation of the input signals and reducing the silicon area and the number-coupling capacitors per channel. The individual isolation channels can operate in excess of 40 Mbps using digital phase-shift-keying modulation. Modulation of the input signal is used to increase immunity to errors at low-input data rates. The device can tolerate ground bounces of 1 V/μs and isolate more than 800 V (refer to Fig. 10.1). The device uses $N + 1$ capacitors for N channels as opposed to $2 N$ of the internally differential implementations presented in the previous sections, thus minimizing the coupling silicon area and increasing reliability.

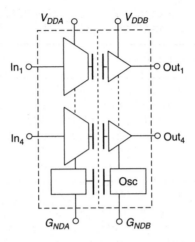

FIGURE 10.19 Digital phase-shift-modulated isolation buffer system architecture. Here, only two channel are shown for simplicity. The device is a four-channel digital isolation amplifier that uses digital modulation of the input signal and capacitive coupling. The digital modulating clock is powered by the receiver side (right, circuit B) and transmitted capacitively to the transmitter (left, circuit A).

10.2.3.1 Digital Phase-Shift–Modulated Isolation Buffer System Overview

Figure 10.20 is a detailed schematic of one of the four digital phase-shift–modulated isolation channels named *isoCap3sc*. The specification for each channel required a data rate of 40 Mbps, military range temperatures, and input signal rise/fall time between 10 and 1.5 ns. The required isolation was at least 100 V in continuous mode. The device is designed to withstand ground bouncing of more than 1 V/μs using a circuit topology able to reject spurious transitions. This feature is obtained by using digital modulation of the input signal before transmission to the receiver through the capacitive isolation interface of Fig. 10.20. The use of modulation increases the switching frequency across the coupling capacitor. Spurious transitions are eliminated if the switching frequency is higher than the maximum allowed ground bouncing. The input signal is buffered and modulated at the transmitter (input) side and communicated to the receiver (output) circuit using capacitive coupling. Each stage's coupling capacitor C_f has a capacitance of 150 fF and has been designed using metal-1 and metal-3 plates. The silicon area used by the capacitor is $175 \times 60 \ \mu m^2$ (Marcus et al., 2006; Culurciello et al., 2007).

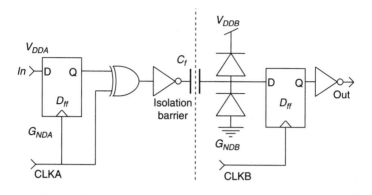

FIGURE 10.20 The *isoCap3sc* digitally modulated isolation channel. Digital phase-shift modulation is employed to communicate a digital signal from transmitter circuit A to receiver circuit B through an isolation barrier.

Notice that our first isolation device (Culurciello et al., 2005b; Culurciello et al., 2005c) used two capacitors per channels, and *isoCap3sc* only uses one capacitor, thus saving precious silicon area. In addition, and the previous device provided nonzero bit-error-rates (BER) at low input rates. We measured BER of $6 \cdot 10^{-7}$ for a 10-kHz, input. The device reported in this chapter, thanks to the digital modulation of the input signal, did not suffer from bit errors at low frequencies.

A 180° phase-shift-keying modulation is performed by XOR-ing the input signal with the transmitter clock CLKA. The transmitter clock is obtained through capacitive coupling from the receiver side, where the global clock is generated. We used a 13-stage ring oscillator at the output side to produce an approximately 200-MHz digital clock signal that modulates the input signal. A D-type flip-flop circuit synchronizes the modulation and demodulation to avoid spurious transitions of the output due to transmission delay. The input flip-flop operates on the rising clock edge. The demodulator is a flip-flop synchronized to the falling edge of the clock CLKB. The output of the *isoCap3sc* isolation channel is the terminal *Out* in Fig. 10.20. At the receiver side, protection diodes located at the receiver input node enforce that the voltage at the floating nodes always drifts to one of the supplies, to prevent damage in case the ground-bouncing rate is much faster than the modulation rate (Marcus et al., 2006; Culurciello et al., 2007).

The final prototype of the capacitive isolation buffer in Fig. 10.19 (named *isoCap3sc*) is organized as an array of four independent isolation channels *isoCap3sc* in one single chip. All channels share the same input and output power supplies. The inputs of each receiver directly after the isolation barrier are protected from high-voltage swings by protection diodes connected to their power supply. Output pads are buffered with digital inverters to be able to drive a 25-pF capacitive load (Marcus et al., 2006; Culurciello et al., 2007).

10.2.3.2 Digital Phase-Shift–Modulated Isolation Buffer Results and Measurements

We simulated the isolation buffer *isoCap3* at the design corners for temperature, and transistors characteristics. We simulated with a temperature range of from −55°C to +125°C for typical, fast, and slow transistors. We conducted the measurements at 25 Mbps. The circuit performed correctly in all settings. We also tested the circuits with a power supply of 3.3 V ±10% of the nominal value at 40 Mbps. In both cases, the circuit was operational. All simulations were conducted with all four inputs tied together and the output connected to a 25-pF capacitance, corresponding to a worse-case scenario for power consumption.

The measured supply current of four channels in parallel was 1.5-mA at low data rates and 3.3-V power supply. This consumption is attributed to the ring oscillator operating at the receiver's side and generating the global modulation clock. The consumption rose to 4-mA with a 10-MHz input and 16 mA with a 40-MHz input. The majority of the power consumption was due to the output drivers, designed to drive capacitances up to 50 pF. A plot of the measured power consumption using four channels in parallel is given in Fig. 10.21.

No crosstalk between channels was observed. All these measurement were conducted with the isolation chip driving 2 ft of coaxial cable and a 25-pF load (oscilloscope load). Fig. 10.22 shows the output of one isolation channel (bottom trace) when driven with a 40-MHz input (top trace).

Operation while providing isolation was verified experimentally, with the circuit operating with an input square wave of 30 MHz, $V_{DDB} = 3.3$ V and $V_{GNDB-GNDA} = 25$ V.

A picture of the fabricated SOS isolator amplifier is given in Fig. 10.23. The die has 12 bonding pads: the left 6 are (top to bottom) the input supply, 4 data inputs, and the transmitter's ground; the right 6 are the output, supply, 4 data outputs, and the receiver's ground.

Finally, we report that each channel *isoCap3sc* uses 230×60 μm² of silicon area, as opposed to the 230×140 μm² used in our first implementation (Culurciello et al., 2005b; Culurciello et al., 2005c). Taking into account that one channel, *isoCap3sc* is used to transmit the global clock from receiver to transmitter, the four-channel device presented in this chapter uses approximately half of the silicon area of our previous devices.

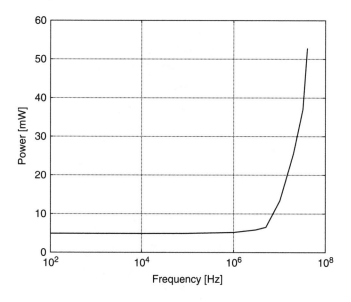

FIGURE 10.21 Power consumption of the digitally modulated isolation amplifier versus input frequency. The device was operated with four channels in parallel driven by the same input.

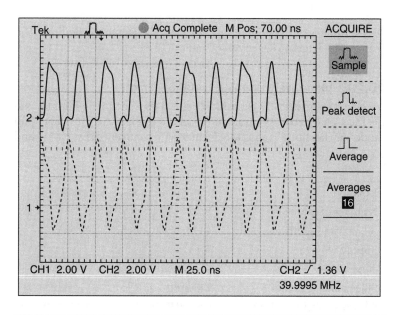

FIGURE 10.22 SOS digitally modulated isolation channel input (top trace) and output (bottom) operating at 40 MHz.

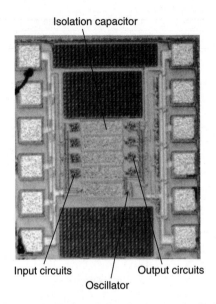

Isolation capacitor

Input circuits Output circuits
Oscillator

FIGURE 10.23 Micrograph of the fabricated SOS isolation amplifier with digital modulation. The die area is 0.6×0.9 mm^2, while each channel is 230×60 μm^2. This prototype did not contain a charge pump.

A summary of the digitally modulated SOS isolation circuit performance characteristics is given in Table 10.4.

10.2.4 Inductively Coupled Isolated Circuits

An electromagnetically isolation circuit uses a magnetic field to couple the signal between a transmitter and a receiver circuit. The circuit model can be thought as a transformer with a current drive circuit and

Isolation	100 V continuous, 800 V peak
Number of channels	Four
Signal bandwidth	40 MHz/channel
Supply voltage	3.3 V
current	4 mA @ 10 MHz, 16 mA @ 40 MHz
Silicon area (channel)	230×60 μm^2
Silicon area (entire chip)	0.6×0.9 mm^2

TABLE 10.4 Summary of the digitally modulated SOS isolation circuit

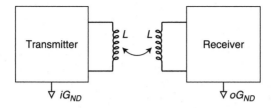

FIGURE 10.24 Principles of operation of a transformer-based electromagnetically coupled isolation circuit. A transmitter circuit (Sec. 10.2.4.1) is coupled to a receiver circuit (Sec. 10.2.4.6) using a transformer. Each coil of the transformer with inductance L.

a receiving circuit. Fig. 10.24 shows the transmitter and receiver circuit with different ground potentials. If the transmitter activates the primary by generating a current through the primary coil, an induction current will be induced in the secondary coil.

This current can be sensed by the receiver circuit and interpreted as a signal. By simply turning on or off the primary current, a secondary current can be transmitted using the magnetic field between two isolated circuits. Notice that this technique is similar to a digital communication system using amplitude shift keying. This isolation system can be implemented by using an oscillator circuit labeled *transmitter* (Sec. 10.2.4.1) and a *receiver* circuit (Sec. 10.2.4.6).

10.2.4.1 Transmitter Circuit: LC–Tank Oscillator

The transmitter circuit for the electromagnetically coupled isolator is an LC–tank oscillator (Lee, 2000; Razavi, 2000). The coils used in the oscillator form the primary coils of the transformer that couples the magnetic field between transmitter and receiver circuit. The LC–tank oscillator is a classic circuit used particularly in narrow-band (GSM, TDMA) and wide-band (CDMA) cellular communication systems for its low-phase noise. The oscillator general output waveform $V_O(t)$ can be expressed by Eq. (10.11).

$$V_O(t) = A\sin(\omega_0 t + \phi(t)) \qquad (10.11)$$

A typical LC–tank circuit is given in Fig. 10.25. This type of oscillator makes use of both an inductance L and a capacitance C. These passive components create an energy tank for the electric field. The oscillation is generated by passing the electric field energy from one passive component to the other. The inductance and capacitance (L, C) trap the electric field inside them for about half a period interval and

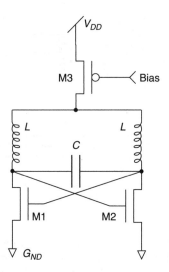

FIGURE 10.25 LC–tank oscillator circuit. The circuit produces a high-quality low-distortion sine wave by periodically converting the electric field stored in capacitors C with the magnetic field stored in inductors L. The transistors bias the DC operational point and actively restore the losses of the oscillator.

transfer the energy of the electric field from one to the other with a time constant of $\tau = 2\pi\sqrt{(LC)}$.

The energy transfer is performed by a couple of transistors (M1, M2) connected in positive feedback with the passive components. M1 and M2 counterbalance energy losses by supplying a charge to the capacitive tank C, in AC operation. As the energy is increasing in one of the coils, the energy stored in the other coil is decreasing proportionally. The capacitor C acts like a charge tank that is redistributed in the form of a current to either one of the coils. Transistor M3 biases the DC operational point of the whole circuit and supplies the bias AC current to restore losses.

This oscillator also supports a mirrored topology using PMOS active devices. PMOS are generally less frequently used because of the lower transconductance of P-type transistors.

10.2.4.2 LC–tank Oscillator Model

Figure 10.26 is an AC model of the oscillator. This model is valid for a great variety of electronic oscillators (Craninckx and Steyaert, 1995). The feedback network is a two-pole lossy system with active restoration. To generate oscillation, the Barkhausen condition must

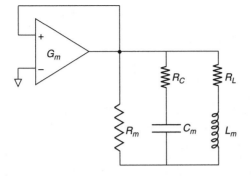

FIGURE 10.26 General model of an oscillator circuit.

occur. From this condition, the frequency at which the imaginary part of the system gain is zero defines the oscillation frequency of the circuit, which is given by Eq. (10.12). The factor $2L$ is due to the series of inductors appearing in the small signal model.

$$f_0 = \frac{1}{2\pi\sqrt{2LC}} \tag{10.12}$$

This is a simple result of linear circuit theory. On the other hand, the amplitude V_{osc} of the oscillation is regulated by the nonlinear effects of the circuit, mainly clipping effect of the active devices due to the limited power supply. Therefore, linear theory will not be able to calculate analytically the amplitude of the oscillations.

The lumped parameters of the model are

R_m = parallel total resistance (mainly the MOSFET drain resistance r_d)

C_m = parallel total capacitance, sum of the explicit capacitance and the MOSFETs gate C_i and output C_o capacitances

L_m = series of the two inductors ($2L$) plus wiring inductance

R_L = parasitic resistance of the inductor L

R_C = parasitic resistance of the capacitor C

G_m = active device transconductance ($G_m = g_m/2$ in the VCO case of Fig. 10.25)

We can define an effective resistance that takes into account of all the parasitic resistances in the circuit and the explicit load resistance of the oscillator. The equivalent effective resistance R_{eff} can be calculated (Craninckx and Steyaert, 1995) using Eq. (10.13).

$$R_{eff} = R_C + R_L + \frac{1}{R_m (\omega_0 C)^2} \tag{10.13}$$

For the oscillator to work properly, a negative resistance must be obtained from the transistor operation. A minimum transconductance G_m can be defined in Eq. (10.14).

$$G_m = R_{eff} \, (\omega_0 C)^2 \qquad (10.14)$$

Given that the active elements in the LC–tank oscillator provide a transconductance higher than G_m, the oscillator can generate resonance and counterbalance the oscillation damping produced by the resistive elements.

We will show in Sec. 10.2.4.3 that the formula for the effective resistance and also of the minimum transconductance G_m clearly specifies all the important design parameters for an LC–tank oscillator and therefore the transmitter circuit of the isolator.

10.2.4.3 Effective Resistance and Minimum Transconductance

The transmitter oscillator frequency depends on parameters L and C [Eq. (10.12)]. It appears that the designer can have at least one degree of freedom in the choice of these values. On the other hand, if the design is to be integrated on a die with no external components, size limitations and cost really make this integrated LC–tank oscillator circuit usable only at high frequencies, of at least about 1 GHz. LC–tank oscillators operating at low frequencies use a significant silicon area, as the size of the passive components L and C increases.

As an example, for 1-GHz operational frequency, L can be 4 nH and C can be 3 pF. These can have reasonable values, despite the fact that inductances of more than 2 nH occupy a very large area on the die (Burghartz et al., 1998). And the capacitance C cannot be increased arbitrarily, because it is directly proportional to the power consumption P_{osc}, since it is charged and discharged twice at each cycle. The power consumption P_{osc} can be computed with Eq. (10.15), where V_{osc} is the oscillation amplitude.

$$P_{osc} = 2C V_{osc}^2 f_0 = \sqrt{\frac{C}{L}} \frac{V_{osc}^2}{\pi} \qquad (10.15)$$

Equation (10.15) shows that to lower the power, consumption C must be small and L large. On the other hand, a large inductance occupies a wide silicon area and the added turns increase its resistance, which in turn lowers the quality factor and raises power consumption again.

Power consumption is also one design constraint. An optocoupler isolation circuit can use 70 mA (Fairchild FOD817) at the transmitter. On the other hand, our application needs to be able to power the circuit from the input signal, and therefore we set a target budget of a few milliamperes for the transmitter circuit. The size of active devices will be limited by the maximum supply power.

This power limitation is ultimately related to the effective resistance and, in particular, the size of capacitor C. This is first because R_{eff} and also G_m is proportional to the parasitic resistances of the circuit components. The parasitic resistance can be changed with fabrication processes or with postprocessing of the die to add a thicker metal layer. The minimum transconductance also depends on the operational frequency ω_0 and the capacitance C because the MOSFETs have to charge and discharge the capacitance twice in a time period. The frequency can be increased to several gigahertz, and the capacitance and inductance value will be lower. If low-power operation is to be maintained, the active device will have to exhibit lower transconductance, since G_m is proportional to the supply current (think of the MOSFET's drain current in the linear region). To lower G_m we have no choice but to lower the value of C and thus, keeping L constant, increase the oscillator frequency. This is one of the most important considerations in designing an LC–tank oscillator. This is the reason why some of our initial prototypes did not oscillate. In fact, the frequency was set to a low value (1 GHz) and the required MOSFET transconductance was too high, as will be explained in Sec.10.2.4.5.

10.2.4.4 Transformer Design

The inductor layout is influenced by the number of turns, which is directly proportional to the final inductance value (Burghartz et al., 1998). As an example, the inductance value in nanohertz for a type-4 in SOS is given by Eq. (10.16) (Peregrine, 2008b).

$$L = 0.35N^2 - 1.34N + 2.45[nH] \qquad (10.16)$$

A plot of this function is given in Fig. 10.27. In addition, the width of the inductor metal is proportional to the parasitic shunt resistance. The wider the metal layer, the lower the resistance. But the wider the metal, the more area will be used, keeping constant the number of turns and inductance. The SOS process features a metal-thick layer to reduce parasitic shunt resistance. The use of this layer cuts the shunt resistance to about one-half (Peregrine, 2008b). In addition, all the metal layers should be used to lower the parasitic shunt resistance. Note that this is not possible if we want to build a transformer using two pairs of superimposed integrated coils. Fig. 10.27 reports the quality factor of the inductor $Q = \omega L/R_L$ at 1 GHz. The shunt resistance was calculated on a single metal layer (metal-1 or metal-2) using the SOS parasitic resistance of 0.06 $\Omega/square$ (Peregrine, 2008b).

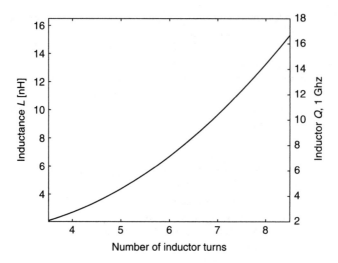

FIGURE 10.27   Inductance and quality factor vs number of turns.

The length of the inductor l_{ind} metal can be calculated with Eq. 10.17.

$$l_{ind} = \sum_{k=0}^{2N} (Di + 2kS) \qquad (10.17)$$

In Eq. (10.17), D_i is the diameter of the hollow center of the inductor, S is the spacing from one metal line to the next, and N is the total number of turns.

As an example, we designed two nominal 2-nH inductors in the SOS process. One is fabricated using the first metal layer (metal-1), the second using the third metal thick layer (metal-3 or metal thick). Both inductors are 200 μm wide. A summary of the characteristics follows.

M1: 20-μm width, 2-μm spacing, *1850*-μm length, $R_L = 5.5\ \Omega$, with 0.06 $\Omega/square$

M3: 17 μm width, 3-μm spacing, *1850*-μm length, $R_L = 3.2\ \Omega$, with 0.03 $\Omega/square$

A variation of these inductors, both 3.5 turns, 130 μm wide for a 1-nH design has the following parameter:

M1: 10.5-μm width, 2-μm spacing

M3: 7.5-μm width, 5-μm spacing

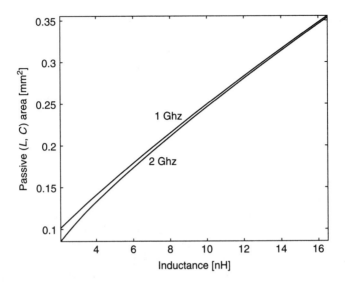

FIGURE 10.28 Passive area vs inductance.

Figure 10.28 plots the silicon area consumption of both inductors and capacitor for an LC-tank oscillator in SOS at 1 GHz and 2 GHz. The inductors are type 4 SOS (Peregrine, 2008b). The area of the two inductors A_L and the capacitor A_C is given by Eq. (10.18).

$$A_L = 2\{D_i + 2[NS_t + (N-1)S_m]\}^2$$

$$A_C = \frac{C_{tr}}{C_{ua}}$$

(10.18)

S_t and S_m are, respectively, the thickness of the metal and the spacing between two metal lines. C_{tr} is the desired design capacitance, and C_{ua} is the nominal capacitance, per unit area.

When the capacitance is reduced, the inductance value increases to maintain oscillation at the desired frequency. The inductances occupy a significant portion of the design. As can be seen in Fig. 10.28, the area of the capacitor is only a small fraction of the total passive area. In fact, the area consumption is linear with the inductance. The area consumption for frequencies higher than 2 GHz was not plotted because the behavior is virtually identical to the 2 GHz plotted in Fig. 10.28. This is another clear indication that the inductor size dominates the area consumption. Note that for frequencies higher than 2 GHz, the inductor size can be further reduced because the product LC gets smaller.

10.2.4.5 Active Devices and Transconductance

The transconductance of the NMOS transistor pair in the LC–tank oscillator is important to start and maintain circuit oscillation, as stated at the end of Sec. 10.2.4.2. NMOS transistors are generally used instead of PMOS because of the higher transconductance due to the mobility of the electrons. The transconductance g_m of the device is defined in Eq. (10.19).

$$g_m = \frac{\partial I_{DS}}{\partial V_{GS}} \bigg|_{V_{DS}, V_{GS}=const} \tag{10.19}$$

The transconductance g_m is proportional to the drain current, which is proportional to the product of the carrier mobility and the gate oxide unit capacitance $\mu_0 C_{ox}/2$ (measured in [μA/V^2]). We carefully examined this quantity in various processes offered by MOSIS, Ca (MOSIS, 1999). The search results are summarized in Table 10.5, providing a list of process, process code, and MOSIS run identifier, and finally the transconductance of NMOS and PMOS transistors. The TSMC processes provided the highest MOS transconductance, and the SOS process provided one of the lowest values. This means that for a given effective resistance of the oscillator, and therefore for a given transconductance, TSMC MOS will be smaller in size, even more than the nominal reduction of features size. Also, it shows that SOS is penalized in this respect, because by requiring larger devices, the silicon area consumption will also be higher.

Because MOS transconductance is proportional to the bias current, the power consumption is affected by the size of the MOSFETs. This is also a disadvantage for using the SOS technology.

Process	Code	MOSIS Run	N, $(U_o C_{ox}/2)$ [μA/V^2]	P, $(U_o C_{ox}/2)$ [μA/V^2]
AMI05	SCN05	T33A	56.3	−18.2
AMI15	SCN15	T33Z	35.1	−11.6
TSMSC0.35	SCN035	T33V	92.4	−33.1
TSMC0.25	SCN025	T33W	122.1	−24.9
TSMC0.18	SCN018	T32L	169.2	−36.1
SOS05	SOI05	T09B	36.1	−21.1
HP05	SCN05	T31D	74.3	−25.4
HP0.35	SCN035	T05Y	74.9	−19.6

TABLE 10.5 Processes, MOSIS run code, N- and P-type transistor transconductance

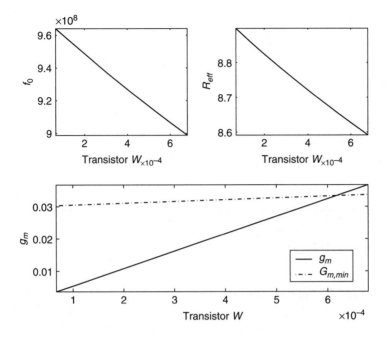

FIGURE 10.29 g_m, R_{eff}, f_0 as functions of transistor width.

Figure 10.29 shows the required minimum transconductance for an LC–tank oscillator designed in SOS for f_0 equal to 1 GHz with L equal to 2 nH, C equal to 6.7 pF. The subplots are a function of the transistor width W. The two topmost plots show the change in oscillation frequency as bigger devices are employed. In fact, the gate capacitance of the MOSFETs is significant and contributes to the total shunt capacitance. The effective resistance R_{eff} is influenced by the oscillation frequency.

The bottom plot of Fig. 10.29 shows the minimum transconductance required from the active devices to balance the shunt resistance of the oscillator. This parasitic resistance is due to the MOSFETs (an average of 250 Ω) and the coil resistance (the inductor layout is in the first metal layer). This plot shows that the active devices width must be such that the MOSFETs' transconductance is higher than the minimum required G_{mmin}. This occurs at and above about 620 μm with the parameters specified above. The transconductance of the transistors is modeled using a velocity-saturated Eq. (10.20), where E_{sat} is the electric field at saturation, or 1.5 MV/m in silicon.

$$g_m = \frac{1}{2} v_0 C_{ox} W \cdot E_{sat} \tag{10.20}$$

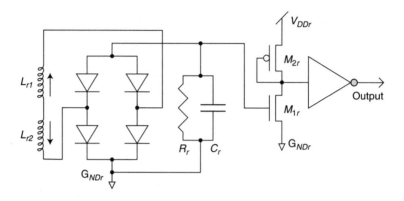

FIGURE 10.30 Receiver circuit. A full-wave rectifier and a RC filter demodulates the amplitude of the received signal. A digital buffer is used as the output stage.

10.2.4.6 Receiver Circuit

The output circuit reconstructs the original input signal by sensing the current on the secondary coil. A schematic caption of the receiver circuit is given in Fig. 10.30. The main operational mechanism is amplitude demodulation of the carrier generated by the input oscillator. The coils in the receiver and the coils on the transmitter (Fig. 10.25) form a coupled transformer with a ratio of 1:1. Whereas the voltage on the transmitter coils is out of phase because the oscillator has two identical branches shifted by 180°, the voltage of the receiving coils is in phase, so that their sum would provide a built-in amplification by 2.

The two coils coupled to the transmitter are connected in phase and their nodes constitute the input of a fully rectifying diode bridge. Their voltage sums because they are axis symmetrically coupled to the transmitter coils, which are 180° out of phase. The bridge was loaded with a resonant capacitance according to the input resonance. This configuration maximizes the output voltage at the receiver because the output loading at resonance is only the parasitic resistance of the coil R_L. The receiving capacitance C_r is also in parallel with a resistor R_r that dumps the oscillation quickly as the transmitter is turned off. This guarantees high throughput and data rates. The output of this envelope detector circuit is then fed into a cascade of one follower biased to detect low thresholds, and an inverter to create a current buffer and interface to the outside world.

The circuit takes advantage of the multithreshold MOSFET available in the SOS process. The diode bridge is made of low-threshold NMOS transistors. This lower threshold makes it possible to demodulate very-low received signals and more effectively couple the secondary coil voltage. Transistor M_{1r} is also a low-threshold NMOS that can detect very-low demodulated signals. This is important, as the amount of voltage coupled to the receiving coils cannot be estimated with precision due to edge effects and nonideality of the coupling transformer.

10.2.4.7 Electromagnetic Isolator: Conclusions

The design of an electromagnetic isolator circuit in SOS is not feasible with the set of parameters and the power budget discussed. The transistor channel length reported in Fig. 10.29 is about 0.6 mm, and this number has to be doubled to 1.2 mm to guarantee operation. From simulations, we measured operation currents of more than 100 mA at 3.3 V for the transmitter circuit, making it undesirable because of excessive power consumption.

A version of the transmitting circuit operating at higher frequencies of 2 GHz or more can alleviate the problem, but the increase in carrier frequency also raises power consumption. An analysis of such a transmitter would reveal the trade-offs; it may be material for future work.

10.3 Three-Dimensional Circuits in SOS

Three-dimensional VLSI fabrication technologies are extremely attractive for their impact on the density and integration of sensory arrays, sensory computation and (communication systems Lee et al., 2000; Lei et al., 2003). As an example, high-density low-power systems for complex visual processing would significantly benefit from 3D VLSI technologies by combining large image sensor arrays with stacked and interconnected processing and communication layers. This would overcome the restriction of standard 2D VLSI processes in terms of type and number of parallel sensing and processing units that can be placed on a single die or connected across chip boundaries. An example of a 3D image sensor is given in Fig. 10.31. The photosensor array is placed at the top of a three-die system to obtain high fill factor and resolution. Image processing and communication circuits are placed in the remaining two dies to minimize noise, crosstalk, and improve the density of computation. The stacking of the three dies achieves high component density without compromising image quality.

Capacitive coupling have has been employed in multichip modules to transfer data signals between multiple dies (Salzman and Knight, 1994; Salzman et al., 1995; Gabara and Fischer, 1997). The coupling capacitors consist of two metal plates residing in separate dies and separated by a dielectric. The dies are aligned to form the coupling

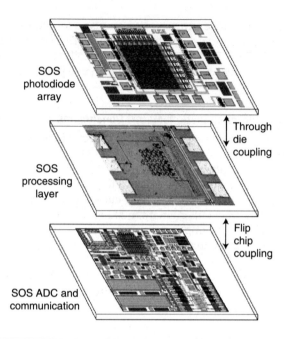

SOS
photodiode
array

Through
die
coupling

SOS
processing
layer

Flip
chip
coupling

SOS ADC and
communication

FIGURE 10.31 Capacitive coupling multichip module for data and power is an enabling technology for 3D VLSI fabrication. Here, an image sensor is obtained by stacking three dies.

capacitance between the metal plates, which are generally bonding pads. Capacitive coupling has been used successfully only to transfer signals (Mick et al., 2002a; Mick et al., 2002b) while still requiring electrical connections for both dies in order to transfer the required power supply. These physical connections are generally obtained using a ball grid array, wire bonds, or probes, all imposing mechanical and cost limitations on the number and density of data signal connections to the package.

10.3.1 Three-Dimensional Interdie Capacitive Data Communication and Power Transfer

We designed a multichip module that uses bondless capacitive coupling to provide both bidirectional communication and also exchange the power supply between two separate dies. A prototype was fabricated on the 0.5 μm SOS FC process. The prototype consists of a transmitting circuit that is packaged and bonded to a conventional

DIP package. This bottom die contains a charge pump that uses a portion of the capacitive coupled connections to provide power to the receiver die. In addition, the transmitter circuit uses capacitive coupling to transmit digital data to the receiver die. The receiver die contains the rest of the circuitry of the charge pump to recover and regulate the supply voltage. The received power supply is used to power a digital buffer that activates the reverse (bidirectional) digital communication from the receiver die back to the transmitter die.

The main advantage of capacitive coupling is its implicit simplicity. The transfer of power as well as the signal between two dies is very desirable for the simplification of packaging. Installation of a die in the package would just require aligning the die to the package's coupling metal plates and using an adhesive. The alignment is unproblematic due to the transparency of the sapphire substrate. This is in contrast to the prevalent bump bond flip-chip techniques, where the yield of the multichip module is proportional to the number of bonds that have to be physically connected (Harper, 2000). In addition, stacking of two or more dies in a multichip module can be achieved with no physical connection by using capacitive coupling of both data signals and power supply.

This technique can be used to test dies both after manufacturing to determine basic functionality and also to avoid low yield before assembling them into a multichip module (Harper, 2000). Instead of requiring physical connection to a test board with capacitive coupling packaging, the die under test needs only to be aligned with the test board for the duration of the test and can be placed and removed at high speed, thus providing very high throughput for postfabrication testing purposes.

Capacitive coupling can provide high data throughput (Mick et al., 2002a) while minimizing noise and parasitic inductance typical of wire bonds. The reduction of the undesired inductance of the bonds can provide higher signal bandwidths, whereas the short interconnections by capacitive coupling reduce antenna effect noise and thermal noise. The combination of short interconnects and reduced parasitics relaxes the required current from the interconnect driver circuits, thus reducing the communication power consumption (M.Secareanu and Friedman, 2000).

10.3.1.1 Three-Dimensional System Overview

The prototype consists of two SOS dies, one acting as transmitter and one as receiver. The transmitter die (bottom die) is placed and bonded into a common dual-in-line package. The receiver die (top die) is flipped and aligned on top of the transmitter die, so that the required capacitive connection are formed between the bonding pads metal sheet of both dies (Fig. 10.32). Figure 10.33 shows a mock of the

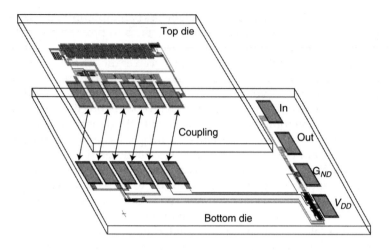

FIGURE 10.32 Capacitive coupling multichip module for data and power. Coupling is performed through capacitance obtained facing bonding pads (*coupling*). The transmitter and receiver circuits are actual layouts of the fabricated multichip module. In, Out, V_{DD}, and G_{ND} are bonding pads for the transmitter circuit.

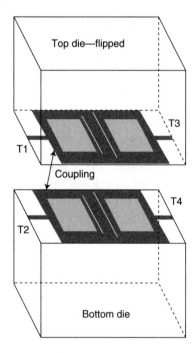

FIGURE 10.33 Coupling is performed through capacitance obtained facing bonding pads (coupling) of two separate SOS dies. The coupling pad area is 90 μm × 90 μm.

500 μm

Bottom die   Top die

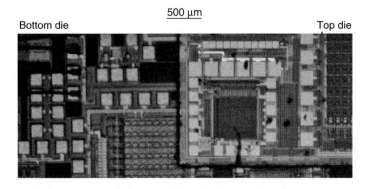

FIGURE 10.34 Micrograph of the assembled multichip module. The bottom (transmitter) die is bonded to a package through bonding wires visible on the left side of the figure. The top (receiver) circuit is on the right side and is flipped so that its bonding pads are facing the bottom circuit's pads. This forms the capacitor for coupling the signal and power across the two dies.

alignments of the pads between the transmitter bottom circuit (pads T2 and T4) and the receiver circuit (pads T1 and T3). The alignment of the dies was performed manually under a microscope. The precision of the alignment was 15 μm, and the dies have been bonded together using a layer of transparent varnish. There were six pads aligning, four of which are used by the charge pump and two more for bidirectional communication of two signals.

The circuits in Fig. 10.32 are actual layout of the fabricated circuit, while Fig. 10.34 is a micrograph of the multichip module. The bottom (transmitter) die is bonded to a package through bonding wires visible on the left side of the figure. The top (receiver) circuit is on the right side and is flipped so that its bonding pads are facing the bottom circuit's pads. This forms the capacitor for coupling the signal and power across the two dies. The distance between the metal plates of the capacitive coupling in air is 8 fF, with a plates distance of 10 μm (3 + 3 μm for the bond to the passivation step and an estimated 4 μm for the varnish layer).

Figure 10.35 is a schematic caption of the integrated circuit, the transmitter circuit is enclosed with in a dotted box, and the circuit outside the box is the receiver circuit. The bidirectional communication circuit is composed of an input pad (In) at the receiver circuit, which is bonded to the package for external stimulation. The input pad is connected capacitively to the receiver circuit, which uses an inverter to buffer the digital signal and drive the return coupling capacitance.

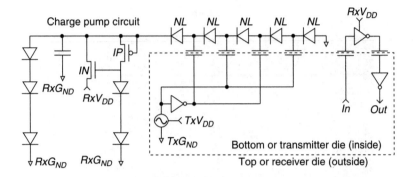

FIGURE 10.35 Schematic of the SOS 3D circuit: a charge pump is used to transfer the power using capacitive coupling between two dies. The transmitter is the schematic inside the dashed box and the receiver schematic is outside the box. In, Out, TxV_{DD}, TxG_{ND} are external connections for the transmitter circuit (refer to In, Out, V_{DD}, G_{ND} in Fig. 10.32).

The signal is the buffered again and output to a pad of the transmitter circuit (Out). The rest of the transmitter and receiver circuits is the charge pump circuit.

The charge pump is based on the Dickson (Dickson, 1976) charge pump design and is composed of 4 stages. An 11-stage ring oscillator at the output side produces a 350 MHz digital clock signal that controls the pump (represented by an oscillator symbol in Fig. 10.35. The output of the oscillator is buffered to drive the isolation capacitances. The Dickson charge pump operates by pumping charge along the diode chain (4 stages, therefore five diodes) as the capacitors are successively charged and discharged during each clock cycle. The capacitors are obtained by using coupling between pads, exactly like the signal capacitors. Notice that the pump design is identical to the one used in the SOS isolation circuits, and precisely to the circuit in Fig. 10.8. Refer to the isolation circuit sections for more details on the charge pump design and characteristics.

10.3.1.2 Three-Dimensional Circuits
Results and Measurements

We successfully tested the functionality of the communication link and power transfer at from 1 Hz to 15 MHz (100 MHz reliably from simulations). Figs. 10.36 and 10.37 show screen frames of the simulations verifying functionality of the prototype with an input square wave, respectively, at 15 MHz and 100 MHz. Fig. 10.38 shows a screen shot from the oscilloscope. For each screen the top waveform is the input signal at the transmitter, and the bottom waveform is the output signal from the receiver. The top left is at 1-V supply and 1-MHz

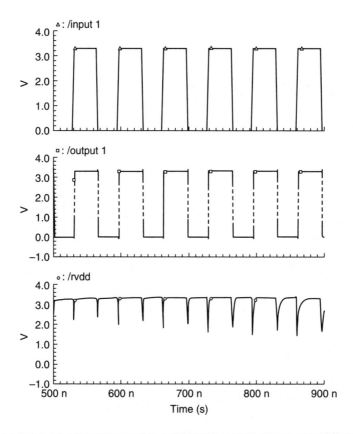

FIGURE 10.36 Simulation verifying functionality of the prototype with an input square wave at 15 MHz. The top trace is the input and the middle trace is the output signal. The bottom trace is the received voltage (rxV_{DD} in Fig. 10.35) output of the charge pump at the receiver side.

input. The top right is at 3.3 V and 1 kHz, the bottom left at 1 MHz and the bottom right at 15 MHz input frequency. The current drawn was 9 mA at 3.3-V supply from 1 kHz to 15 MHz. The current was 3 mA at 1-V supply and 1 MHz operation. Notice the close match between simulations and measured data, due to the fidelity of the SOS models above threshold (as reported in Chap. 1). The square wave output of the 15-MHz trace in Fig. 10.38 appears corrupted due to impedance mismatch with the prototype's package.

The capacitive vias used in our prototype occupy an area of 90 × 90 μm^2 for a plate separation of approximately 10 μm. By keeping the coupling capacitance constant, this area can be reduced to 30 × 30 μm^2 with a plate separation of 1 μm, and to 9 × 9 μm^2 with a

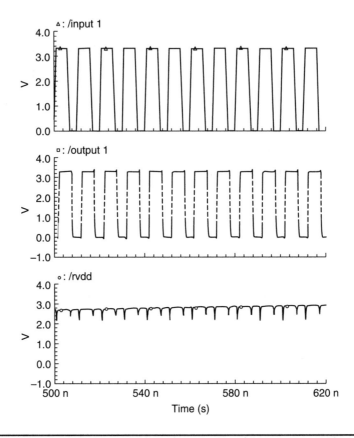

FIGURE 10.37 Simulation verifying functionality of the prototype with an input square wave at 100 MHz. The top trace is the input and the middle trace is the output signal. The bottom trace is the received voltage (rxV_{DD} in Fig. 10.35) output of the charge pump at the receiver side.

0.1-μm separation. Typical inductive vias occupy an area between 100×100 μm^2 (Mizoguchi et al., 2004) and 300×300 μm^2 (Chong and Xie, 2005). Inductive vias of 50×50 μm^2 have been proposed, but no demonstration of communication was given (Mick et al., 2002a). Three-dimensional galvanic vias occupy an area of 5×5 μm^2 in the Defense Advanced Research, Projects Agency-sponsored 3D-SOI process from MIT Lincoln Laboratory (MIT Lincoln Laboratory, 2005). The size of the vias are compared in Fig. 10.39. Through-die galvanic vias obtained in 3D processes obtain very high density of vertical connections but require expensive fabrication technologies. Capacitive coupling requires less silicon area than inductive coupling, especially when plates are close to each other. Plate spacing of 1 μm or

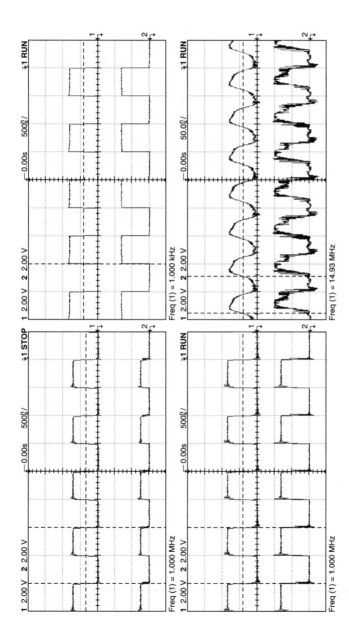

FIGURE 10.38 Oscilloscope traces verifying functionality of the prototype. The top left trace is obtained with 1-V power supply and 1-MHz input square wave. The top right has 3.3 V and 1 kHz input. The bottom left trace has 3.3 V and 1 MHz input. The bottom right trace has 3.3 V and 15-MHz input.

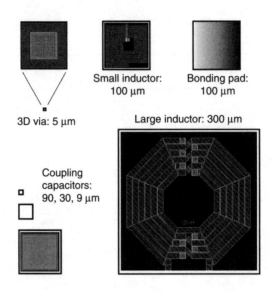

FIGURE 10.39 Comparison between layout sizes of inductive, capacitive and galvanic vias for data communication. The large inductor (10 nH) and the 3D via is in the MIT LL 3D-SOI process. A magnified (20 x) 3D via layout is also displayed. The capacitive vias and the small inductor (1 nH) are SOS layouts. The solid square is the size of a bonding pad, reported here as a frame of reference. Sizes are in (linear) micron.

less can be obtained easily by etching the passivation layer of facing dies.

Capacitive vias can be optimized for power and signal transfer. When transferring data, low-capacitance vias are desirable because less power consumption is required to drive them. On the other hand, the capacitive vias used to couple power should be obtained with large coupling capacitances so that power transfer is maximized. This in turn requires more silicon area. The required silicon area for a given power can be obtained from Fig. 10.10.

It is important to notice that capacitive coupling provides electrical insulation between dies (Culurciello et al., 2005a; Culurciello et al., 2005b). This feature is advantageous for the design of sensitive instruments that needs to be decoupled from noise sources. In addition, insulation is often required in body implants and biomedical circuits for the safety of patients and users.

In summary, capacitive coupling of data has been demonstrated practical and advantageous in 3D assemblies and die packaging. Capacitive coupling of power is particularly suitable for low-power sensors and sensory front ends. Insulated sensory systems running on low-power budgets can also take advantage of the technology to de-couple circuitry.

10.3.2 Three-Dimensional Integrated Sensors in SOS

In this section, we present a 3D integrated system capable of thermal actuation and digital temperature measurement using a bandgap reference. Fig. 10.40 shows the organization of the prototype system. The bottom die transmits power to the top die using a four-stage Dickson charge pump and capacitive coupling (Culurciello et al., 2005a). The top die contains a bandgap reference and two digital Voltage-Controlled Oscillator (VCO) converters. These circuits convert a temperature dependent (V_{ref}) and a temperature independent (Temp) signal from the bandgap reference into a square-wave digital signal. These two signals are communicated back to the bottom die again using capacitive coupling. Notice that the top die has no galvanic connection to the bottom die or the external world. Capacitive coupling and data and power allow the top die to be electrically floating (Culurciello and Andreou, 2005).

The system contains a thermal heater implemented as a shunt silicon resistor of 100 Ω between the power supplies of the top die. The heater shows the capabilities of the systems in actuation and is a fundamental components of chemicals and gas sensors. Having no direct galvanic connection to the power supply, the heater efficiency is much

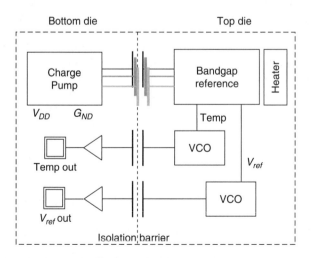

FIGURE 10.40 System architecture.

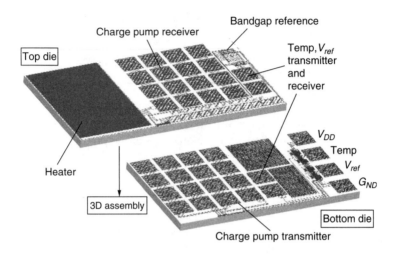

FIGURE 10.41   Integration of the fabricated 3D integrated thermal actuator with digital temperature sensor. The bottom die transmits power to the top die. The top die contains a bandgap reference and two digital VCO converters. Notice that the die has no galvanic connection to the bottom die.

higher as no thermal loss occurs via the terminal's metal connections. The temperature measurement systems uses a VCO to converts the V_{ref} signal of the bandgap reference to the frequency of a digital square wave. An identical VCO is used to convert the reference signal *Temp*. These two signals can be used together to verify the functionality of the bandgap reference and also obtain a digital reading of the temperature of the top die.

Figure 10.41 shows the three-dimensional arrangement of the system. The bottom die couples with the bottom-facing top die by means of 18 bonding pads. The top die is flipped and aligned on top of the bottom die so that the required capacitive connection are formed between the bonding pads metal plates of both dies. Sixteen pads are used to pump charge to the top die and supply current to its circuits. Two pads are used to communicate back the temperature and reference signals to the bottom die. The bottom die is placed and bonded into a common dual-in-line package. The alignment of the dies was performed manually under a microscope. The precision of the alignment was less than 10 μm, and the dies have been bonded together using a layer of transparent varnish.

Figure 10.42 is a micrograph of the fabricated and assembled multichip module. The bottom die is bonded to a package through bonding wires visible on the right side of the figure. The top circuit, visible on the right side of the figure, is flipped so that its bonding pads are

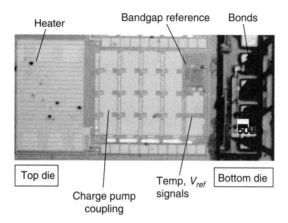

FIGURE 10.42 Micrograph of the assembled multichip module. The bottom die is bonded to a package through bonding wires visible on the right side of the figure. The top die, visible on the left side of the figure, is flipped so that its bonding pads are facing the bottom circuit's pads. This forms the capacitor for coupling the signal (*Temp, V_ref*) and power across the two dies (using a charge pump).

facing the bottom circuit's pads. This forms the capacitor for coupling the signal and power across the two dies. The capacitive coupling in air is about 8 fF, with a plate distance of 10 μm (3 + 3 μm for the bond to passivation step and an estimated 4 μm for the varnish layer).

10.3.2.1 Three-Dimensional Integrated Sensor Components

The core circuitry of the 3D system is a CMOS bandgap reference with a sub-1-V output and implemented on the top die. This circuit has been described in Sec. 4.3.

The charge pump is based on the Dickson (Dickson, 1976) charge pump design and consists of by 4 stages. An 11 stage ring oscillator at the transmitter side produces a 350-MHz digital clock signal that controls the pump. The output of the oscillator is buffered to drive the isolation capacitances. The Dickson charge pump operates by pumping charge to the top die (Culurciello et al., 2005a; Culurciello and Andreou, 2005). The 16 coupling capacitors are organized to pump during the 2 phases of the clock cycles, in groups of 8 capacitors per phase.

The VCO circuit is presented in Fig. 10.43. The circuit is a self-reset asynchronous oscillator that converts an input voltage into the frequency of a square digital wave. The core of the oscillator is an integrator based on capacitor discharge. The input transistor converts

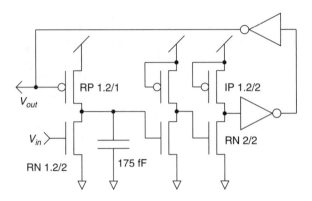

FIGURE 10.43 VCO used to convert an analog signal into a digital clock with varying frequency.

an input voltage into a nonlinear current that discharges the capacitor. When the capacitor is discharged, a feedback loop composed of four inverters provides a delayed reset signal to restart the integration. Notice that we take advantage of the multiple threshold devices in the SOS process to design two of the inverters in the feedback loop. These self-bias inverters use an intrinsic transistor to self-bias. These first stages also provide a delay before communicating the reset signal to the integrator.

10.3.2.2 Three-Dimensional Integrated Sensors Results and Measurements

We measured the performance of the bandgap circuit by evaluating the temperature measurement capabilities and the stability of the output V_{ref} with a power supply of 2 V. Fig. 10.44 shows the dependance of signals *Temp* and V_{ref} with temperature. The signal *Temp* is derived from the V_{be} voltage of the bandgap circuit and it is linearly and inversely proportional to temperature. The voltage V_{ref} was designed to be approximately 950 mV. Notice that the V_{ref} signal is constant in the temperature range 20°C to 80°C. The V_{ref} ripple is approximately 6%, a very good result given the low reference voltage. Figure 10.45 represents the frequency of the square wave of signal *Temp* as a function of temperature. This is the digital reading of the temperature. Notice the linearity of increase of the output frequency with temperature. A linear model of the frequency dependance predicts a conversion factor of 3.5 kHz/°C.

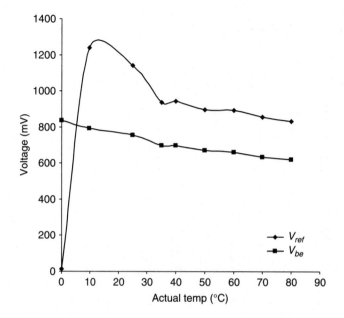

FIGURE 10.44 Voltage reference output V_{ref} and *Temp* as a function of the temperature.

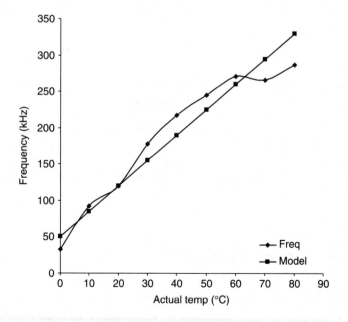

FIGURE 10.45 Temp signal frequency versus temperature.

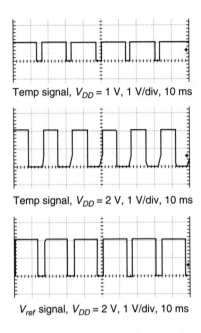

Temp signal, V_{DD} = 1 V, 1 V/div, 10 ms

Temp signal, V_{DD} = 2 V, 1 V/div, 10 ms

V_{ref} signal, V_{DD} = 2 V, 1 V/div, 10 ms

FIGURE 10.46 Oscilloscope traces verifying functionality of the prototype. The top and middle traces show the temperature digital signal *Temp* obtained respectively with a 1- and 2-V power supply. The bottom trace is the reference voltage V_{ref} signal with a 1-V power supply.

We successfully tested the functionality of the data communication and power transfer of the 3D integrated system. Fig. 10.46 shows screen frames from the oscilloscope verifying functionality of the prototype. The top and middle traces show the temperature digital signal *Temp* obtained respectively with a 1- and 2-V power supply and at a temperature of 25°C. The bottom trace is the reference voltage V_{ref} signal with a 1-V power supply. The current drawn was 2 mA with a 3.3-V supply and 1 mA with a 1-V supply.

10.4 Summary and Conclusions

The SOS process, with its insulating substrate, allows us to fabricate isolation circuits in a single die. Monolithic isolator circuits are beneficial because of the reduction of costs in the packaging production step and for a reduce power consumption due to circuit integration. We developed and tested multiple SOS digital isolator buffers. We reported with high performance and show that they are capable of high-voltage isolation. The intended application for the isolator circuit of this chapter is to decouple two regions of a spacecraft, a control unit, and high-voltage solar panels. Other typical applications are in harsh industrial environments, transportation, medical and life-critical systems. The intended isolation range is at least 100 V in continuous mode.

In addition, with the same circuit component used for the design of the digital isolator circuit, we were able to demonstrate the exchange of power and data between two SOS dies coupled capacitively through bonding pads. This technique can be used to package multichip ensembles of sensors and processors without using galvanic connections. In addition, it can be used to advance the 3D fabrication technology of multichip modules.

We demonstrated that capacitive coupling is capable of transferring power to the receiver die by employing a charge pump. We also show that communication can be achieved between the two dies by reading two digital signals coupled across the dies. One signal is the temperature reading of a thermometer circuit residing on the top die. A bandgap reference implements the temperature sensors and also provide, a stable temperature-independent voltage that is also transmitted to the bottom die.

References

Abshire, P. 2001. *Sensory Information Processing Under Physical Constraints*. PhD thesis, Johns Hopkins University, Baltimore, MD.

AMIs 1999. American Microsystems Inc. URL: http://www.amis.com/foundry/.

Anderson, C., Petrovick, J., Keaty, J., and al. 2001. Physical design of a fourth-generation POWER GHz microprocessor. In *Solid-State Circuits Conference, 2001. Digest of Technical Papers. ISSCC. 2001 IEEE International*, pp. 232–233. IEEE.

Andreou, A., Meitzler, R., Strohbehn, K., and Boahen, K. 1995. Analog VLSI neuromorphic image acquisition and pre-processing systems. *Neural Networks* 8(7-8):1323–1347.

Andreou, A. G., Kalayjian, Z. K., Apsel, A., Pouliquen, P., Athale, R. A., Simonis, G., and Reedy, R. 2001. Silicon on sapphire CMOS for optoelectronic microsystems. *IEEE Circuits and Systems Magazine* 1(3):22–30.

Apsel, A. 2002. *Optoelectronic Receivers in Silicon on Sapphire CMOS: Architecture and Design for Efficient Parallel Interconnects*. PhD thesis, Johns Hopkins University, Baltimore, MD.

Apsel, A. and Andreou, A. 2000. Quality of data reconstruction using stochastic encoding and an integrating receiver. In *Proceedings of the 43th Midwest Symposium on Circuits and Systems*, pp. 183–186, Ames, MI. Best Student Paper Award.

Apsel, A. and Andreou, A. 2001. A 5-mW, gigabit/s silicon on sapphire CMOS optical receiver. *IEE Electronics Letters* 37(19):1186–1188.

Apsel, A. and Andreou, A. 2005. A low power SOS CMOS optoelectronic receiver using low and high threshold devices. *IEEE Transactions on Circuits and Systems I* 52(2).

Apsel, A., Culurciello, E., Andreou, A., and Aliberti, K. 2003. Thin film pin photodiodes for optoelectronic silicon on sapphire CMOS. In *IEEE International Symposium on Circuits and Systems, ISCAS '03*, vol. 4, pp. 908–911.

Apsel, A., Fu, Z., and Andreou, A. 2004. A 2.5-milliwatt SOS CMOS optical receiver for chip-to-chip interconnect. *IEEE Journal of Lightwave Technology* 22(9).

Axon Instruments OpusXpress model 2008. http://www.moleculardevices.com/pages/instruments/opusxpress.html.

Ayers, S., Gillis, K., Lindau, M., and Minch, B. 2007. Design of a CMOS potentiostat circuit for electrochemical detector arrays. *IEEE Transactions on Circuits and Systems* 54(4):736–743.

Baker, R. 2005. *CMOS Circuit Design, Layout and Simulation*. New York: Wiley Interscience.

Baker, R. J., Li, H. W., and Boyce, D. E. 1998. *CMOS Circuit Design, Layout and Simulation*. New York: IEEE Press.

Banba H. et al. (1999). A CMOS bandgap reference circuit with sub-1-V operation. *IEEE Journal of Solid-State Circuits* 34:670–674.

Bandyopadhyay, A., Mulliken, G., Cauwenberghs, G., and Thakor, N. 2002. VLSI potentiostat array for distributed electrochemical neural recording. In *Proceedings of the IEEE International Symposium on Circuits and Systems (ISCAS'2002)*, pp. II–740–II–743, Phoenix, AZ.

Barbaro, M., Burgi, P.-Y., Mortara, A., Nussbaum, P., and Heitger, F. 2002. A 100×100 pixel silicon retina for gradient extraction with steering filter capabilities and temporal output coding. *IEEE Journal of Solid-State Circuits* 37(2):160–172.

Bo, X., Grudowski, P., Adams, V., and al. 2006. Optimization of dual-ESL stressor geometry effects for high performance 65nm SOI transistors. In *SOI Conference, 2006 IEEE International* pp. 19–20.

Boahen, K. 2005. Neuromorphic microchips. *Scientific American* 292:56–64.

Boahen, K. A. 2000. Point-to-point connectivity between neuromorphic chips using address events. *IEEE Transactions on Circuits and Systems—II: Analog and Digital Signal Processing* 47(5):416–434.

Brouk, I., Alameh, K., and Nemirovsky, Y. 2007. Design and characterization of CMOS/SOI image sensors. *IEEE Transactions on Electron Devices* 54(3):468–475.

Bucher, M., Lallement, C., and Enz, C. 1996a. An efficient parameter extraction methodology for the EKV MOST model. In *ICMTS, IEEE International Conference on Microelectronic Test Structures*, Trent, Italy, pp. 145–150.

Bucher, M., Lallement, C., Enz, C., and Krummenacher, F. 1996b. Accurate MOS modelling for analog circuit simulation using the EKV model. In *ISCAS '96. Proceedings of the 1996 International Symposium on Circuits and Systems* pp. 703–706.

Buchholtz, T., Aipperspach, G., D.T., Phan, N., Storino, S., Strom, J., and Williams, R. 2000. A 660-MHz 64b SOI processor with Cu interconnects. In *Solid-State Circuits Conference, 2000. Digest of Technical Papers. ISSCC. 2000 IEEE International* pp. 88–89. IEEE.

Burghartz, J. N., Edelstein, D. C., Soyuer, M., Ainspan, H. A., and Jenkins, K. A. 1998. Low-noise voltage controlled oscillators using enhanced LC tanks. *IEEE Journal of Solid-State Circuits* 33(12):2028–2034.

Chan, A., Man, T., Jin, H., Yuen, K., Lee, W., and Chan, M. 2004. SOI flash memory scaling limit and design consideration based on 2-D analytical modeling. *IEEE Transactions on Electron Devices* 51:2054–2060.

Choa, Y., Takaob, H., and Sawada, K. 2007. High speed SOI CMOS image sensor with pinned photodiode on handle wafer. *Microelectronics Journal* 38(1):102–107.

Chong, K. and Xie, Y. 2005. High-performance on-chip transformers. *IEEE Electron Device Letters* 26:557–559.

Cohen, M., Edwards, R., Cauwenberghs, G., Vorontsov, M., and Carhart, G. 1999. AdOpt: Analog VLSI stochastic optimization for adaptive optics. In *Proceedings of the International Joint Conference on Neural Networks (IJCNN'99)*, vol. 4, pp. 2343–2346, Washington DC.

Colinge, J.-P. 1997. *Silicon-on-Insulator Technology: Materials to VLSI*, 2nd ed. Boston: Kluwer Academic Publishers.

Craninckx, J. and Steyaert, M. 1995. Low-noise voltage controlled oscillators using enhanced LC tanks. *IEEE Transactions on Circuits and Systems II*, 2(14):794–805.

Cristoloveanu, S. and Li, S. S. 1995. *Electrical Characterization of Silicon-on-Insulator Materials and Devices*. Boston: Kluwer Academic Publishers.

Culurciello, E. 2007. Three-dimensional phototransistors in 3D silicon-on-insulator technology. *Electronics Letters* 43(7):418–420.

Culurciello, E. and Andreou, A. 2003. A comparative study of access topologies for chip-level address-event communication channels. *IEEE Transactions On Neural Networks* 14:1266–1277. Special Issue on Hardware Implementations.

Culurciello, E. and Andreou, A. 2005. Capacitive coupling of data and power for 3D silicon-on-insulator VLSI. In *IEEE International Symposium on Circuits and Systems, ISCAS*, vol. 4, pp. 4142–4145, Kobe, Japan.

Culurciello, E. and Andreou, A. 2006. CMOS image sensors for sensor networks. *Analog Integrated Circuits and Signal Processing* 49(1):39–51.

Culurciello, E., Andreou, A., and Pouliquen, P. 2002. Modeling hot-electrons effects in silicon-on-sapphire MOSFETs. In *ISCAS '02. Proceedings of the 2002 International Symposium on Circuits and Systems Phoenix, AZ*, pp. 569–572.

Culurciello, E. and Andreou, A. G. 2004. ALOHA CMOS imager. In *IEEE International Symposium on Circuits and Systems ISCAS '04*, Vol. 4.

Culurciello, E. and Etienne-Cummings, R. 2004. Second generation of high dynamic range, arbitrated digital imager. In *IEEE International Symposium on Circuits and Systems ISCAS '04*, vol. 4, pp. IV– 828–831, Vancouver, BC, Canada.

Culurciello, E., Etienne-Cummings, R., and Boahen, K. 2001. Arbitrated address event representation digital image sensor. In *IEEE International Solid-State Circuits Conference. Digest of Technical Papers. ISSCC. 2001*, pp. 92–93.

Culurciello, E., Etienne-Cummings, R., and Boahen, K. 2003. A biomorphic digital image sensor. *IEEE Journal of Solid-State Circuits* 38(2):281–294.

Culurciello, E., Pouliquen, P., and Andreou, A. 2005a. An isolation charge pump fabricated in silicon on sapphire CMOS technology. *IEE Electronics Letters* 41(10):520–592.

Culurciello, E., Pouliquen, P., and Andreou, A. 2007. A digital isolation amplifier in silicon-on-sapphire CMOS. *Electronics Letters* 43(8):451–452.

Culurciello, E., Pouliquen, P., Andreou, A., Strohbehn, K., and Jaskulek, S. 2005b. A monolithic digital galvanic isolation buffer fabricated in silicon on sapphire CMOS. *IEE Electronics Letters* 41(9):526–528.

Culurciello, E., Pouliquen, P., Andreou, A., Strohbehn, K., and Jaskulek, S. 2005c. A monolithic isolation amplifier in silicon-on-insulator CMOS. In *IEEE International Symposium on Circuits and Systems, ISCAS*, vol. 1, pp. 137–140, Kobe, Japan.

Culurciello, E. and Weerakoon, P. 2007. Three-dimensional photodetectors in 3D silicon-on-insulator technology. *IEEE Electron Device Letters* 28(2):117–119.

Dickson, J. 1976. On-chip high-voltage generation in NMOS integrated circuits using an improved voltage multiplier technique. *IEEE Journal of Solid-State Circuits* 11(6):374–378.

Enz, C., Krummenacher, F., and Vittoz, E. 1995. An analytical MOS transistor model valid in all regions of operation and dedicated to low-voltage and low-current applications. *Journal on Analog Integrated Circuits and Signal Processsing* pp. 83–114.

Enz, C. and Temes, G. 1996. Circuits tehcniques for reducing the effect of op-amp imperfections: autozeroing, correlated double-sampling, and chopper stabilization. *IEEE Proceedings* 84(11):969–982.

EPFL (2004). URL: http://legwww.epfl.ch/ekv/.

Faramarzpour, N., Deen, M., and Shiranic, S. 2006. Signal and noise modeling and analysis of complementary metal-oxide semiconductor active pixel sensors. *Journal of Vaccum Science Technology* 24(3):879.

Fish, A., Yadid-Pecht, O., and Culurciello, E. 2007. Responsivity of gated photodiode in sos technology. In *IEEE Conference on Sensors* Atlanta, GA.

Fossum, E. 1995. Digital camera system on a chip. *IEEE Micro* 18(3):8–15.

Fossum, E. 1997. CMOS image sensors: Electronic camera-on-a-chip. *IEEE Transactions on Electron Devices* 44(10):1689–1698.

Fu, Z. and Culurciello, E. 2006. An ultra-low power silicon-on-sapphire ADC for energy-scavenging sensors. In *IEEE International Symposium on Circuits and Systems ISCAS '06*, Kos, Greece.

Fu, Z., Weerakoon, P., and Culurciello, E. 2006. A nano-watt silicon-on-sapphire ADC using 2C-1C capacitor chain. *IEE Electronics Letters* 42(6):526–528.

Gabara, T. J. and Fischer, W. C. 1997. Capacitive coupling and quantized feedback applied to conventional CMOS technology. *IEEE Journal of Solid-State Circuits* 32:419–427.

Gambini, S. and Rabaey, J. 2007. Low-power successive approximation converter with 0.5V supply in 90 nm CMOS. *IEEE Journal of Solid State Circuits* 42(11):2348–2356.

Garcia, G., Reedy, R., and Burgener, M. 1988. High-quality CMOS in thin (100nm) silicon on sapphire. *IEEE Electron Device Letters* 9(1):32–34.

Geiger, R. L., Allen, P. E., and Strader, N. R. 1990. *VLSI, Design Techniques for Analog and Digital Circuits*. New York: McGraw-Hill.

Genov, R., Stanacevic, M., Naware, M., Cauwenberghs, G., and Thakor, N. 2006. 16-channel integrated potentiostat for distributed neurochemical sensing. *IEEE Transactions on Circuits and Systems* 53(11):2371–2376.

Gomila, G., Pennetta, C., Reggiani, L., Ferrari, G., and Bertuccio, G. 2004. Shot noise in linear macroscopic resistors. *Physical Review letters* 92(22):226601-1.

Gray, P., Hurst, P., Lewis, S., and Meyer, R. 2001. *Analysis and Design of Analog Integrated Circuits*. New York: Wiley Interscience.

Gupta, M. 2000. *Handbook of Photonic*. New York: CRC Press.

Hamill, O., Marty, A., Neher, E., Sakmann, B., and Sigworth, F. 1981. Improved patch-clamp technique for high-resolution current recording from cells and cell-free membrane patches. *European Journal of Physiology* 391:85–100.

Harper, C. A. 2000. *Electronic Packaging and Interconnection Handbook*. New York: McGraw-Hill Professional.

Harrison, R.R. and Charles, C. 2003. A low-power low-noise CMOS amplifier for neural recording applications. *IEEE Journal of Solid-State Circuits* 38:958–965.

Harrison, R., Watkins, P., Kier, R., Lovejoy, R., Black, D., Greger, B., and Solzbacher, F. 2007. A low-power integrated circuit for a wireless 100-electrode neural recording system. *IEEE Journal of Solid-State Circuits* pp. 123–133.

Hasler, P. 2005. Floating-gate devices, circuits, and systems. In *IEEE International Database Engineering and Application Symposium (IDEAS05)* vol. 1, pp. 482–487.

Hastings, A. 2005. *Art of Analog Layout*. Upper Saddle River, NJ: Pearson.

Howes, R., Redman, White, W., Nicols, K. G., Murray S. J., and Mole, P. J. 1990. Modelling and simulation of silicon-on-sapphire MOSFETs for analogue circuit design. In *ESSDERC '90*. ESSDERC.

Interpoint Corporation, 2005. MGA series product datasheet.

Johns, D. and Martin, K. 1997. *Analog Integrated Circuit Design*. New York: Wiley

Johnson, J. 1928. Thermal agitation of electricity in conductors. *Physical Review* 32(7):97–109.

Kalayjian, Z. 2000. *Optoelectronic Vision and Image Processing*. PhD thesis, Johns Hopkins University, Baltimore, MD.

Kanekawa, N., Kojima, Y., Yukutake, S., Nemoto, M., Iwasaki, T., Takamiand, K., Tekeuchi, Y., and Shima, Y. Y. Y. 2000. An analog front-end LSI with on-chip isolator for V.90 56 kbps modems. In *IEEE Custom Integrated Circuits Conference*, pp. 327–330.

Kansy, R. 1980. Response of a correlated double sampling circuit to 1/f noise. *IEEE Journal of Solid-State Circuits* 15(3):373–375.

Karlsson, P. and Jeppson, K. 1992. An efficient parameter extraction algorithm for MOS transistor models. *IEEE Transactions on Electronic Devices* 39(9):2070–2076.

Kleinfelder, S., Lim, S., Liu, X., and Gamal, A. E. 2001. A 10,000 frames/s CMOS digital pixel sensor. *IEEE Journal of Solid-State Circuits* 36(12):2049–2059.

Krymski, A., Blerkom, D. V., Andersson, A., Block, N., Mansoorian, B., and Fossum, E. 1999. A high-speed, 500 frames/s, 1024 × 1024 CMOS active pixel sensor. In *Proceedings of the Symposium on VLSI Circuits*, Kyoto, Japan, pp. 137–138.

Kucewicz, W., Bulgheroni, A., and M.Caccia (2004). Fully depleted monolithic active pixel sensor in SOI technology. In *IEEE Nuclear Science Symposium Conference* vol. 2, pp. 1227–1230.

Kuhn, S., Kleiner, M., Thewes, R., and Weber, W. 1995. Vertical signal transmission in three-dimensional integrated circuits by capacitive coupling. In *ISCAS '95. Proceedings of the 1995 International Symposium on Circuits and Systems* vol. 1, pp. 37–40.

Kuhn, W., Boyd, R., Shumaker, R., M.M.Mojarradi, and Li, H. 2001. An RF-based IEEE 1394 ground isolator designed in a silicon-on-insulator process. In *IEEE Midwest Symposium on Circuits and Systems* volume 2, pp. 764–767, Where, InMars.

Kuo, J. B. and Su, K. W. 1998. *CMOS VLSI Engineering Silicon-on-Insulator (SOI)*. New York: Springer.

Kuttner, F. 2002. A 1.2V 10b 20Msamples/s non-binary successive approximation ADC in 0.13 μm CMOS. In *Proceedings of the IEEE International Solid State Circuit Conference*, volume 1, pp. 176–177.

Laiwalla, F., Fu, Z. M., Culurciello, E., Sigworth, F. J., and Klemic, K. 2005. A CMOS integrated patch clamp amplifier. In *Connecticut Symposium on Microelectronics and Optoelectronics*, New Haven, CT. CMOC.

Laiwalla, F., Klemic, K., Sigworth, F., and Culurciello, E. 2006a. An integrated patch-clamp amplifier in silicon-on-sapphire CMOS. *IEEE Transactions on Circuits and Systems, TCAS-I, Special Issue on Life Science and Applications* 53(11):2364–2370.

Laiwalla, F., Klemic, K., Sigworth, F., and Culurciello, E. 2006b. An integrated patch-clamp amplifier in silicon-on-sapphire CMOS. In *IEEE International Symposium on Circuits and Systems, ISCAS '06*, pp. 4054–4057, Kos, Greece.

Lambda. 2005. Lambda pss/psd series product datasheet, San Diego, CA.

Lee, K. W., Nakamura, T., Ono, T., Yamada, Y., Mizukusa, Y., Hashimoto, H., Park, K. T., Kurino, H., and Koyanagi, M. 2000. Three-dimensional shared memory fabricated using wafer stacking technology. In *IEEE International Electron Devices Meeting* pp. 165–168, San Francisco, CA.

Lee, T. H. 2000. Oscillator phase noise: A tutorial. *IEEE Journal of Solid-State Circuits* 45(5):326–336.

Lei, X., Liu, C. C., Kim, H. S., Kim, S. K., and Tiwari, S. 2003. Three-dimensional integration: technology, use, and issues for mixed-signal applications. *IEEE Transactions on Electron Devices* 50:601–609.

Lundstrom, M. 2000. *Fundamentals of Carrier Transport*. New York: Cambridge University Press.

Marcus, G., Strohben, K., Jaskulek, S., Andreou, A., and Culurciello, E. 2006. A monolithic isolation amplifier in silicon-on-insulator CMOS: Testing and applications. *Analog Integrated Circuits and Signal Processing* 49(1):63–70.

Marwick, M. and Andreou, A. 2007. A UV photodetector with internal gain fabricated in silicon on sapphire CMOS. In *IEEE SENSORS 2007 Conference*, pp. , Atlanta, GA.

McIlrath, L. 2001. A low-power low-noise ultrawide-dynamic-range CMOS imager with pixel-parallel A/D conversion. *IEEE Journal of Solid-State Circuits* 36:846–853. Issue 5.

Mead, C. and Mahowald, M. A. 1988. *A silicon model of early visual processing*. Neural Networks 1(1):91–97.

Mead, C. A. 1989. *Analog VLSI and Neural Systems*. Reading, MA: Addison-Wesley.

Megahed, M., Burgener, M., Cable, J., and al. 1998. UTSi(R) CMOS technology for system-on-chip solution. In *1998 Topical Meeting on Silicon Monolithic Integrated Circuits in RF Systems, 1998. Digest of Papers*, pp. 94–99, Ann Arbor, MI.

Mick, S. E., Wilson, J. M., and Franzon, P. 2002a. 4 Gbps AC coupled interconnection. In *IEEE Custom Integrated Circuits Conference* pp. 133–140, Hong Kong.

Mick, S. E., Wilson, J. M., and Franzon, P. 2002b. Packaging technology for AC coupled interconnection. In *IEEE Flip-Chip Conference*.

MIT Lincoln Labs (2005). MITLL low power FDSOI CMOS process design guide. http://www.ll.mit.edu/, Boston, MA.

Mizoguchi, D., Yusof, Y. B., Miura, N., Sakura, T., and Kuroda, T. 2004. A 1.2-Gb/s/pin wireless superconnect based on inductive inter-chip signaling (IIS). In *IEEE International Solid-State Circuits Conference* vol. 1, pp. 142–144 Yokohama, Japan.

Martin, M. N. and Andreou, A. A. 1998. Floating-gate logic (FGL) for low voltage digital systems. In *Proceedings of the Second International Workshop on Design of Mixed-Mode Integrated Circuits and Applications* pp. 125–128, Guanajuato, Mexico.

Moini, A. 1994. Design of a VLSI motion detector based upon the insect visual system. Master's thesis, The University of Adelaide, Australia.

Molecular Devices Electrophysiology Instruments (2008). http://www.moleculardevices.com/.

Montezapour, S. and Lee, E. K. F. 2000. A 1-V, 8-bit successive approximation ADC in standard CMOS process. *IEEE Journal of Solid State Circuits* 35(4):642–646.

MOSIS (1999). The MOSIS service. http://www.mosis.org.

Murari, K., Thakor, N., Stanacevic, M., and Cauwenberghs, G. 2004. Wide-range, picoampere-sensitivity multichannel VLSI potentiostat for neurotransmitter sensing. In *Proceedings of the 26th Annual International Conference IEEE Engineering in Medicine and Biology Society (EMBS'2004)*, San Francisco, CA.

Nanion Technologies 2008. http://www.nanion.de/.

Neher, E. and Sakmann, B., eds. 1995. *Whole-Cell Recording*, Chapter 7. Plenum Press. A Marty and E Neher.

Nemirowsky, Y., Brouk, I., and Jacobson, C. 2001. 1/f noise in CMOS transistors for analog applications. *IEEE Transactions on Electronic Devices* 48(5):921–927.

Pain, B., Yang, G., Oritz, M., McCarty, K., Hancock, B., Heynssens, J., Cunningham, T., Wrigley, C., and Ho, C. 2000. A single-chip programmable digital CMOS imager with enhanced low-light detection capability. In *13th International Conference on VLSI Design* pp. 342–347, Calcutta, India.

Pan, J., Topol, A., Shao, I., and al. 2007. Novel approach to reduce source/drain series and contact resistance in high-performance UTSOI CMOS devices using selective electrodeless CoWP or CoB processing. *IEEE Electron Device Letters* 28(8):691–693.

Pandey, S. and White, M. H. 2001. An integrated planar patch-clamp system. In *International Semiconductor Device Research Symposium* pp. 170–173.

Park, J. and Culurciello, E. 2008a. High-speed back-illuminated image sensor in silicon-on-sapphire. In *IEEE International Symposium on Circuits and Systems, ISCAS '08*, Seattle, WA.

Park, J. and Culurciello, E. 2008b. Phototransistor image sensor in silicon on sapphire. In *IEEE International Symposium on Circuits and Systems, ISCAS '08*, Seattle, WA.

Park, S., Kim, Y., Ko, Y., Kim, K., Kim, I., Kang, H., Yu, J., and Suh, K. 1999. A 0.25-μm, 600-MHz, 1.5-V, fully depleted SOI CMOS 64-bit microprocessor. *Solid-State, IEEE Journal of Circuits* 34(11):1436–1445.

Peregrine Semiconductor 2003. 0.5-μm FC Design Manual (52/0005). http://www.peregrine-semi.com/.

Peregrine 2008a. *0.5-μm FC Design Manual*. Peregrine Semiconductor Inc., San Diego, CA, 52/0005 ed.

Peregrine 2008b. *Peregrine UTSi 0.5-μm RF Spice Models*. Peregrine Semiconductor Inc., San Diego, CA, 53/0016 edition. http://www.peregrine-semi.com/.

Pham, D., Aipperspach, T., Boerstler, D., and al. 2006. Overview of the architecture, circuit design, and physical implementation of a first-generation cell processor. *IEEE Journal of Solid-State Circuits* 41(1):179–196.

Pozar, D. 1998. *Microwave Engineering, 2nd ed*. New York: Wiley.

Promitzer, G. 2001. 12-Bit low-power fully differential switched capacitor noncalibrating successive approximation ADC with 1MS/s. *IEEE Journal of Solid State Circuits* 36(7):1138–1143.

Rabaey, J. M. 1996. *Digital Integrated Circuits*. Upper Saddle River, NJ: Pearson.

Razavi, B. 2000. *Design of Analog CMOS Integrated Circuits*. New York: McGraw-Hill.

Reedy, R. 1982. Characterization of defect reduction and aluminum redistribution in silicon implanted SOS films. *Journal of Crystal Growth* 58(1):53–59.

Reedy, R., Sigmon, T., and Christel, L. 1983. Suppressing Al outdiffusion in implantation amorphized and recrystallized silicon on sapphire films. *Applied Physics Letters* 8(42):707–709.

Reggiani, L., ed. (1985). *Hot Electron Transport in Semiconductors*. New York: Springer-Verlag.

Roden, D. 2004. Drug-induced prolongation of the QT interval. *New England Journal of Medicine* 350:1013–1022.

Rozeau, O. and al. 2000. Impact of floating body and BS-tied architectures on SOI MOSFET's radio-frequency performances. In *IEEE International SOI conference* pp. 124–125.

Sakmann, B. and Neher, E., eds. 1995. *Electronic Design of the Patch-Clamp*, Chapter 1. Plenum Press. F.J Sigworth.

Salzman, D. and Knight, T. K. Jr. 1994. Capacitive coupling solves the known good die problem. In *IEEE Multi-Chip Module Conference* pp. 95–100.

Salzman, D., Knight, T., Jr., and Franzon, P. 1995. Application of capacitive coupling to switch fabrics. In *IEEE Multi-Chip Module Conference*, pp. 195–199, Santa Cruz, CA.

Sarpeshkar, R., Delbruk, T., and Mead, C. 1993. "white noise in mos transistors and resistors". *IEEE Circuits and Devices Magazine* pp. 23–29.

Schanz, M., Nitta, C., Bubmann, A., Hosticka, B., and Wertheimer, R. 2000. A high-dynamic-range CMOS image sensor for automotive applications. *IEEE Journal of Solid-State Circuits* 35:932–938.

Schrey, O., Hauschild, R., Hosticka, B., Lurgel, U., and Schwarz, M. 1999. A locally adaptive CMOS image sensor with 90dB dynamic range. In *IEEE International Solid-State Circuits Conferences* pp. 310–311. IEEE.

Scott, M., Boser, B., and Pister, K. 2003. An ultralow-energy ADC for smart dust. *IEEE Journal of Solid State Circuits* 36(7):1123–1129.

Secareanu, M. R. and Friedman, E. G. 2000. Low power digital CMOS buffer systems for driving highly capacitive interconnect lines. In *IEEE Midwest Symposium on Circuits and Systems* vol. 1, pp. 362–365.

Serrano-Gotarredona, T., Linares-Barranco, G., and Andreou, A. 1999. Very wide range tunable CMOS/Bipolar current mirrors with voltage clamped imput. *IEEE Transactions on Circuits and Systems, Part I: Fundamental Theory and Applications* 46(11):1398–1407.

Sigworth, F. and Klemic, K. 2005. Microchip technology in ion-channel research. *IEEE Transactions On Nanobioscience* 4(1):121–127.

Sigworth, F. J. 2003. Life's transistors. *Nature* 423:21–22.

Sinencio, E. S. and Andreou, A. G. 1998. *Low-Voltage Low-Power Integrated Circuits and Systems*. New York: IEEE Press.

Singh, J. 1994. *Semiconductor Devices: An Introduction*. New York, McGraw-Hill.

Sodini, C. and Howe, C. 1996. *Microelectronics: An Integrated Approach*. Upper Saddle River, NJ: Prentice Hall.

Srinivasan, V., Graham, D., and Hasler, P. 2005. Floating-gates transistors for precision analog circuit design: an overview. In *48th IEEE Midwest Symposium on Circuits and Systems*, vol. 1, pp. 71–74.

Stanacevic, M., Murari, K., Rege, A., Cauwenberghs, G., and Thakor, N. 2007. VLSI potentiostat array with oversampling gain modulation for wide-range neurotransmitter sensing. *IEEE Transactions on Biomedical Circuits and Systems* 1(1): 63–72.

Stanojevic, Z., Ioannou, D., Loncar, B., and Osmokrovic, P. 1997. Design of a SOI memory cell. In *21st International Conference on Microelectronics*, vol. 1, pp. 297–300.

Stuber, M., Megahed, M., Lee, L., and al. 1998. SOI CMOS with high-performance passive components for analog, RF and mixed-signal Design. In *SOI Conference, 1998 IEEE International* pp. 99–100.

Sze, S. M. 1981. *Physics of Semiconductor Devices*. New York: Wiley-Interscience.

Sze, S. M. 1990. *High-Speed Semiconductor Devices*. New York: Wiley-Interscience.

Tecella Electrophysiology Instruments (2008). http://www.tecella.com/.

Tian, Fowler, B., and Gamal, A. E. 2001. Analysis of temporal noise in CMOS photodiode active pixel sensor. *IEEE Journal of Solid-State Circuits* 36(1):92–101.

Uehara, A., Kagawa, K., Tokuda, T., Ohta, J., and Nunoshita, M. 2003. Back-illuminated pulse-frequency modulated photosensor using silicon-on-sapphire technology developed for use as epi-retinal prosthesis device. *Electronics Letters* 39(15):1102–1104.

Uryu, Y. and Asano, T. 2002. CMOS image sensor using SOI-MOS/photodiode composite photodetector device. *Japan Journal of Applied Physics* 41(4):2620–2624.

Verma, N. and Chandrakasan, A. 2007. An ultra-low energy 12-bit rate-resolution scalable SAR ADC for wireless sensor nodes. *IEEE Journal of Solid State Circuits* 42(6):1196–1205.

Vittoz, E. 1994. Low-power design: Ways to approach the limits. In *IEEE International Solid-State Circuits Conference, ISSCC*, pp. 14–18, San Francisco, CA.

VPT, Inc. 2005. 2800s series product datasheet, Everett, WA DVSA.

Waaben, S. 1975. High performance optocoupler circuits. In *IEEE International Solid-State Circuits Conference* vol. XVIII, pp. 30–31, San Francisco, Cali.

Walden, R. H. 1999. Analog-to-digital converter survey and analysis. *IEEE Journal of Selected Areas in Communications* 17(4):539– 550.

Weerakoon, P. and Culurciello, E. 2007a. An integrated patch-clamp amplifier for high-density whole-cell recordings. In *IEEE International Symposium on Circuits and Systems, ISCAS '07*, pp. 1205–1209, New Orleans, LA.

Weerakoon, P. and Culurciello, E. 2007b. Vertically integrated three-dimensional SOI photodetectors. In *IEEE International Symposium on Circuits and Systems, ISCAS '07*, pp. 2498–2501, New Orleans, LA.

Weerakoon, P., Klemic, K., Sigworth, F., and Culurciello, E. 2008a. An integrated patch-clamp amplifier for high-throughput planar patch-clamp systems. In *IEEE International Symposium on Circuits and Systems, ISCAS '08*, Seattle, WA.

Weerakoon, P., Klemic, K., Sigworth, F., and Culurciello, E. 2008b. Integrated patch-clamp biosensor for high-density screening of cell conductance. *Electronics Letters* 44(2):81–82.

Weerakoon, P., Klemic, K., Sigworth, F., and Culurciello, E. 2008c. An integrated patch-clamp potentiostat with electrode compensation. *IEEE Transactions on Biomedical Circuits and Systems TBCAS* 0(0).

Wise, K. D., Anderson, D., Hetke, J., Kipke, D., and Najafi, K. 2004. Wireless implantable microsystems: High-density electronic interfaces to the nervous system. *IEEE Proceedings* 92(1):76–97.

Wise, K. D. 1984. A micromachined integrated sensor with on-chip self-test capability. Invited Paper, Digest, at the *Solid-State Sensors Conference*, pp. 12–16, Hilton Head, Island, SC.

Wise, K. D., Angell, J., and Starr, A. 1970. An integrated circuit approach to extracellular microelectrodes. *IEEE Transactions on Biomedical Engineering* 17:238–247.

Xu, J., Mick, S., Wilson, J., Luo, L., Chandrasekar, K., Erickson, E., and Franzon, P. 2004. AC coupled interconnect for dense 3-D ICs. *IEEE Transactions on Nuclear Science* 51:2156–2160.

Yang K. and Andreou, A. 1994. The multiple input floating gate MOS differential amplifier: An analog computational building block. In *Proceedings of the 1994 International Syposium on Circuits and Systems* volume 5, pp. 37–40, London.

Yang, D., Gamal, A. E., Fowler, B., and Tian, H. 1999. A 640×512 CMOS image sensor with ultrawide dynamic range floating-point pixel-level ADC. *IEEE Journal of Solid-State Circuits*, 34:1821–1833.

Yang, H. and Sarpeshkar, R. 2005. A time-based engergy-efficient analog-to-digital converter. *IEEE Journal of Solid State Circuits* 40(8):1590–1601.

Yang, W. 1994. A wide-dynamic range, low-power photosensor array. In *IEEE International Solid-State Circuits Conference, 1994. Digest of Technical Papers. ISSCC. 1994*, pp. 230–231, San Francisco, CA.

Zheng, X., Wrigley, C., Yang, G., and Pain, B. 2000. High responsivity CMOS imager pixel implemented in SOI technology. In *IEEE International SOI Conference* pp. 138–139.

Zhou, Z., Pain, B., and Fossum, E. R. 1997. CMOS active pixel sensor with on-chip successive approximation analog-to-digital converter. *IEEE Transactions on Electrical Devices* 44(10):1759–1763.

Index

Index